# *Sketching Theoretical Biology*

# *Sketching Theoretical Biology*

*Toward a Theoretical Biology*

*C.H. Waddington*
*editor*

*Volume 2*

Routledge
Taylor & Francis Group
LONDON AND NEW YORK

First published 1969 by Transaction Publishers

2 Park Square, Milton Park, Abingdon, Oxfordshire OX14 4RN
711 Third Avenue, New York, NY 10017

*Routledge is an imprint of the Taylor & Francis Group, an informa business*

First issued in hardback 2017



Library of Congress Catalog Number: 2009043982

Library of Congress Cataloging-in-Publication Data

Toward a theoretical biology. 2, Sketches.
Sketching theoretical biology : toward a theoretical biology, volume 2 /
C.H. Waddington, editor.
p. cm.
Includes index.
ISBN 978-0-202-36319-6
1. Biology--Congresses. I. Waddington, C. H. (Conrad Hal), 1905-1975. II. Title.

QH301.T68 2009
570--dc22

2009043982

ISBN 13: 978-0-202-36319-6 (pbk)
ISBN 13: 978-1-138-53260-1 (hbk)

# Preface

In 1965 the Executive Committee of the International Union of Biological Sciences asked me to arrange a series of small Symposia at which biologists, mathematicians, theoretical physicists and others would explore the possibility of formulating, at least in broad outline, the structure of a discipline of General Theoretical Biology. It was felt that the time is approaching when such a subject might merit a recognized place within the field of biology, as Theoretical Physics has done for so many years among the physical sciences. The first Symposium was held at the Villa Serbelloni, Lake Como, at the kind invitation of the Rockefeller Foundation, from 28 August to 3 September 1966. From this meeting, there resulted a book of essays, published under the title *Towards a Theoretical Biology, I: Prolegomena* (Edinburgh University Press, and Aldine Publishing Company, Chicago, 1968). The present volume has issued from the second Symposium, which was held from 3 to 12 August 1967.

During our first meeting we were asking ourselves the entirely preliminary question – is there, or ought there to be, such a subject as General Theoretical Biology ? Tentatively, we concluded that there might be. It would have to be quite different in character from Theoretical Physics. It would not seek for universal and eternal laws. Instead, it would accept that living systems are particular examples of some kind of 'organized complexity', and its task would be to ask 'what kind of complexity ?', and 'what are the principles of its organization ?' In this, the second Symposium, I think we began to find meaningful ways of formulating these questions, and to sketch out some general lines along which there is hope of answering them.

The purpose of this volume is not to record the very latest thoughts about Theoretical Biology, but rather to bring together a number of elements which, we thought, would be useful in the eventual construction of a coherent and comprehensive Theory of Biology. Some participants felt that the major contributions they have available at present had already been adequately expressed. I have therefore included a few already-published articles. (I have, with the authors' concurrence, considerably shortened the paper by *Gmitro* and *Scriven*, in which the basic ideas had been developed more thoroughly than is necessary for our purpose here.)

Further, this book emerges from conversations at a Symposium, which have a more informal character than the orthodox Scientific Paper, carefully tailored to the conventional format of scientific publication ; and I offer the reader such non- (but I hope not sub-) standard items as Michael Arbib's Lecture Notes, my autobiographical Note, and the poems of Mary Reynolds.

Once again, we were fortunate enough to be the guests of the Rockefeller Foundation, and to enjoy the beautiful surroundings of the Villa Serbelloni and the hospitality of Dr and Mrs John Marshall, who understand so well how to create the best possible conditions for informal but intense discussions, so that a great deal of communication can be achieved within a comparatively few days. The I U B S, and in particular myself and all the participants at the meeting, would like to express our great gratitude for this most generous support. I am also very grateful to one of our fellow-guests, Mrs Mary Reynolds, for permission to print two poems she wrote during our stay at the Villa.

C. H. WADDINGTON
*University of Edinburgh*

## Contents

# Sketch of the Second Serbelloni Symposium

C. H. Waddington
University of Edinburgh

What we are trying to do at these Symposia is to bring into being a not yet existent academic discipline of THEORETICAL BIOLOGY. This should exist in the same sort of way that Theoretical Physics does today. That is to say, not, of course, as a closed body of doctrine with all the questions neatly answered. A few decades ago some biologists wrote as though they thought that physics was in this condition, and had in fact revealed the fundamental laws of physics and the basic constituents of matter. It is doubtful if physicists themselves ever thought this, and they certainly do not at the present time. They have, of course, many profound theorems such as quantum mechanics, but as we heard from David Bohm, it is by no means generally accepted as the final word of physics. Again the physicists know many 'fundamental particles', but now that they are looking inside the nucleus they are turning out new fundamental particles at the rate of several a year. It is very clear that they do not yet know *the* fundamental constituents of matter and it is becoming more and more probable that there are no such things, but that the material universe is open-ended to investigation in both directions towards the very small (sub-nuclear) and towards the very large (cosmological). The recognition of this largely removes the point of trying to distinguish between vitalist and mechanist theories of biology, where vitalism is defined as the notion that the objectively observable behaviour of living systems demands the postulation of entities not contemplated by the laws of physics. Since we do not know in full what entities are demanded by the laws of physics, the distinction is more or less inapplicable. The question to ask about biological theory is not whether it is vitalist or mechanist, but whether it is a useful scientific theory or not. Many useful biological theories cannot yet be expounded in terms of conventional physics, e.g. conditioned reflex learning, and genetics could not be until about 1960 at earliest.

The question for our Symposium therefore is not to produce a coherent body of theory, but rather to decide what are the type of questions which in biology might correspond to the kind of questions discussed by theoretical physicists in their field. Then, of course, we can best elucidate the nature of the questions by attempting some provisional answers of them.

In practice we have attempted to discover these fundamental questions by approaching biology from the outside. We have done this from two directions: firstly from general metaphysics and secondly from considering the origin of life from non-living systems.

The metaphysical approach was little in evidence in the first Symposium, but much more so in the second, where it was strongly represented by the quantum physicist David Bohm. He urged us to begin by considering the nature of the basic entities that science is about. His thought followed what I should consider a very Whiteheadian line. The basic entities are events or experiences involving something going on and a percipient who discerns particular specifiable characteristics in what is proceeding. Bohm was very concerned with how we divide this unitary experience up into two sections, one which we consider objective, and the other subjective. He made the point that we are pushed by our language, which is heavily functional, to call things we can talk about easily objective and things we can't talk about easily subjective. In his view many 'subjective' concepts (which tend to be thought of in pairs of opposites such as harmony and conflict, beauty and ugliness, etc.) should be thought of as much more objective than they usually are, since they are also inherent in the crude basic experiences and it is only convention which leads us to take them out of the 'real world' and put them into ourselves. For instance, at one point I asked him whether 'noise' can be defined without reference to a biological function. He answered 'Not yet, but I think it should be.'

Coming from these very general questions nearer to biology we find that one of our main means of discerning specific characteristics in ongoing processes is a visual sense, i.e. the seeing apparatus, the eye. Here biology immediately brings us up against two very interesting and challenging points. In the first place, as Richard Gregory pointed out, the eye does not directly provide us with nearly as much information as we succeed in some peculiar way in extracting from it. For instance, the image on the retina is only two-dimensional, yet we can use it as though it were three-dimensional. Putting his point in a crude way for brevity, you have to set up a computer in the eye-brain system which could carry a working model of the three-dimensional world with only occasional checking, and the checking involves making the best bet you can of what the two-dimensional retinal image indicates. It may be that the necessity for this computer-like system to interpret visual signals is the basis for the whole development of intelligence and conceptual thought. Moreover, here we meet for the first time a type of ambiguity or indeterminacy not unlike the physicist's indeterminacy of velocity and position of a particle—an uncertainty which we shall keep on coming across in other contexts in many aspects of fundamental biology.

A second point about our perceptive system, which I have been making for many years, is that natural selection must have tailored our perceptive system, which is a product of evolution, so that it is appropriate to perceive the kinds of things that it pays off to perceive—or rather the kind of things it paid our mammalian ancestors to perceive. Relation between this point and David Bohm's questions about how we divide the primitive experience into subjective and objective sections was not pursued, but clearly needs thinking about. Moreover, here we have another of those circular or even knotted intellectual situations which are so characteristic of biology. We perceive only what it is good for us (evolutionarily) to perceive, but when we look at the process of evolution we realize that we could equally well say that our perceptive apparatus has progressed in evolution along the lines on which it was already best at functioning. We know what it is good for us to know, and it is good for us to know what we know.

The other approach to biology from outside, namely the problem of the origin of life and the criteria for recognizing life, also led us rather directly to the true subtleties of the evolutionary process. At the first Symposium, Cairns-Smith and Pattee had both given us examples of non-biological systems which exhibit the hereditary transmission of mutable information, Cairns-Smith in terms of crystal dislocations and Pattee of polymerization processes. Pattee in particular raised some very general theoretical points in this connection. He argued that a hereditary process is equivalent to a classification process, i.e. a choosing between sharply defined alternatives on inadequate evidence. He suggested that an enzyme choosing between possible substrate molecules is the simplest model of this. He went on to argue that any such classification process must depend on something over and above the basic quantum mechanical laws, and in particular demands the action of a non-holonomic constraint, which in the crudest terms may be thought of as a structure whose material integrity has a decay time very much longer than that of the process with which one is concerned. He also raised the question whether any quantum mechanical chemical system can possibly have the reliability required for biological hereditary transmission (the shade of Wigner rose at this point). But it is not clear to me whether the reliability need after all be so high. Cells we now know have quite a lot of repair enzymes for dealing with the results of unreliability; and in any case a very large number of cells die, both amongst micro-organisms and during the development of higher organisms.

We reached general agreement, however, that the mere hereditary transmission of mutable genetic information is not really good enough to qualify as life. We need something that can really undergo evolution by the process of natural selection, and that involves something which interacts with the environment sufficiently for natural selection to be effective. Life requires not only transmission of a memory store, subject to occasional mistakes, which will presumably be encoded in some rather inert substance such as DNA, but also the production of something much more actively connected with the surrounding world which will take the form of more reactive substances, such as proteins. Life which is to engage in evolution requires in fact not only a genotype but a phenotype. The process by which the genotype becomes worked out into the phenotype can be referred to as 'epigenesis' ('metabolism', which thirty or more years ago was considered by most leading biologists as the discriminating characteristic of biology, 'local reversals of the Second Law' is no more than a short time-scale section of epigenetics).

▶ Following this line of thought there are three fundamental problems for theoretical biology:

i. What is involved in replication, i.e. the hereditary transmission of specificity (we will go back to this old-fashioned term instead of using the current jargon 'information' because though hereditary specificity is merely information while it is being transmitted, as soon as it becomes active in determining the character of development it becomes instructions or 'algorithms').

ii. We need a theory of epigenesis which is sufficiently powerful and flexible to encompass the development of higher organisms as well as, say, the production of a single enzyme in a bacterium.

iii. How exactly does evolution work? And again we need an inclusive theory such that the simplest systems can be regarded as degenerate cases.

(i) The logical structure of the process of replication is by no means as simple as might be thought at first sight, or as one would think after looking at such model systems as a Xerox copying machine or Penrose's ingenious toys. At the first Symposium the 'Party Game Model' presented by Michie and Longuet-Higgins brought out some of the implications. At the second Symposium Michael Arbib expounded a computer programme for building a machine that would replicate itself—a synthesis of the approaches of Turing and von Neumann.

(ii) We actually spent more time on the metabolism-epigenetics problem. If we are to set up an inclusive theory in this field in which simple cases are degenerate examples, this theory must deal with situations in which systems containing large numbers (say $10^5$) genes develop along time-trajectories which are to some extent self-stabilizing, and which lead to a relatively small number of sharply separated quasi-stable (or adult) states.

There were several approaches.

(a) Elsasser has been rather obscurely worried about this problem for a long time. His original formulation in terms of an 'increase in information during development' was not very acceptable to most of the rest of us, who felt that information was not an applicable concept in this field, where we are really dealing with algorithms. However, Elsasser rephrased his basic uneasiness in terms of an 'immense number indeterminacy'. If $10^5$ elements interact with one another, the total number of equations required to express the properties of the system are measured in such immense numbers that they are certainly in practice uncomputable, and, Elsasser argues, in principle uncomputable for the reason of what he calls the 'immense number phenomenon'. Thus the system he claims has a basic indeterminacy in it, of a kind which presents the same intellectual challenge as the usual quantum physicist's indeterminacy, although it is different in nature.

(b) The second approach is to develop a theoretical system in which one can describe these processes in terms of topological structures in a multi-dimensional function space. At the second Symposium we did not carry this much beyond the stage we had got at the first Symposium, where René Thom and Waddington had discussed such ideas as chreods, catastrophes, and the like.

(c) Another way of dealing with regularities emerging in multi-dimensional interacting systems is Gibbsian statistical mechanics. One of the highlights of the second Symposium was a convergence on to this type of mathematics from several sides. Ed Kerner, studying as a theoretical physicist looking at the problems of general ecology and interrelations between species, had made one of the crucial steps in applying Gibbs' ensemble theory to processes governed by non-linear feedback equations. Brian Goodwin had applied methods derived from Kerner to the problems of metabolism and epigenetics in cells and had emphasized the importance of oscillatory phenomena that are almost bound to arise. He defined indeed a series of concepts parallel to those of temperature, free energy, etc., but characterizing the oscillatory phenomena of cell behaviour and therefore qualified by the adjective 'talandic'. Jack Cowan, studying interactions between neurones and precisely defined anatomical regions of the human visual cortex, had found himself calling on essentially similar types of mathematics. Art Iberall had also come by a rather different route to a realization of the basic importance of the limit cycles of non-linear oscillators in the organization of many metabolic processes in animals. It seemed to emerge, since nearly all biological processes are controlled by feedback loops of one kind or another and occur as factors in enormously complex aggregates of interacting processes, that the properties of non-linear oscillators and the mathematics of the Gibbs ensemble form one of the most important methods by which theory can approach some of the most fundamental aspects of biology.

Another aspect of epigenetics to which we devoted some time was morphogenesis—the development of specific geometrical forms. Waddington outlined a scheme for classifying the types of processes by which biological forms can be brought into being. This has already been published in his *New Patterns in Genetics and Development* and will not be repeated in these notes. Scriven provided an extremely interesting discussion of the formation of regular morphological structures resulting from the breakdown of uniformity in originally homogeneous systems in which chemical or physical processes are actively proceeding. Examples are the arising of hexagonal cells subdividing large areas from which convection currents are being generated, the break-up of jets of liquid into separate drops, and so on. This was a generalization of Turing's fundamental work on the production of form within originally uniform systems. It would seem also to be related to the discussion of 'catastrophes' by René Thom, but unfortunately Thom was not present this year to take part in the discussion.

(iii) Another major topic for discussion was the theory of evolution and the adequacy of the present biological orthodoxy on this subject. Of course, nobody doubts that the introduction of Mendelism into evolutionary thinking resulted in very great progress, for instance in the realization that enormous stores of genetic variability are hidden in apparently uniform-looking populations, in understanding the genetic structure of special groups such as social insects, polyploid plants, and the like. Maynard Smith gave a very good account of some of these major advances and he expressed himself as in general satisfied with what he referred to as the theory of neo-Darwinism. Waddington, however, expressed the view that the term neo-Darwinism should really be reserved for the strictly theoretical work of mathematicians who have dealt with evolution in recent times. He argued that the most widely accepted mathematical formulations, those of Haldane and Fisher, are seriously defective in several ways—the criticisms apply to a much less extent to the much more subtle and profound work of Sewall Wright, though even he does not escape them entirely. According to Waddington, the basic error in the current formulation is to attach coefficients of selective value to genotypes, whereas in fact it is the phenotype which is acted on in natural selection. This point again brings us from another angle face to face with what looks like an 'indeterminacy' situation. As we have seen above, Elsasser produced arguments based on the immense number phenomenon, that the relation between the genotype and the phenotype may be, in principle, indeterminate. This may or may not be justified, but in any case we have to realize that natural selection operates without attempting to resolve the relation: the phenotype provides the criterion which is used to decide at what rate the genotype will be transmitted to later generations.

▸ Waddington also pointed to three other questions which seem intellectually challenging to theorists. Two of these are concerned with the concept of random mutation, and the third with the concept of evolutionary value or evolutionary progress.

1. The effect of a gene mutation on the phenotype is determined by the interaction of the mutant gene with all the other genes and with the environment during epigenesis. Thus, if the epigenetic system has certain stabilities and instabilities built into it—as is obviously the case—the effect of random changes in genes will not be random by the time they are worked out into phenotypes.

2. If we are looking for an inclusive theory of evolution it must be able to deal with the evolution of higher organisms. In these the characters on which natural selection acts will often be such characteristics as the ability to run or fly fast, to eat certain kinds of food, etc. Two further orders of complexity are introduced. In the first place, a character such as the ability to run fast is affected by a very large number of genes. Waddington suggested that the contribution of an individual gene mutation to such an ability is not much more direct than the contribution of a particular random shaped pebble with the engineering qualities of a concrete member of which this pebble forms part of the aggregate. But secondly, running fast may be only one of the possible ways in which members of the species can score highly in the natural selection stakes. It might be equally effective to develop a greater ability to fight off the predator, rather than to escape from him by flight. The situation is therefore radically different from one in which there is a known fixed endpoint to be reached, and we ask whether the process of random search is an adequate method of reaching it. This is the paradigm situation envisaged by many mathematicians who approach the theory of evolution. According to this argument, it is basically mistaken and inapplicable. A somewhat better model (though still very imperfect) would be to consider

the natural selection within one generation as equivalent to one trick in a card game such as bridge. The gene pool of the population corresponds to a hand of cards which have indeed been produced by random processes of mutation. The question at each particular round of the game is whether the species has got a suitable card to play—has it got some small trumps when it needs them, so that it does not have to commit its high cards, and on other occasions does it have the necessary high card to get by ? The process of natural selection must in fact be envisaged in terms of games playing strategy.

3. The current theory of natural selection is in terms of the leaving of offspring in the next generation. It has little to say about the long-term prospects of a species which may find itself having to meet the challenge of a new Ice Age, new predator, new disease, and so on. Waddington argued that here again an adequate formulation would need to be in terms of an extended and improved games theory. Can we find a way of defining in the first place what is a good bridge hand, or a good set-up on the chess board ? The 'evolutionary potential' of a species cannot be defined in terms of the number of offspring it will leave in the next generation, but must be considered as something comparable to a hand of cards which will prove useful whatever cards the opponents may hold.

The problem of 'randomness' was approached by Pattee from another angle. 'Simple physical systems are perfectly reliable (electrons do not make mistakes). At some level of organization we speak of error. What is this level ?' It seems, in fact, to be some sort of intermediate level. The movement of individual molecules in a gas is perfectly determinate if we look at it in detail. It becomes 'random Brownian' movement or 'noise' when we consider the movements within a small volume of gas, but there is again nothing random about the overall gas pressure exerted on the walls of the container. Part of Waddington's argument above is that a great deal of evolution takes place at the level corresponding to the gas laws, so that the randomness of the mutation (cf. Brownian movements of the molecules) is not basically relevant. Again, the molecular biologists studying changes of nucleotide sequences in DNA can also not remain content with regarding them merely as random but would search for their deterministic mechanisms (cf. following in detail the collisions and flight path of a molecule in a gas).

At this point Bohm brought us back to fundamental physics and metaphysics, with the argument that all realms of existence contain some things which are fortuitous or contingent, and others which are necessary or organized. This he claims applies to all levels, including physics. Fundamental physical entities such as quanta arise as organized entities from a lower level of fortuitous elements. This arising of the necessary or ordered from the fortuitous is what he regards as creation. His paradigm is a child starting to draw a random line, which may express something of his internal state; then he sees he has drawn a square and he takes that into himself and begins to express it again by putting into his drawing definite corners and straightening the lines and so on. On a more down-to-earth biological level, consider a genetically variable population, members of which may try all sorts of ways of earning a living in the face of the environment surrounding them. When one way turns out to pay off well in terms of natural selection (e.g. escaping predators by flight rather than fighting them), then there will be natural selection for that particular solution of the problem. Then the population will find itself being drawn into a fast-running type of organization.

## A Personal Overview

A. S. Iberall
General Technical Services Inc., Pennsylvania

It is possible to argue that the physicists succeeded in the aim of the conference: to consider and determine whether a theoretical biology might be brought into existence in the same sense that a theoretical physics has. However, despite their self-satisfaction, it is far from certain that the representatives of the other disciplines, particularly the biologists, would accept that conclusion. This personal summary, threading the proceedings in overview, may bind the diversity of physical positions that were expressed.

*Bohm* sounded a keynote of the troubled metaphysical foundations of modern physics. Partly pure professionalese, his views irritated the non-physicists, and, at the beginning, a substantial number of the physicists. Yet by the end of the conference his insistence that the physicists consider the faulty foundations of quantum mechanics, the foundations of thinking about space-time and of order itself, finally began to stroke sufficient chords among them into resonance. This arose because of the striking parallels between the quantization of elementary physical systems—in nuclear, atomic, molecular physics; and in the biological systems—in the molecular biology of the formation of life; in its evolution—in the internal cellular processes, in the cell, the internal organ systems, the total organism, and in the species in the ecosystem. It really wasn't until positions about the quantization at these other levels had emerged in the conference that any meaningful dialectic began with Bohm's point of view.

Consider, for example, one such exchange. Gradually a (vertical) hierarchy of organizational levels began to appear at the conference; and a (horizontal) heterarchy of diverse elements. One might see at most a finite matrix of this sort; or, by extension, an indefinite denumerable growth which might proliferate emanatively in time. Yet Bohm took exception to this. His point (in this interpretation) was essentially that ordering was non-denumerable. It is the abstraction provided by people that establishes autonomous hierarchical order.
However, this ordering, from a 'prime ministering' or 'executing' function (using social system elements as analogue) down to the lowest 'functionaries', represents the orders of the hierarchy, while there is a flux of state identifications upward, which represents information. Such ordering information, however, represents timeless processes. It is the horizontal heterarchial order that gives

the time order (i.e. time is given by the processes of different variety).

The law of the hierarchy is to govern the hierarchy, and this must flow from the timeless order (e.g. consider the ordering of the king and the duke, etc.).

Now the concept of function cannot be described or defined in terms of function. It must have an aim which is not function. It is for this reason that a functionless (and thus timeless) order must exist. However, function and its opposite, malfunction, lead to harmony and conflict, as objective realities. 'Survival' involves harmonization (e.g. the continued existence of cars is harmonized by the traffic laws). Conflict is dissipative. Every view of function tends to maintain harmony and order.

Ultimately, it is the law of constancy of numbers, of a one-ness other than function, that establishes the relations of order, and it is the time space that determines the hierarchical space.

While no certain idea emerged whether Bohm had any significant positive idea for establishing an abstract structure for quantum mechanics (and no reader could judge this further from these remarks), nevertheless the attack he mounted on the problem of quantization is one of a number of paths that may have some meaning. (The reader may judge for himself from Bohm in July 1966 *Rev. Mod. Phys.*) His attack—it appears in this overview—is that time and space are not to be equated, but, instead, that 'timely' properties emerge from hierarchical ordering itself. Beyond this, Bohm's own remarks on his notion of order will have to speak for themselves.

It is certainly true that Bastin and Lieber were also concerned with such global questions, in seeking a line to view the physical-material universe, and thence, biological processes.

A similar message was received (1967) in a New York Academy of Science meeting ('Interdisciplinary Perspectives of Time', R. Fischer, ed.; *Ann. N.Y. Acad. Sci., 138*, 367, 1967), particularly as speaker after speaker made his distinctions between physical time in its many faces (astronomical, radioactive, geological, etc.), and biological time (e.g. physiological, psychological).

This appears to me then to afford the transition to my contribution, which began from a very specialized problem. I had started, as a classical physicist (of a 'systems' persuasion), to try to clarify a few topics in the physics of regulation and control of the macroscopic biological system. Experimentally, instead of finding the small amplitude dynamics of linear regulation and control theory (for example, the variations in regulated metabolic power), I found characteristics much more similar to so-called bang-bang control theory. In a number of related studies I found the system loaded from one end to another with large-amplitude oscillations. 'Limit cycles !' I said. 'The system must be described by spectroscopy.' Thus, purely on empirical grounds, I was led to the same path that emerged in physics, which started from the chemical 'field' spectroscopy of individual atomistic species, and continued to the physical 'temporal' spectroscopy of atoms and molecules. My contribution, thus, is to offer the same path for a theoretical biology that developed as a theoretical physics, namely, that the gross 'material' system is made up of temporally quantized atoms in the case of simple solids, or of atomistic biochemical chains in the case of the biological system, whose everbeating cooperative effort represents the foundation for the characteristics of the overall system.

It is quite pointed of Bohm and others to have asked the question as to what is the foundation for quantization, since this has meaning at many different levels. It is important also to recognize that the ordering relations which may give the meaning at different levels may result in differences in detail and connection at various levels, yet possess great similarities. In the case of the gross biological system, I have chosen to insist that the spectroscopy is due to non-linear limit cycles at the biochemical level, largely.

It was most fortunate (for me) that Warren McCulloch had put me in touch with Goodwin a few years ago (and through whose good offices I have been able to attend this conference). It was clear that Goodwin had taken on for himself the same task at the 'atomistic' biological level, namely at the level of the biological cell, that I had taken for the system as a whole. His advantage was that whereas my chains (I prefer chains to loops or networks to emphasize its chained causal nature without stressing the fixity of the lumped element-to-element connection of networks) were much more complex thermal-mechanical-hydraulic-electrical-chemical systems sweeping cooperatively throughout the system—say as represented in particular by any endocrine system response—his chains were shorter enzyme-linked chains nearly available by a more direct spectroscopic separation. It is to his credit that his quick appreciation of the non-linear exposition of its dynamics became the most direct path to sink it into biological consciousness. This is shown by the acceptance of his book, *Temporal Processes in Cells,* and by the speed with which it has inspired other investigators. In these efforts, Goodwin and my colleagues have therefore been the 'practical' translators of from whence the foundations for a theoretical biology should come.

It was pleasing to Goodwin and me (in the important sense of its fittingness) to find in the remarks of Scriven that the modern chemical engineer was capable of following up the non-linear mechanics of the biological system with his unit processes, transport processes, and reaction systems. It is not surprising, for it was in the works and examples of such people as Onsager, Kirkwood, Eyring, etc., that people like myself learned enough about non-equilibrium mechanics to begin to tackle real systems.

Then Pattee, immersed at the lowest biological level question, the origin of life, better than anyone squared the circle. It turns out that here, too, at the lowest level of biochemical synthesis, the question is the non-linear dynamics of the chain. It is not sufficient to obtain a replication mechanism. A tactical polymerization by which aggraded forms may be created, by condensation reactions, is simply not sufficient. These will limit, reach a static, steady state. There must be a degradation step; and, further, it must be a step which produces a reliable gain from the energetic (free energy) changes. This requires specific catalysis ('No hereditary process can take place without uncatalyzed reactions').

From my point of view, Pattee's remarks are directed at the steps of a non-linear chain that can produce cyclic certainty from a DC potential bath of chemicals. It is not the DNA, RNA geometric specificity that provides the steps. They provide a geometric milieu over which the decisive steps can take place. Specific catalytic reactions are the escapement-like processes that make the non-linear clock run. (See, for example, Pattee *et al*, *Natural Automata and Useful Simulations*, Sparton, 1966.) The timing phases are possibly the many unit process steps involved in protein synthesis. Thus, Pattee's story is not the completion of the mystery of non-linear limits cycles—the spectroscopy of life formation and maintenance of the molecular biological level—but only its sketchy formal beginning. Room is still left for Nobel prizes at this level. (Arbib's contribution was to emphasize that there is a mathematical apparatus for doing the kind of thing that the complex biological system must do in self-replication. That is, he attempts an elegantly simple proof of the von Neumann result of whether a machine can be built that can reproduce machines as complex as itself.) We can only hope that as investigators like Pattee can elucidate more of the experimental problems of dynamic synthesis, that Arbib's algebraic semi-groups will be able to keep up, and perhaps sometime even surpass the experimental problems in predictive and prognostic value.

There is little doubt that the mathematical-computational complexity will be great. ('Isomorphic' mathematical theories will be quite important.)

While Kornacker directed his attention pointedly at the nervous system elements, he was really casting light on the one intermediate element which is needed to clarify the non-linear mechanisms of both the gross biological system and the microscopic cellular biological system; that is, the membrane. At present, his structure is still formal.

Guided by the statistical mechanics of thermal fluctuations and the formation of thermodynamic averages, he applies these ideas to cellular transport, specifically to electrical excitability of nerve membrane. The basic need, he shows, is to come up with a negative resistance within the membrane model. He finally proposes that a diode rectifier operating in a particular direction against a concentration gradient (sodium is his proposed gradient source) is the necessary and sufficient condition for a negative resistance.

While the details of his own particular problem are exceedingly clever, it is the broader implication that is really more interesting. His modelling contains some fertile seeds for the necessary instability to get the biological membrane to act as an active source for transport—whether at Goodwin's level of the cellular non-linear performance, or my level of the system response of such a system, say, as red cell-capillary interaction in the vascular beds.

Beyond that, Cowan suggests that the next few steps in understanding the electrical activity in the nervous system are coming along quite well. In the path in which Hebb has been so influential, a rough view exists of the relation of cortex and the reticular formation-thalamic link. Cowan has been developing a modelling of the cortico-thalamic nets, proposed as an improvement on the crude McCulloch-Pitts descriptions of processes in neural arcs. Thus, the high-frequency, tens of millisecond, responses in the brain are gradually coming into perspective.

For the gross nature of behaviour, McCulloch and I have recently sketched out a description of what a non-linear patterned model of man has to look like (copies were made available at the symposium, '1967 Behavioral Model of Man—His Chains Revealed', *NASA Contractors Report*, CR-858, July 1967)*. The model, physiological-psychological, is based on a central concept, homeokinesis, which echoes Waddington's homeorhesis. Waddington's thoughts long pre-date ours (apologies are in order for not having stumbled on his concept earlier). However, vanity and religion make me prefer or feel more comfortable with our concept (yet), for the ground we covered with it.

What is clearly missing is the 'theoretical' intermediate structure that covers the behavioural spectrum from 0·1 second to a few days. That it may be dominated

*The paper is now available in *Currents in Modern Biology, 1* (1968), 337.

by endocrine system and neuro-endocrine system responses is quite probable. However, we did not succeed at the conference in covering this ground. Although not represented at the conference, there are 'systems' scientists in endocrinology who have begun the laborious task of outlining the dynamic networks that govern their response. (Illustrative is the modelling of Yates and Urquhart. For example, Yates' most recent model of the adreno-cortical system was presented at an October 1967 conference on 'Hormonal Control Systems in Health and Disease', San Diego.)

However, the biologist claims there are two global questions which these statements of operating mechanisms do not answer:

1. What is the adaptive nature of the individual in his milieu ?
2. What is the operative and adaptive nature of the species in their milieu ?

The first question—which is where Waddington's and Maynard Smith's interests lie, and the 'real' reason for calling this conference—can be examined by default. Let us first pursue the second.

Fortunately, Kerner furnishes the beginnings of an answer. (I was very fortunate to attend a N.Y. Acad. Sci. conference—see *Ann. N.Y. Acad. Sci., 96,* 975, 1962—in which he presented a primary paper on the subject.) Following Volterra dynamics for interacting species, Kerner embeds this in the statistical 'mechanics' of Gibbs' ensemble theory. By this step, he reminds us that—if we can find the interparticulate forces—we can treat all kinds of ensembles, including biological, by this technique. I must confess that his paper inspired me, and in fact I think that I *have* furnished an answer for the 'effective' characterization of the inter-particle forces in biological spectroscopy. Just as in theoretical physics—if the energetics presented by the spectroscopy of the underlying atomistic particles can be identified, then in summation, thermodynamic functions can be set up (e.g. the free energy—see for example Landau, Lifshitz, *Statistical Physics,* 1958) and the equations of change can be derived. This technique avoids the description of the detailed 'force' laws (actually it embeds them in a quantized structure which is quite close). Thus, I believe a structure now exists to describe both the individual quantization and the population changes. Kerner has continued with his problem of the application to competition among species.

Thus, we approach the question Waddington wants answered: 'You physical fellows have had your fun, and I've paid for it, now where is the answer to my question ? What is the theoretical basis for evolution ?' (This is a case in which I hope I am only putting my words in his mouth !)

Maynard Smith said that the fitness of an individual species is governed by or related to the number of its children. We will go along with this. The fitness of a species depends upon its ability to achieve a non-decaying number existence in the ecology. It is not a question of brain size or abilities. It is a question whether, for changing ecologies, the genetic content, epigenetically enfolding, has enough survival value to produce a yield slightly greater than one-for-one. This is not a deterministic question, but an emanative evolutionary one. Thus one cannot state the course of future evolution. (Today's species may die; yesterday's certainly have died. This is well illustrated in Gaylord's *Life in the Past,* which shows the growing, peaking, and decline in many phyla.)

However, this again is a quantization problem, i.e. the formation of a stable cycle within a quasi-static changing milieu. The theoretical foundation here is whether the epigenetically produced branchings—of phenotype—are exceedingly rich for changing environmental conditions. This must be at the basis of a science of the theory of evolution. We physicists could start to examine such questions, but the hint is better directed to the geneticist. He knows so many more details than we; we would only fumble. However, he has to absorb from us the static problem, Arbib's theory of automata, and the dynamic problem—Bohm's, Pattee's, Goodwin's, our spatial and temporal quantization by dynamics.

As Pattee indicated on the lowest level, the genotype is the instruction for a working cycle, the phenotype is the enzyme that makes it carry over. In our terms, to illustrate non-linear quantization—using the theory of the (very good) clock as example (very good will mean highly reliable in hereditary terms)—the timing phase (more generally now the spatial phase with a rate governing reaction) is given by a linear isochronous element—pendulum, mass-spring, atomistic element. While this will create a cycling phase, it either will not sustain, or will die down; thus an escapement which injects 'impulsive energy' (the enzyme), carries it over. A variety of phenotypic escapements might do. This governs the direction of evolution.

The contribution of 'practical' Richard Gregory, at this point, is very impressive. We mathematical physical types, prepared to lay out the mathematical and electrical characteristics of the nervous system, and of the emergent patterns of behaviour, may sometimes be casual about one link—the sensory link—and its implied informational data processing. For the species to stay out of trouble—normally to have children and not get eaten—remote 'at-a-distance' processing is desirable. The remote sensory detector (why it arose in the genotype is Waddington's and Maynard Smith's business), the eye, proliferated the whole nervous system, says Gregory. The clue to its origin, likely, is the pervasive dark-light variation, that most common signal. Did this start the higher development? Agreeing with Gregory, we are inclined to say, 'Yes'.

Thus, the fundamental pieces fall into line—obviously not in any well developed sense, but in outline. At this point now it can be (and is being) followed by competent workers clarifying the pieces. At what level description will subsequently break down is not for us to say. However, this era puts the foundation of the living system at every level into quantized limit cycles and their adaptive, emanative evolution.

Some incidental tidbits that concerned us also arose at the conference. In conducting a personal dialectic with Bohm, a thought struck me on a feature common to his presentation and mine, on the nature of creativity. Creativity in humans, to me, meant the establishment of novel emergent cyclic patterns in behaviour. To Bohm, the next instant of time itself was the only 'creative' emanative process. These, I suddenly saw, were the same. The 'creation' of time is the only continuing monotonous thing that is 'created'. In this sense, the 'creation' of man (as a linear chain) is the only act of creation of man to man. Kornacker and I worked this over to the conclusion:

*The creation of time order is to the creativity of dynamic process as the hereditary (i.e. reproductive) sequence is to the creativity of evolution.*

We presented this as one unifying theme.

Kornacker posed one other problem—of the definition of obscenity—which I attempted to dismiss, until he forced a meaningful dialectic. In my terms, a person whose 'body' image (of both internal and external cyclic events, i.e. his entire 'superego') is well composed and placid, will regard any presentation put before him which is too complex in structure and 'jittery' with regard to his 'ego' image (by definition, one's own ego image is stable, all else is moving—when this is upset, the individual is in trouble), as 'obscene'. To a person whose image is multi-dimensional and darting in time structure, almost no image can be obscene. This is the essence of the concept.

By this means we succeeded further in seeing ourselves, seeing others, and seeing the temporal-spatial problem of ideas, and creativity, and discovery in science, and the central problem of the conference.

Finally, Kornacker proposed, in apt summary, a (hopefully non-obscene) structural picture to unify the conference. I leave that summary to him (see p. 321).

# Some Remarks on the Notion of Order

**David Bohm**
University of London

This conference is concerned with the question of whether the development of biology has now reached the point where a coherent over-all theory of the subject can begin to be formulated. In my view, such a theory would very probably have to involve the notion of *order*, in a way that is more fundamental than that in which order now enters into the theories of physics. Indeed, as I shall try to explain, our physical theories are at present in a state of flux, that may lead to radical changes in them, such that current fundamental ideas, based on measure and metric, may also have to be replaced by new ideas, based on order. So order may well be a fundamental notion that underlies both physics and biology, and permits them to be related in a deep and essential way, by making available a common language and concepted structure for the formulation of both.

In regard to this point, I would like to go even further, by emphasizing that order is something that is more fundamental and more universal than most of what has previously been generally regarded as basic in our thinking. This is because order is common not only to physics and biology, but also to all that we can know and all that we can perceive. Thus, there is the order of events in time, the order of cause and effect, and the manifold topological orders that constitute the essence of what is meant by space (e.g. order of inside and outside, right and left, up and down, open curves and cycles, etc.). Without this vast totality of topological orders, there would be no meaning to *measuring* intervals of time and space, nor even to the idea of continuity or discontinuity of these intervals. And then there are also the directly perceived orders of warm and cold, hard and soft, and shades of colour, as well as the tremendous possibilities for orders in the notes of the scale which are the basic content of music. There is the order of words (both temporal and syntactical) that makes communication possible and the order of feelings that is an inseparable part of the meaning of communications (e.g. pleasure and pain, interest and boredom, etc.). Indeed, wherever one looks, whether outwardly at nature, or inwardly at the thoughts and feelings that are the expressions of the operation of the mind, one finds that the essence of things is always in one kind of order or another. Thus, order may well be the basic factor which unites mind and matter, living and non-living things, etc.

Moreover, the notion of order is evidently more fundamental than other notions, such as, for example, that of relationships and classes, which is now generally regarded as basic in mathematics. To illustrate this by an extreme example, one can point out that to establish a new order of society would evidently be more fundamental than to establish new relationships and classes in society. And more generally, wherever two things are related, they are related by being comprehended within a totality of common or similar orders. So some sort of order, either tacit or explicit, is always a kind of ground or foundation that is logically prior to the notion of relationships.

If order is more fundamental than almost any other notion that we can think of, how then can we hope to define it ? That is to say, how are we to arrive at the essence of order, which must, as we have seen, in some ways transcend the whole field of what can be put into words ?

Of course, we cannot possibly obtain a complete verbal definition of order. Rather, we must begin with the fact that everyone already has a vast totality of tacit and implicit knowledge about order. What we can do with words is to 'point to' certain essential features of this tacit knowledge, and thus to bring out explicitly what is already implicit in the whole structure of our thinking and perception.

With all this in mind, I now propose that a good point of departure into this subject is to consider the notion that order is basically a set of *similar differences.* To illustrate what is meant, consider a geometric curve, which is in some way an *ordered* set of points. To describe this order, let the curve be approximated by linear chords of equal length. Then, intuitively, we can see in a general way that to obtain a regular curve rather than an arbitrary set of points the differences in the chords must be similar.

The simplest curve is a straight line. Here the successive chords differ only in position, being similar (and indeed the same) in direction. The whole curve is determined by the first chord. So we can call it a curve of *first order.*

The next curve is a circle. Here the chords differ both in position and in angle, but successive differences in angle are similar (and indeed equal). So a circle is determined by the first two chords and can be called a curve of *second order.*

The next curve is a spiral. Here the planes determined by successive pairs of chords are different, so that the curve turns into a third dimension. However, the differences of angle between the planes is similar (and indeed the same). Thus, a spiral is a curve of *third order* determined by its first three chords.

Evidently we can in this way define curves of higher and higher order, eventually reaching curves of *infinite order.* Among these would be curves so complex and 'tangled up' that everyone would be inclined to call them 'random'.

The question of the meaning of the term 'randomness' has never been answered very clearly. Very often the quality of randomness has been equated with what is called 'disorder'. But if one thinks for a moment he will see that disorder, in the sense of the total absence of any kind of order whatsoever, is both a logical and a factual impossibility. Thus, if an object moves on what is called a random curve, it always moves in some kind of order. At least, after the motion has taken place, one can describe the order of the curve in question and distinguish it from any other curve which follows a different order. (Such a description might for example be made going through the analysis of similar differences, in the way described above.) Of course, the future order of a random curve would not be predictable (which is just another way of saying that a random curve is of infinite order). But it is evidently wrong to identify the totality of all possible order with nothing more than predictability (for example, the subsequent order in a musical composition is not predictable from what came before any given part, and yet it has a real order, considered as a totality).

It would seem then that it is a source of confusion to equate randomness with disorder, or even to say that disorder can exist in any context whatsoever. No matter what happens, it always has to happen in some kind of order, and what we have to do is to describe and analyse the order rather than to avoid the question by calling it disorder.

We have said that a random curve is one of infinite order. Evidently this is a necessary but not a sufficient condition for randomness. The attempt to get a full definition of the sufficient conditions for randomness leads to some very subtle questions, into which we cannot enter here. It may be pointed out, however, that one requirement on a random order is that it must eventually contain every possible kind of sub-order or partial order. In other words, a random order is, in some sense, *open.* In addition, it must satisfy certain statistical conditions on the partial orders (which allow the concept of probability to be applied). A great deal of work remains to be done in clarifying these questions. But a necessary condition to begin this work is that we cease to use the word 'disorder', which blocks the way to thinking about the *order of randomness,* because it formally and logically denies the existence of any and every kind of order that could possibly be conceived of.

We are now ready to consider the question of the *difference of orders.* First of all, it is evident that two geometrical curves have different similarities of their differences, and that this is their essential difference. To avoid confusion of terminology, let us say that there are 'differences$_2$ in the similarities of the differences$_1$', where the term 'differences$_1$' describes differences *in* a given curve, while 'differences$_2$' denotes differences *between* the two curves.

Two such different curves can then be related. The most elementary relationship is a one-one correspondence of their differences. Such a relationship is a set of similarities$_2$ in the differences$_1$ of the two curves. There can also be similarities$_2$ of the similarities$_1$ in the curves in question, as well as differences$_2$ in the similarities$_1$, etc. In this way we can explicate the full subtlety and complexity of the relationship of different curves.

Here it is important to emphasize that logically speaking one can properly relate only things that are different. Indeed, the Latin root of 'different' means 'to carry apart', while the root of 'related' or 'referent' is 'to carry back together'. That is to say, when things are 'carried apart' they will 'refer' to each other, and this is the essence of their relationship. So what is logically prior to relationship is difference and similarity, leading to order.

This is true perceptually as well as conceptually. Thus, in the corner of the eye, we can perceive that things have changed and are different before we know what it is that is different. As we turn the central part of the retina toward the thing that has changed, our perceptions sharpen up to reveal just what it is that has altered. Very probably the brain is thus enabled, step by step, to register the differences and the similarities, giving rise to a perception of order. This is then related to other orders that are stored in memory. In this way we can see what is happening and recognize it by referring it to the order of what is already known.

In discussing the subject of relationship, it is necessary to understand that two kinds of differences are always involved. First there are the constitutive differences, which determine the essence of the order of whatever we are talking about (in the case of the geometrical curve these are the differences in the chords). Then there are the distinctive differences, which determine and define how one order can be distinguished from another, and yet refer to the other through their mutual relationships.

Of course, the constitutive differences can always be related (or referred) to distinctive differences in another order. For example, the chords that constitute a curve can be related to another set of curves constituting a coordinate system or a reference frame (e.g. by giving the coordinates of each part of the various chords). So now we can focus on the distinctive differences between the chords by calling attention to the differences between their coordinates, slopes, etc. But note first of all that to do this we have had to introduce the notion of constitutive differences in the coordinate curves themselves. In other words, underlying each analysis of distinctive differences there must be a set of constitutive differences *somewhere,* to which the distinctive differences can be referred. Thus, we never totally remove the need for constitutive differences, but at best only transfer the constitutive differences to another part of our conceptual structure. But secondly, granting this, we do not in general even reduce a given set of constitutive differences to distinctive differences by relating them to another referential order. Thus, with regard to the geometrical curve, mathematicians have seen the need to define and work out the intrinsic properties of the curve, i.e. those properties that are independent of the coordinate frame to which it is referred (also called the invariant features of the curve). So when one relates constitutive differences to distinctive differences, this is in general merely a descriptive process rather than an explanation of one in terms of the other.

What is needed in this field is to see clearly the difference between constitutive differences and distinctive differences, and then to see how these two kinds of differences are related. Indeed (as has already been pointed out), what is constitutive difference at one level corresponds to a related set of distinctive differences at another level, and so on in principle without limit. For this reason, orders tend to develop into indefinitely extending hierarchies.

To a certain extent, such hierarchies of order are introduced by us as a result of our analysis (e.g. geometrical curves are referred to coordinate frames which can in turn be referred to other frames of finer mesh, etc.). But, more generally, we find that there are certain natural hierarchies of order which reflect not mainly our particular procedures of analysis, but rather the existence of a real structure.

Structure can best be described as a constitutive order of constitutive orders (constitutive order being the result of a set of similar constitutive differences). To illustrate what this means, let us consider the structure of a house. One begins with the bricks, which are similar in size and shape but different in position and orientation. The similarity of these differences of the bricks leads to the order of the wall. The wall in turn becomes an element of a higher order, in such a way that the similar differences in the walls make the rooms. Likewise, the similar differences of the rooms make the house, those of the houses the streets, those of the streets the city, etc.

It is clear that the principle of structure is universal. Thus, the elementary particles are ordered to make the atoms, the atoms are ordered to make molecules the molecules make micro-objects, and so on to the planets, stars, galaxies, galaxies of galaxies. For living matter the molecules are ordered to make the components of cells. These are ordered to make the cells, these the organs, these the organisms, these the societies of organisms. And something similar goes on in perception and thinking. Indeed, even our most abstract concepts form structures in this way (e.g. a set of ordered classes makes a class of higher order, and in turn can be the beginning of a class of yet higher order, and so on without limit).

One of the most characteristic features of structure is that partial constitutive orders can be abstracted from it in such a way that distinctive differences and relationships show themselves naturally. For example, one can abstract a pair of parallel walls, whose distinctive differences and relationships are indicated by another pair of parallel walls that connect the end points of the first pair. In all structures one finds a very rich set of such cross-references of every kind. These cross-references are both inferences from the structure in question and indications that what we are dealing with is in fact a unified totality of structure rather than an arbitrary and fortuitous array of elements. So when we discover a set of data with very rich cross-references in it we try to find a structure, and if we succeed in doing this, we test our assumed structure by observing further cross-references in the data, often of new and hitherto unsuspected kinds.

To carry out this kind of inquiry adequately we need a language that describes order and structure properly. In my view we do not at present have such a language. Evidently the common language is inadequate, because its terms referring to order are extremely vague and confused. Indeed, it is to remedy this situation that I have called attention to the need to consider similar differences, different similarities, the hierarchy of orders, the constitutive and distinctive orders, etc.

One might then ask whether we could not describe orders and structures properly with the aid of mathematical language. However, I do not think existing forms of mathematics are really adequate for this purpose either. To be sure, something along this line is being done in *topology* and in information theory. But in both subjects, what is absent is an adequate notion of order. Indeed, the general mathematical notion of order is now formulated basically in terms of certain relationships. Thus, in what is called a lattice one defines an ordering relation, symbolized by $>$, which has, in essence, the same qualities as the notion of 'greater than'. This relation is assumed to be asymmetric, transitive, and reflexive, and further assumptions are made to allow for the many 'strands' of a partially ordered system.

## Some remarks on the notion of order

The main difficulty with this notion is that it does not readily allow for the basically hierarchical possibilities of order, which lead to structure as an order of orders. In essence, it is order at one level only. As such, it is only a special case of the general descriptions of order questions, as similar differences and different similarities of the differences. Moreover, it does not permit a clear expression of the difference between constitutive differences and distinctive differences. Indeed, the relation > really refers only to distinctive differences. For it tacitly supposes that each element is considered separately as fully constituted, so that the order of elements is *external* to the elements themselves, and refers only to the way in which the elements are distinct from each other and yet related On the other hand, when for example we considered the differences in the chords that make up a curve, we were regarding the order as basic to what the curve is, and not as some purely external descriptive property that was being applied to show how a set of distinct but related points happened to be aggregated in such a way as to be distributed along a given curve.

What is needed is to develop a new mathematics of order and structure. This requires an extensive study, in which one slowly and carefully 'feels one's way' into the subject. It cannot properly be done solely by applying existing mathematics, because the latter does not have the right general structure. To this end it is necessary to formulate a new set of mathematical axioms which treat order and structure as bare concepts. After all, our present axioms do not explicitly define the basic elements in our mathematical thinking (for example, in geometry points and lines are taken as purely abstract words, defined only tacitly by the ways in which they are used). So there is no reason why we cannot introduce new axioms, in which the notions of order and structure, defined only tacitly and not explicitly, are taken as the fundamental points of departure for our thinking.

Indeed, I have been doing some preliminary work on this question. The main difficulty seems to be to develop a new structure of mathematical symbolism that takes into account the hierarchical potentialities of order and that does not tacitly commit one to the view that the world is composed of separate 'elements' whose orders and relationships are external to what these 'elements' are. In addition, it is necessary that the symbolism explicitly differentiates between constitutive differences and distinctive differences so that it will permit the expression of how these two kinds of differences are related in a vast set of cross-references of one aspect of structure to another.

Thus far, both in general terms and with reference to mathematics, we have been considering order and structure largely as static. But in reality, to do this is to abstract from a process of movement and development in which each order and each structure is always becoming different. What is essential to process is not merely that there is a change of order and structure, but that the differences are similar, so that the changes are themselves ordered. In other words, process is an order of change. And, needless to say, even the orders of change can themselves be ordered to form a larger hierarchy of process, which is an order of orders of change.

Consider, for example, the laws of motion in physics. To simplify this discussion we will consider changes taking place in equal but short intervals of time rather than trying to use the infinitesimal calculus. Now, in free space an object moves at uniform speed in a straight line. This means that successive differences in position define segments that are similar (and indeed equal) in magnitude and direction. Thus, the law of motion is just an assertion of similar differences, implying a certain linear order of change. In the presence of forces it is necessary only to go on to the second differences, i.e. to differences in successive segments, whose similarities define the acceleration of the body. Thus, Newton's laws of motion may be stated by saying that similar differences in the applied force always lead to similar differences in the acceleration.

Evidently the laws of physics are expressed in terms of a very simple kind of order, i.e. the order of mechanical motion of a body. In biology we may express the growth of an organism (the phenotype) in terms of a very rich and complex set of similar differences and different similarities in the changes of its various aspects and features. And as we go on to consider the growth and development of intelligent responses of the higher animals and man, we find yet higher orders of similar differences and different similarities.

The need to proceed in this way illustrates a very characteristic feature of process, i.e. that *the breaks or changes in the order of a given process can themselves be the basis of a higher order of process.* (This is the temporal counterpart of how the walls, which are the ordered set of bricks, are themselves the basis of the rooms, which are formed by the ordered set of walls.) Such a possibility can best be seen in music. Thus, there may be a short set of notes in a given order. This order changes, then changes again and again. But all the changes of order form a yet higher order, which constitutes a part of the *development* of the over-all theme. Each order of development itself changes in an ordered way to form a still higher order of development. And to the possibilities of going on with this process there is in principle no limit.

It seems clear that in biological and psychological processes something similar is involved. Thus, we see how on one level the DNA and the RNA function is an ordered way to build and maintain the cell. On the next level, the cells function so as to maintain the organs, the organisms, the society of organisms, etc. Here we are emphasizing not merely the order of orders of static structure, but also the order of orders of dynamic function which is needed for maintaining coherent growth and life itself. A similar emphasis is needed in psychological processes, which reveal an even greater richness and variety of orders of orders of function.

The study of order leads to particularly interesting questions when we consider the process of evolution. Of course, to a certain extent, evolution can be considered to be a set of changes within a given order of process, from one possibility to another. But I would like to suggest that there is another kind of evolution, which is the coming into being of a new and higher order of process. Thus, in music there can be variation on a given theme. But then there can be a basic change of order of the whole theme. And then there can be something yet more—an ordered series of such changes in this theme. This latter order is not only new relative to what was there before, but it is also evidently of a higher order. Likewise, we can think of the evolutionary process by considering not merely a set of variations on a particular kind of structure of organism, but also the coming into being of new orders, along with an ordering of the changes of order in the whole process.

For example, in a certain sense, intelligence enables man to order his physical actions in new ways. Over a short period of time, man's physical actions are, on the whole, not very different from those of the higher animals. But in the higher animals these actions vary in a way that tends to have no particular *intrinsic order* (i.e. an order not imposed mainly by the environment). However, in man, intelligence reveals itself through the ordering of these physical orders in a new and different way. Something similar is also to be observed at lower levels of the evolutionary process. Thus, when cells work together to form an organism, the main new factor is that certain variations of the behaviour of individual cells previously determined fortuitously by the environment are now ordered intrinsically in the over-all functioning of the organism. Indeed, the evolutionary process can be said, in a certain way, to be leading to ever higher degrees of intrinsic determination of the order of lower orders' actions, so that, at least in this sense, it has a kind of direction of development. Perhaps life could be regarded as an early stage in this process, and intelligence as a later stage. In this connection it may well be that even inanimate matter is evolving (e.g.

there may have been a time before which electrons, protons, and neutrons did not exist). Thus, the difference between life and non-life (and between different levels of intelligence) is perhaps not in the process of evolution itself, but rather in the degree and kind of intrinsic order of order which has thus far resulted from the process of evolution.

In discussing the evolutionary process I want to emphasize that the dynamic feature of the process of *ordering* has to be taken as prior to the more static feature of *order.* This process contains two inseparable aspects—the dissolution of an older order and the creation of a new order. Thus, if individual organisms did not die (i.e. undergo a dissolution of their orders) it would be impossible for new orders to come into being in the successive generations. But what is of crucial significance here is not merely the replacement of one order by another. Rather, it is that each change of order (whether in the actions of the individual organism or in the nature of the successive generations of a species) is itself capable of entering into a yet higher order of changes of order. Therefore creation is not just the death of the old organism and the birth of the new one. Rather, its essence is that in it there is scope for the coming into being of ever higher orders of order. And just as order is itself logically and existentially prior to relationships and classes, so the process of ordering is logically and existentially prior to the orders which result from it, and which are created and dissolved in this process.

At this point we come naturally to the question of mechanism. For we are led to ask whether what has been said about the complexity and subtlety of biological, psychological, and evolutionary ordering refers to the basic constitutive differences that make up the order of natural processes, or whether it refers merely to the distinctive differences in these processes, that are perhaps convenient and useful as descriptions, but that have no really fundamental significance.

When one asks such a question one is usually referring tacitly to the argument that on the level treated by physics the whole world is in reality constituted out of some basic kinds of elementary particles, which move mechanically according to certain laws, in a way that is determined by their inertia and by the forces of interaction between them. To be sure, it may well turn out that known particles, such as electrons, protons, and neutrons, are not truly elementary and that they are in fact constituted out of some as yet unknown 'really elementary' particles of a finer nature. Whatever the truth might be in this regard, however, one can assume that the 'really elementary' particles move in a completely determined mechanical way, given in essence by Newton's laws or by some variation of these. This means that the order of motion is assumed to be nothing more than an automorphism, i.e. a movement in which there is a limited field of possible states, and in which each change corresponds to going from one state to another, within this limited field. Thus, according to Newtonian mechanics, the state of a system of particles is completely specified by the position and velocities of each particle at a given moment of time. The totality of the available states is taken to be the whole set of possible positions and velocities open to the various particles. And the movement consists of a process in which each particle goes from one position and velocity to another position and velocity. The law of movement then determines the order in which this automorphism takes place, for each possible state of the system.

If one accepts such an assumption, it follows that the whole order of behaviour of any system of particles is in reality determined completely by the mechanical order of movement of the constituent particles. To be sure, one may find it convenient to group these particles into systems, such as atoms, molecules, cells, organs, organisms, etc. Because the particles interact with each other, the systems can display a sort of 'collective behaviour' in which they 'work together' in a general over-all way as a kind of relatively stable unit on a higher level. As a result, one can simplify things by abstracting from the basic laws a suitable partial treatment of the order in which these systems move. In effect, this partial treatment ignores the complexities of the deeper structure of the systems in question and approximates them as single systems. But in a more fundamental sense one must regard these systems as in essence nothing more than convenient abstractions for establishing distinctive differences that are useful in the description of the general behaviour of particles when they happen to be aggregated into groups. On the other hand, the fundamental constitutive differences would be only in the particles themselves and in their movements.

The whole comparison of the evolution of the order of orders of movement would then be a sort of figure of speech, like the 'average man' of economics. One knows that the 'average man' is nothing more than a purely conceptual idealization, while the individual man and his groups or aggregates are what actually exist. Similarly, one might say that the hierarchy of orders and orders of orders is, in this point of view, a purely conceptual abstraction, while what really exist are the particles and their groups or aggregates.

In biology, such a point of view is exemplified in the assumption that the *entire* behaviour of cells and organisms can be explained, more or less, mechanically in terms of the properties of molecules, such as DNA, RNA, amino acids, proteins, etc. Of course, one realizes that these are made of smaller particles, such as atoms, which are in turn made of electrons, protons, neutrons, etc. But one assumes that the mechanical laws determining the motions of the fundamental particles are such that through the interactions of these particles there arise groupings, systems, or aggregates (such as DNA molecules) which can be treated in a simplified way as relatively stable units, having certain essentially mechanical properties that are known or that can be discovered by further experiment and observation. And, of course, if this assumption is correct, it follows that the whole development of life, intelligence, society, etc., can in principle eventually be explained by referring it to an ever more complete knowledge of the properties of these basic molecules.

Of course, all these conclusions will follow only if it is true that the basic constitutive order of the universe is indeed that of the supposedly fundamental particles and their supposedly mechanical motions. These conclusions would, however, all fall to the ground if it turned out that natural processes cannot, in general, be reduced to mere automorphisms of mechanical order, and that they contain a really creative movement, in which there appear new orders and orders of orders. Therefore the question of whether the basic laws of physics are in fact mechanical or not is of the utmost potential significance in biology.

Now the fact is that physics has, in the past 50 years or so, been making gigantic strides away from mechanism. In my view, the main steps in this direction have taken place in statistical mechanics (to a relatively small extent) and in quantum theory (where the step is really clear and decisive).

As is well known, statistical mechanics has been able to explain the thermodynamic properties of matter in bulk by means of a statistical treatment of the movements of large aggregates of atoms. But in doing this it has been led into certain very confused questions, having to do with the statistical interpretation of the concept of entropy. Usually the increase of entropy (which takes place irreversibly) has been equated with an increase of what is called 'disorder'. As I have already indicated, however, there can be no such thing as 'disorder'. Therefore all efforts to proceed along this line are bound to end up in confusion. And, indeed, one does discover that proofs of the increase of entropy always encounter contradictions and paradoxes (such as those involved in Boltzmann's H theorem). More and more subtle analyses are made to resolve these paradoxes but one generally finds that these attempts transfer the difficulties to other parts of the theory, where they are rather hard to see (like sweeping the dust under the carpet).

In my view, these contradictions and paradoxes arise because one begins with classical physics, which works solely in terms of a simple mechanical kind of movement generating particle orbits that are curves of *second order* (i.e. curves determined by the differences in two successive steps). On the other hand, the notion of entropy is inseparable from that of probability, which is in turn inseparable from that of randomness. And as has been seen earlier, a random curve (a typical case of which is the orbit of a particle in Brownian motion) is a curve of infinite order which cannot be determined by the differences in two successive steps. So there is an unresolvable conflict between the order of classical mechanics and the order of randomness implied by the concept of entropy. By calling randomness 'disorder' one fails to notice that it is in reality some kind of order, so that one is able to overlook the conflict between the two entirely different kinds of order, and thus one falls into confusion.

This confusion is further confounded by a presently current tacit assumption that in all fundamental laws (having to do with basic constitutive differences rather than distinctive differences) the movement must be either a simple curve of second order (determined by two successive steps) or a random curve of infinite order (treated by the laws of probability). In trying to explain entropy in this way we are led to impose both these extremes at the same time, and thus to attempt the impossible. But the conflict can be avoided if we admit the new concept that the basic laws of physics may involve curves of all orders, from the first or second to infinity. The increase of entropy can be explained as a *change of order* of the orbital curve, from one of lower to one of higher order. This is evidently not a mechanical process. So thermodynamics takes us out of the domain of mechanics. (But, of course, we will not be able to get very far in our inquiry into these processes until we develop a new mathematics of order, just as Newton could not get very far in mechanics until he developed the new mathematics of the differential calculus.)

One of the reasons why this question is hard to put clearly is that, in a random curve, the statistical properties become simpler, and thus in a sense can be said to change from a higher order to a lower one (e.g. as the gas molecules move at random, they produce a practically uniform over-all density of matter). Thus, when the entropy increases, the detailed motions of the individual particles undergo an increase of order, to a state that is less symmetric than before (since a low order of order implies a high degree of symmetry), while statistically the system as a whole moves toward a more symmetric state. The tendency to identify the change of order with nothing more than the statistical large scale properties lies behind the conclusion that the flow of heat leads only to a state of greater symmetry. This tendency is made particularly inevitable by the assumption that the individual molecules move toward disorder, i.e. no order at al . Clearly, from this assumption, it follows that the change of order can only be in the statistical properties. But once we recognize that the individual particles also have an order, which is always tending to increase (with a corresponding decrease of symmetry), then we can avoid the paradoxes that arise when the individual particles are assumed to become 'disordered'.

In the light of the above discussion, it becomes clear that the so-called reversibility of the basic laws of physics is only a simplifying abstraction. That is, if we simplify the real orbit by treating it as a curve of fixed order, then the laws will allow any given motion to be carried out in a reverse order. But if the real orbit is characterized by a continual and generally unidirectional change of order (from one of higher symmetry to one of lower symmetry), then the motion is not really reversible.

To a certain extent, a similar kind of irreversibility may prevail in living matter. For the evolution of life resembles the random curve in that it is a process of at least potentially infinite order. But it is different, in that the random curve merely goes through all the possibilities on a given level, in such a way as to give rise to a long-run tendency to statistical symmetry in its over-all structure. On the other hand, living matter tends to evolve hierarchically in an ever-increasing totality of orders of orders, so that, in the long run, it is always going to higher levels of over-all structure rather than to statistical symmetry. So, even though the increase of entropy involves a change from a lower to a higher order of order on the microlevel, this is very different from the change of order involved in the evolution of life.

In discussing the problems of statistical mechanics we have seen how notions have been introduced into physics that tend to move it away from a mechanistic point of view. However, it is only when we come to quantum theory that we see to full extent how far modern physics has departed from its earlier basically mechanical foundations. This departure involves three new aspects:

1. Process is discrete rather than continuous. Thus, electrons are said to 'jump' from one 'orbit' or quantum state to another without passing through intermediate states. What characterizes the 'jump' is the change of what is called the 'action variable' by an integral number of units (one unit of action being measured by Planck's constant $h$). However, the continuous change of action that characterizes classical mechanics is recovered with the aid of the correspondence principle, i.e. in the limit where the change of action contains many units, the discrete

changes can be approximated as continuous and are in correspondence with those prescribed by classical laws.

2. The constitutive order of this discrete process is determined by laws of probability, and not by the classical orbit of curves of second order. However, the correspondence principle enables us to recover classical laws, with the aid of the further statement that in the limit where many discrete steps are taken the probabilistic (random) process of quantum law leads, on the average, to the usual orbits of classical physics or a good approximation (i.e. through the statistical law of large numbers).

3. It is found that electrons, etc., can behave under certain conditions like particles and under other conditions like waves. This phenomenon is often referred to as the 'wave-particle duality'. What it means is that not merely have detailed classical laws broken down, but that the *whole order of movement conceived as the displacement of particles from one place to another has been transcended* (e.g. even the Brownian motion curves of infinite order are inadequate to describe the phenomena of the wave-particle duality).

Now, in classical physics it was known that there is a domain of phenomena (electromagnetic fields) which move in an entirely different order, i.e. that of wave motion. But what has been discovered is that, in processes involving only a few quanta of action, phenomena are observed which suggest that in some sense the same entity follows both the particle order of motion and the wave order of motion. However, a further analysis of the situation reveals that there is an inherent vagueness or lack of complete definition in these orders, which is completely foreign to classical concepts. When the particle order is relatively well defined, the wave order becomes correspondingly vague and vice versa. In other words, there is a reciprocal relationship (implied by the so-called uncertainty principle) between the sharpness of definition of the two opposing orders of motion, i.e. wave and particle. Different experimental conditions then determine which of these aspects is sharper and which is less sharp.

Bohr has formulated this behaviour in terms of what he calls the principle of complementarity. To come to this principle, we first note that while wave and particle orders are both necessary, and therefore complement each other in the full description of the phenomena, they also contradict each other when both are completely defined. In this quantum theory this contradiction is, however, avoided with the aid of an assumption of the kind described above, of inherent vagueness of the orders of wave and particle motions. Bohr assumes that this vagueness is not just characteristic of current theories of quantum phenomena, but that it represents a general principle applying universally. And this assumption is, in essence, the principle of complementarity.

While it appears to be possible to formulate Bohr's view in a logically consistent way, I am inclined to favour another approach, suggested by a comparison of the situation to that obtaining in the statistical explanation of entropy, where we are faced with a contradiction between the one extreme of the simple Newtonian order of motion and the other of an infinite random order. We may perhaps surmise that as there is a whole spectrum of new kinds of order between these two extremes, so there is a whole spectrum of new kinds of order between particle motion and wave motion. I have in fact made some progress toward formulating such orders. But because this work is as yet incomplete, I shall not refer to it further in this talk. Rather, I shall try to stay, as far as possible, within the framework of inherent vagueness of orders that is generally accepted by physicists today for the description of the wave-particle dualism.

Even if one remains within this framework, it is clear that the quantum theory implies a genuinely new order in physical law, radically different from the older order of classical mechanics. One may fail to notice this when one thinks of detecting electrons, protons, etc., by their tracks in a cloud chamber. From these tracks one tends to suppose that one practically 'sees' a little ball moving through the chamber, leaving a set of droplets in its path. But actually one sees only the droplets. The 'little ball' is purely an inference. In the classical domain of processes involving many units of action, this inference is well confirmed as a good approximation. But in the quantum domain one discovers that what we have been thinking of as a 'little ball' also moves in an order resembling that of a wave. So it cannot *really* be a 'little ball'. Therefore, in the interests of clarity of thought it would be best if we would drop the idea of the 'little ball' forever (except as a simplification that is approximately valid in the classical domain). The electron is almost infinitely more complex and subtle in its full behaviour than is any implication of such a model.

Some notion of the subtlety of possibilities open to the electron can be obtained with the aid of a more modern form of quantum mechanics, called 'field theory'. In this theory the starting point is the idea of 'quantum state'. This is defined mathematically with the aid of what is called the wave function. Although this function is rather abstract, certain of its properties can be visualized through the notion of order. For example, the places where the wave function is zero define surfaces. If one analyses these functions in some detail, one sees that different quantum states correspond to different orders in which these surfaces are placed.

Thus, a state of well-defined momentum corresponds to a set of planar surfaces in a simple linear order. On the other hand, a state of well-defined angular momentum corresponds to a set of planar surfaces in a cyclical order, going around a common axis. And, more generally, each quantum state is reflected in some such order.

Now, in quantum field theory we have the further basic factor that a transition is described as the annihilation of an existing quantum state and the creation of a new state. But since each state corresponds to a certain order, as described by its wave function, this implies that all movement is being treated as a change of order. And this change of order is in certain crucial ways similar to what happens in biology, where development of the species proceeds through the death (annihilation) of one organism and the creation (birth) of another.

While we do not wish to suggest that the analogy between electrons and living beings is complete, we do wish to emphasize that it goes far enough to show that physics has really totally abandoned its earlier mechanical basis. Its subject matter already, in certain ways, is far more similar to that of biology than it is to that of Newtonian mechanics. It does seem odd, therefore, that just when physics is thus moving away from mechanism, biology and psychology are moving closer to it. If this trend continues, it may well be that scientists will be regarding living and intelligent beings as mechanical, while they suppose that inanimate matter is too complex and subtle to fit into the limited categories of mechanism. But of course, in the long run, such a point of view cannot stand up to critical analysis. For since DNA and other molecules studied by the biologist are constituted of electrons, protons, neutrons, etc., it follows that they too are capable of behaving in a far more complex and subtle way than can be described in terms of the mechanical concepts.

However, probably because biologists are (tacitly if not explicitly) guided by notions resembling the correspondence principle, they are often able to suppose that these quantum mechanical subtleties are of no significance for systems as large as those studied in the inquiry into the genetic structure of cells. Nevertheless, it must be emphasized that, from the point of view of quantum physics, current experiments in this field are extremely crude. Therefore they can be counted on to reveal only the grosser features of what is happening, while finer points that may be of crucial importance can very easily be overlooked.

Even in physics, similar situations have arisen. Thus, one finds in the phenomena of superconductivity and superfluidity how quantum properties can be significant even on the scale directly observable with the naked eye (a scale that is obviously much larger than that involved in the behaviour of DNA molecules). What is striking here is the appearance, even on the large scale, of a kind of stability of certain states of motion, which is made possible only by the fact that action is discrete and quantized. Some physicists (notably Schrödinger) have suggested that the genetic process is made stable because the relevant molecules (such as DNA) are also in well-defined quantum states. On the other hand, without taking such quantum properties into account, it is practically certain (as emphasized by Pattee) that we cannot understand the stability of transmission of genetic characteristics. And if this is the case, it is extremely likely that we will miss certain key aspects of the process when we ignore the quantum properties of the molecules, by treating them as if they were nothing but large-scale classical objects.

With regard to this point, it is important to note that the whole subject is beset with some very serious problems. Thus, the current quantum theory treats the change of quantum states of molecules basically as a random process (of infinite order) with probabilities determined by certain mathematical quantities which are called 'matrix elements' and which can in principle be calculated by solving Schrödinger's equation for the relevant system. Now, on the basis of certain fairly reasonable assumptions about these matrix elements various physicists (notably Wigner) have demonstrated that the probability that a molecule (such as DNA) could engage in a self-replicating process without a basic change of structure is essentially nil.

In a way, this brings us to the same conclusion as that which tends to be held intuitively by biologists, i.e. that in the DNA molecules, quantum mechanical subtleties will be lost in the effects of random motions, so that the system can be treated essentially classically. However, if one reflects on this for a moment, one will see that this leaves us with the serious problem that the underlying physical laws almost certainly imply an unstable behaviour in the replication process, rather than the observed stable behaviour. Does this not lead us to the notion that (as suggested by Wigner), in some way, the laws of physics may themselves be incomplete?

This brings us to the further question of how universal the current laws of physics actually are. With regard to this question, we know even now that our notion of the nature of the so-called 'elementary' particles is extremely inadequate. Thus, it has been found that these particles can be created, destroyed, and transformed into each other. In addition, a large family of similar, but unstable, particles has been discovered which has been classified phenomenologically in a complex set of interrelated orders. It already appears to many physicists as if the beginnings of an entirely new order of natural law were being revealed, in which the particles would be like the flowers on a carpet pattern, while there would be something as yet new and unknown, which corresponds to the woven structure of cords that constitute the carpet. So when we analyse the world as if it were made of particles, this might be similar to analysing a carpet as if it were made of the flowers, that can be abstracted from its patterns. Such an analysis might give certain correct results and yet be very misleading when applied too broadly. Similarly, it may well be that the whole structure of physics is inadequate and misleading when extended too far into the processes of living matter.

Of course, we do not at present know whether the new order that must underlie current physical laws will be significant in biology or not. Nevertheless, the fact that the whole order of basic physical law is currently in the process of radical transformation may perhaps suggest to us that none of the inferences drawn from these laws are iron-clad certainties applying with absolute and unlimited universality. Therefore, even before we know just what form these new laws will take, we may perhaps be led to allow ourselves to question the completely universal validity of some of the basic assumptions underlying current physical theories.

When we look at the quantum theory, we see that the assumption that is on the weakest footing is that of the *random order* into which transitions between successive quantum states are supposed to take place.

First of all, there is the fact that the very idea of randomness has been confused by equating it with the impossible and meaningless concept known as 'disorder'. So the first step is to try to clear up our thinking about the randomness of quantum processes.

As in the discussion of classical statistical mechanics, we are led in doing this to assume that in a series of quantum processes (i.e. 'jumping' from one quantum state to another) there may be an order, defined, for example, by a suitable analysis of similar differences in the steps. In a typical case, the process may have a very high order of order, which would, in certain ways, approximate the quality of randomness. The tendency of the degree of order to increase in a unidirectional way would, as in the classical theory, be the explanation of the increase of entropy. And as happened classically, the process would be reversible, only if one simplified the theory, by abstracting the situation in which the order of order was more or less fixed.

It follows from the above discussion that a series of quantum jumps may not have a completely random order (with probabilities given by the matrix elements computed in accordance with the quantum theory). But does this conclusion not contradict the vast body of experimental fact, which has thus far confirmed the inferences drawn from the quantum theory ?

If one looks carefully at this body of fact, one discovers that the randomness of a series of quantum jumps has actually been tested only in a few limited contexts, and then only to a rather limited degree of approximation. A typical test would be to measure some property of a particle and then, a little later, to measure either the same property or another property of this particle. Such observations as have been made thus far involve rather long time intervals between successive measurements. If one reflects a moment, one can see that if one allows a long time to elapse between successive measurements it becomes difficult to distinguish between a process of high order and a random process. To make such a distinction it is necessary to have a series of successive measurements on the same system that follow each other in as short a time as possible.

New experiments are therefore needed here. Recently, preliminary efforts to set up such experiments have been begun by some of my colleagues of Birkbeck College. However, it must be emphasized that the technical difficulties in the way of such a programme are quite considerable, so that one must not expect quick results in this field.

We see, then, that even in physics, quantum processes may not take place in a completely random order, especially as far as very short intervals of time are concerned. But after all, molecules such as DNA are in a continual process of rapid exchange of quanta of energy with their surroundings, so the possibility clearly exists that the current laws of quantum theory (based on the assumption of randomness in *all* quantum processes, whether rapid or slow) may be leading to seriously wrong inferences when applied without limit in the field of biology.

Of course, unless the situation in living matter differs in some way from that in non-living matter, then, although there may be something non-random in the order of physical processes, the quality will nevertheless tend to be lost over long intervals of time. But we have already admitted that, even in inanimate matter, the order of a process is not fixed, and can change, for example, when the entropy increases. It is evidently possible to go further and to assume that, under certain special conditions prevailing in the development of living matter, the order could undergo a further change, so that certain of these non-random features would be continued indefinitely. Thus, there would arise *a new order of process.* The changes in this new order would themselves tend to be ordered in a yet higher order. This would lead not merely to the indefinite continuation of life, but to its indefinite evolution to an everdeveloping hierarchy of higher orders of structure and function. The situation in physics could then be compared to the pattern of sound or 'background noise' which has momentary fragments of simple and symmetric order but a long run tendency to approximate a random order of sounds that gives an over-all impression of statistical symmetry or uniformity. On the other hand, the situation in biology would then be more like the pattern of sound in music, which is ordered in a hierarchy that is in principle capable of further development without limit.

Because of our scientific training, we may at first find it hard to accept such an idea. For we have been heavily conditioned to the belief that the higher orders of nature are determined completely by the lower order mechanical motions of the particles which constitute the complex structures that we meet in everyday life. But, after all, this is merely a rather poorly tested assumption. As yet, it is in fact impossible to exclude the contrary assumption that in some crucial ways the higher order features of natural laws are as 'fundamental' as are those features referring to the movements of electrons, protons, atoms, D N A molecules, etc. Certainly, to fail even to entertain such notions, at least in a provisional way, would constitute a kind of dogmatism. Such dogmatisms could very easily put 'blinkers' on our field of mental vision that could prevent us even from looking for any phenomena that do not fit into current mechanical hypothesis. In particular what it prevents us from looking for are phenomena capable of showing whether, the essence of natural law is the hierarchy of orders that has been described here or whether it is some fundamental level of mechanical order.

What kinds of experiments and observations does this point of view suggest ? Naturally, one can say little in this regard until we have, to some extent, digested the new structure of ideas that it implies. Nevertheless, one can even now make a few preliminary suggestions.

Firstly ( as is necessary in physics ) one has to inquire very carefully into the randomness of biological evolution, both for the genotype and for the phenotype. Of course, it is recognized by all that natural selection constitutes a non-random feature of the evolutionary process. But the further question is that of whether *all* non-random features are of this type, i.e. essentially external to the basic processes that constitute the living organism itself. Indeed, within such an organism it is now generally assumed that evolution is due entirely to purely random mutations of the genes. The real question here is that of finding whether this process is in fact purely random or not.

One observation that could be relevant would be to trace a series of successive mutations to see if the order of changes is completely random. In the light of what hasbeen said here, it is possible that while a single change (or difference) may be essentially random relative to the previous state of a particular organism, there may be a tendency to establish a series of similar changes (or differences) that would constitute an *internally ordered* process of evolution.

In this regard, it is possible that there is an important difference between an almost stable kind of species, in which mutations tend to be both slow and very nearly random, and a species in the process of transition, in which mutations tend to be fairly rapid and strongly directed in some order, in the way described above. Of course, most species that we can now observe will tend to be almost stable, having passed through their rapid and strongly directed transitional phases long ago. So it is unlikely that a cursory inspection will immediately show up a species that is in an easily observable process of transition ordered in some well-defined direction. Therefore, in most cases that one is likely to meet, the deviations from randomness will very probably be small. However, there may be a few cases of unstable species, in which a series of mutations can be found, which is in an order that is appreciably different from that of randomness.

If such deviation from randomness could be found, this could have very far-reaching consequences. For it would imply that when a given type of change had taken place there is an appreciable tendency in later generations for a series of similar changes to take place. Thus, evolution would tend to get 'committed' to certain general lines of development. But because even order of change can itself change, there could be a higher-order tendency to alter the 'line' of evolution and to start on new lines. Of course, survival in the total natural environment would ultimately decide which lines could be sustained. Nevertheless, an entirely different principle of internal order would determine both the nature of these 'lines' and how they would tend to vary and transform into other 'lines'.

More generally, what is needed, in physics as well as in biology, is to perceive the existing facts anew, in the light of the notion of order, and of a hierarchy of orders of orders (e.g. one may try to see where new orders of order have come into being in evolution). Such perception will evidently tend to lead us to ask new questions in our scientific research. And, as is well known, it is as important to ask the right kind of question as it is to find the answers to the question by observation and experiment. Thinking within a fixed circle of ideas tends to restrict the questions to a limited field. And, if one's questions stay in a limited field, so also do the answers. Thus, nature itself is apparently confirming our assumption that the general framework of current ideas is in principle complete and exhaustive, requiring only detailed development before leading to a full understanding of everything in the universe.

In reality, however, it is only we who thus tether ourselves so that we never wander too far from familiar fields that seem to be safe, secure, and rewarding. Nevertheless, we can, at any moment, cut our tethers, by considering the available phenomena in the light of new general notions, such as those of order. We may fear that we will in this way be led to go out of our familiar rich pastures and to lose ourselves in the unknown immensity of trackless deserts and mountainous wildernesses. But it is not at all unlikely that we can in this way come upon much vaster fertile areas than those which we have known thus far. At least, it seems reasonable that some of us should engage in this kind of exploration, and that not everybody should spend his whole life in extensions of the field of the known.

# Further Remarks on Order

**David Bohm**
University of London

As a result of a very fruitful dialectic resulting from the interchange of points of view at the Bellagio Conference, I now find myself in a position to extend my original remarks on the general notion of order in a considerable number of ways.

▸ *On Metaphysics.* I think the most important aspect of the interchange is the emergence of a common realization that metaphysics is fundamental to every branch of science. Metaphysics is not a well-defined field of study, a single, basic foundation on top of which we erect a towering structure of physics, chemistry, biology, psychology, sociology, and so on; but, rather, something that pervades every field, that conditions each person's thinking in varied and subtle ways, of which we are often not conscious.

Metaphysics is a set of basic assumptions about the general order and structure of existence. Whenever we say '*All* is X' or 'X is *basic*' or 'X is *always* (or *never*) true' we are expressing a metaphysical position. Thus, various ancient Greek philosophers enunciated views, such as '*All* is fire . . . water . . . flux . . . atoms', etc. Scientists later assumed that *all* is a universal mechanism, following Newtonian laws. Then came assumptions, such as '*All* is to be described and predicted by suitable structures of mathematical formulae (such as those of quantum mechanics)'. This Conference heard yet other metaphysical positions exposed, such as: '*All* is to be understood as a structure of automata, expressed in terms of semi-groups' and 'What is *basic* to *all* life is a genetic process, in which changes in the genotype are *always* fortuitously related to the experiences of the phenotype and in which these changes will survive *only* if they are favourable to continued propagation of offspring in the existing environment.'

It seems clear that everybody has got some kind of metaphysics, even if he thinks he hasn't got any. Indeed, the practical 'hard-headed' individual who 'only goes by what he sees' generally has a very dangerous kind of metaphysics, i.e. the kind of which he is unaware (e.g. 'You can *never* change human nature'; 'There must *always* be wars', etc.). Such metaphysics is dangerous because, in it, assumptions and inferences are being mistaken for directly observed facts, with the result that they are effectively riveted in an almost unchangeable way into the structure of thought. What is called for is therefore that each one of us be aware of his metaphysical assumptions, to the extent that this is possible.

One of the best ways of a person becoming aware of his own tacit metaphysical assumptions is to be confronted by several other kinds. His first reaction is often of violent disturbance, as views that are very dear are questioned or thrown to the ground. Nevertheless, if he will 'stay with it', rather than escape into anger and unjustified rejection of contrary ideas, he will discover that this disturbance is very beneficial. For now he becomes aware of the assumptive character of a great many previously unquestioned features of his own thinking. This does not necessarily mean that he will reject these assumptions in favour of those of other people. Rather, what is needed is the conscious criticism of one's own metaphysics, leading to changes where appropriate and, ultimately, to the continual creation of new and different kinds. In this way, metaphysics ceases to be the master of a human being and becomes his servant, helping to give an ever changing and evolving order to his overall thinking.

The proper role of metaphysics is as a metaphor which provides an immediate perceptual grasp of the overall order and structure of one's thoughts. It is therefore a kind of poetry. Some 'hard-headed' individuals may object to bringing such 'poetry' into science. But, just as Molière spoke of the man 'who talked prose all his life without knowing it', so even the practical man is 'speaking poetry all his life without knowing it'. The point that I wish to emphasize here is that all of us will think more clearly when we frankly and openly admit that a lot of 'hard-headed common sense' and 'factual science' is actually a kind of poetry, which is indispensable to our general mental functioning.

▶ *The Metaphysics of Process.* The basic metaphysics that I am now considering is one in which *process* is fundamental. I therefore suggest that one entertains the notion 'All is process'. That is to say, 'There is *no thing* in the universe'. Things, objects, entities, are abstractions of what is relatively constant from a process of movement and transformation. They are like the shapes that children like to see in the clouds (e.g. horses, mountains, buildings). Actually, the clouds are an aspect of the movement of air, the condensation of water vapour, and such. The forms that we see in them have only a certain relative stability. Rocks, trees, people, electrons, atoms, planets, galaxies, are also to be taken as the names of centres or foci of vast processes, extending ultimately over the whole universe. Each such centre or focus refers to some aspect of the overall or total process, which is relatively stable (i.e. which has a certain tendency to 'survive'). Some things (like clouds) last a short time, while others last much longer. But fundamentally (i.e. metaphysically) I am assuming that nothing lasts for ever. This assumption cannot be 'proved' definitively. However, it has evidently never yet been falsified in experience, experiment, or observation (i.e. nobody has ever yet discovered anything that is permanent).

Now, the notion of process is evidently practically empty of content until we can say something about its order. Generally speaking, there are three levels in which this order can be discussed. These are:

(i) Quasi-Equilibrium Process
(ii) Dynamic Process
(iii) Creative Process

As a rule, we tend to begin in a situation close to equilibrium, which enables us to recognize certain relatively static, or constant, features of process. We give these features names, and are thus led to regard them as stable objects or entities (e.g. as we can do when looking at clouds). Then, as we see that these features are changing and transforming, we seek to explain their relative stability in terms of a dynamic process of interaction of some basic entities (e.g. the shapes of the clouds are the results of movements of molecules of air and water). Still later, we come to the notion of a creative process, in which there are no basic objects, entities, or substances, but in which all that is to be observed comes into existence as a certain order, remains relatively stable for some time, and then passes out of existence (e.g. as physics now explains the movement of electrons through annihilation of existing orders and creation of new orders). In the metaphysics of process, creation and transformation of order is always taken to be the deepest and most fundamental account of the laws of a process.

▸ *The notion of structure-function.* Although the notion of creative process is taken here as basic, it is necessary to have a language, in which the relatively stable aspects of process can be given a fairly detailed and precise description. Such a language is needed, in order that our communications concerning process shall be capable of being given a certain relatively unambiguous kind of content. It is here that the notion of structure-function becomes relevant.

The notion of structure has already been explicated in my earlier talk as 'Constitutive order of constitutive orders'. To extend the precise specification of structure to the case in which change and transformation are relevant, we are now led to consider that particular kind of process which is known as *function.*

In general terms, one can say that function is a certain kind of ordered change of structure [1]. Each function can be regarded as having an 'input' consisting of a certain range of possible structures. The result of a particular function is to transform any given kind of input structure into some corresponding 'output structure'. A function of the stomach is to transform undigested protein structures into simpler 'digested' structures. A function of the brain is to 'digest' informational structures in a similar but different way.

The above is an extension of the mathematical notion of function: to each value of the 'independent' variable there is a corresponding value of the 'dependent' variable. But, evidently, even this extension of the notion of mathematical function contains only a part of what is more broadly meant by the word 'function'. For, usually, we think of each function as having a role, an aim, or an end, in something that goes beyond the field of that function. For example, suppose that a man is doing a job in the role of a government 'functionary'. This activity evidently has a part to play in some broader set of functions, which evidently extend up to include the whole government. And the function of the whole government is, in some sense, to help regulate the overall order of society.

What is the function of the overall order of society ? Presumably, it is aimed at creating conditions in which individual human beings can live happy lives. But what then is the function of happiness ? Is it to enable people to be 'better citizens' so that they can properly fulfil the functions in their respective jobs ? If this were the case, we would be going around in a circle, in which no sense could ultimately be made of what is happening. The existence of society would then seem to be the mere result of the arbitrary and fortuitous coupling of functions (e.g. as it is said of some: they eat to live and live to eat).

The above discussion illustrates the general feature of function: that it must ultimately be referred to something beyond the field of function if it is to be properly understood. In the case of society, the ultimate aim of all social functions may be taken as happiness, meaning by this the harmonious ordering of life, both individually and collectively. Of course, others may disagree with this and say that the ultimate aim is to express the glory of the state, or of God, or to realize an ideal. But all these are also beyond the field of function.

A similar set of questions arises in the study of biological evolution. In all forms of life we observe highly coordinated, integrated sets of functions. But to understand these we must appeal to what is beyond the field of function. In earlier days, people attributed all this to the creative action of God. Current metaphysics appeals instead to survival value as the trans-functional feature to which all biological function has ultimately to be referred. But when one inquires more carefully, one finds that it is not possible to specify just what is meant by survival value without bringing in something beyond the field of function. For example, the older Darwinian phrase 'survival of the fittest' could have been taken to mean survival of those organisms that best harmonized with (*fitted* into) the prevailing environment. Without some notion equivalent to harmony, fitness, suitability, viability, or the like, the notion of survival becomes largely empty of content, being indeed a mere tautological statement that those forms of life that continue indefinitely to produce offspring from one generation to the next are the ones that will survive.

Such metaphysical difficulties with the notion of function are evidently not restricted to the study of biological evolution or of the organization of society. Indeed, the very word 'organization' contains a tacit reference to what is beyond the field of function. For example, the difference between a mob and a group of people working together on a constructive job is that the latter is so organized that its functions are harmoniously directed to an end determined beyond the whole field of the functioning group, while the former has no such harmonious ordering of its actions.

Physics has traditionally been a science which was supposed to be free of such trans-functional references. But this freedom was always more apparent than real. Thus, even if we restrict ourselves to Newtonian mechanics, we are inevitably led to ask: 'Why do the various parts of the universe function in just the order implied by these laws and not in some other order?'

That is to say, is the form of Newton's laws a contingency, that depends on something beyond these laws? If so, we are led back on the endless search for some ultimate ground which must eventually bring us, as has been seen, beyond the field of function. And if, instead of God, it is a 'natural necessity' (i.e. something that could not be otherwise, because of what the world *is*), then we are already appealing to something outside the field of function. Moreover, we are doing so in a very unclear way because there is, in fact, no means of knowing that the world *is* such that Newton's laws are inevitable. Indeed, what such a statement amounts to is that this is what we believe to be true in an absolute sense. And, evidently, believing something to be absolutely true is an act of faith that is outside the field of function. In one way or another, therefore, deep inquiry into the foundations of the laws of Newton must sooner or later carry us beyond the field of function.

The relevance of such trans-functional questions was demonstrated rather sharply when the whole Newtonian structure was overturned by subsequent developments in physics in such a way as to show that the approximate validity of Newton's laws is in fact contingent on the satisfaction of certain conditions (smallness of velocity relative to that of light and largeness of action relative to Planck's constant). Moreover, in the actual development of theories of relativity, cosmology, and quantum mechanics, scientists seem to have been forced to make wide use of further trans-functional criteria, such as 'simplicity' or 'elegance', which have indeed played a crucial role in determining both the form and the content of such theories. In addition, because of the use of probabilistic laws in various fields (statistical mechanics and quantum theory), trans-functional notions implicit in the words 'disorder' or 'randomness' have begun to be brought into the very foundations of physics. As far as one can tell, people mean by the word 'disorder' only that what actually happens fails to harmonize with their tacit conventions concerning what is supposed to constitute order.

Indeed, in all sciences, trans-functional notions are being introduced through largely tacit metaphysical assumptions of various kinds. One of the most common of these assumptions is contained in the distinction between what is fortuitous and what is not. Now the word 'fortuitous' includes the notion of 'randomness' but goes beyond this notion in the sense that two different orders, each perfectly determinate in itself, may have no essential relationship between them. For example, if one assumes that the genotype is fortuitously related to the phenotype, this allows causally determined changes in the genes (e.g. by means of certain chemicals). But these changes are then not significantly related to the basic order and structure of the phenotype.

One cannot fail to note here that the definition of the word 'fortuitous' is inseparable from related words, such as 'essential', 'significant', 'relevant'. All of these words bring in notions going beyond the field of function. At present, our notions of this kind are largely tacit and unconscious. So, in effect, we are introducing all sorts of arbitrary (and indeed fortuitous) elements into our thinking by using the word 'fortuitous' in an uncritical way. It is incumbent on all of us to pay very careful attention to what we really mean by such words. In doing this, we may well open the door to fruitful new lines of research.

▶ *Are harmony and conflict merely subjective judgments ?* As one goes into these questions one cannot fail to notice that trans-functional notions all tend ultimately to involve some kind of tacit distinction between harmony and conflict. But here one may begin to feel uneasy, because of the operation of the widespread belief that this distinction is merely a subjective judgment, whereas it is felt that science ought to be based, as far as possible, on objective fact. Thus physics deals with properties, such as mass, length, time, charge, etc., which are supposed to exist 'out there' independently of human beings, while qualities like harmony and conflict, beauty and ugliness, are supposed to exist only in the eye of the beholder.

I regard it as very important to question this division. Actually, the concepts of length, time, mass, charge, have been created by man. A few thousand years ago, nobody felt that these qualities are what is 'out there'. It is true that they are creations that in some way reflect a reality beyond man's thoughts. Nevertheless, man sees them in nature, as he can see beauty or ugliness.

As a result of our history of technological development, we have many technical words describing all sorts of functions, but few words to describe trans-functional qualities, such as harmony and conflict. Because we can so easily talk about function, and find it so hard to talk about what is behind function, we readily slip into the habit of assuming tacitly that only function is relevant in science. It is therefore necessary to develop the non-functional aspects of our language to correct this imbalance.

In order to do this we can begin by considering a few simple cases where harmony and conflict evidently have more than a purely subjective significance. Thus, suppose that we have motor cars moving along a road. These are co-ordinated in some kind of order, in the sense that they do not collide and destroy each other. Now, if this road intersects another road with a similar stream of cars, the two streams will not in general be coordinated. Rather, they will be only fortuitously related, with the result that they will tend to collide and clash, destroying each other in an order that is not harmony but conflict. With the aid of traffic signals, these two streams can be coordinated harmoniously. Similarly, in an organism, the growth rates of cells have to be coordinated harmoniously. When there is a cancer, this coordination breaks down and there is a state of conflict, which ultimately destroys the organism. In all these cases and many other similar ones it can be seen that we are actually able to distinguish harmony and conflict in a factual way, but that because there is no room for such a distinction in our scientific language we do not believe that it has any significant role to play in the development of our basic theories. (In other words, our tacit metaphysics is that harmony and conflict have no metaphysical relevance whatsoever.)

The importance of these considerations to biology is evident. For example, we can see that survival of an organism depends on two levels of harmony: (a) within the organism itself, and (b) in the relationship between the organism and its environment.

At first sight it may seem that to see this doesn't really help the biologist with his actual work. But, in my opinion, this would be a short-sighted view. For a change in metaphysics can make one alert and attentive to possible new lines of development, which may eventually open up new fields of research.

Then again, there is, as I have indicated earlier, a wide field of investigation to which one can be led by inquiring carefully into the meaning of the word 'fortuitous'. For in the understanding of how this notion is to be limited, one discovers one of the main factors that make real creation possible. (It is not the only one, however.)

Here I can make use of the example from the work of Piaget, in his observations on the development of the intelligence in infants and young children. He cites cases in which children learn to draw by first simply moving the pencil in what appears to be a 'random' or 'fortuitous' way which presumably expresses the ever-changing 'internal' state of the child. But if the child is looking at an object (e.g. a rectangle), this internal state will be influenced by such 'impressions' from the outside. As a result, the line ceases to be 'fortuitous'. Instead it begins to show a vague resemblance to the rectangles. At a certain age the child is able to notice this resemblance and to have a persistent interest in it. As he pays attention to what is happening, the internal 'impression' of the rectangle is probably sharpened up, and this 'expresses' itself by an 'improvement' in the resemblance of the line to the rectangle [2]. At first corners are introduced, and later the lines are straightened out, made perpendicular to each other, and so on. And this is potentially only the beginning of an endless movement, in which can emerge, in principle at least, a very great artist. What is crucial here is that the initially fortuitous relationship of the 'inner' state ('expressed' by the variations in the line) and the 'outer' perceptible structures is steadily transformed. This transformation from fortuitous to necessary is typical of a wide range of creative processes. (There may also be an inverse process in which certain orders, previously necessary, become fortuitous.) In such processes a genuinely new order is brought into existence.

This notion of creation is potentially relevant in all science. In biology, it might help to throw light on the origin of life and of intelligence. In addition, it suggests the need for more careful attention to the relationship between genotype and phenotype (already emphasized by Waddington on other grounds). For, evidently, between the genetic material and the whole organism is a vast hierarchy of structure and process on many levels. As the organism grows, and as it maintains itself, there is room on all these levels for transformations from fortuitous relationships to necessary relationships, and vice versa [3]. Careful attention to these transformations may well be a key to understanding what kind of harmony is really involved in the survival of a species.

However, I would like to extend this notion of the fortuitous and the necessary to the whole of existence. In other words, since even the so-called 'elementary' particles of physics are now known to be created, annihilated, and transformed, there is no known factual reason that would even favour the postulation of an ultimate 'bottom level'. Rather, all that is known seems to fit very beautifully into the metaphysics of process, assumed to be ordered hierarchically, in the general way I discussed in my earlier talk.

Each level or order of the hierarchy is assumed to contain necessary aspects and fortuitous aspects. In the overall movement within this process, the fortuitous can transform into the necessary and the necessary into the fortuitous, thus making possible the creation of new orders (along with the ending of old orders). In this way we come to consider an 'open-ended' universe, with neither 'bottom' nor 'top' levels. Indeed, even the level structure is itself always transforming, although this tends to be more nearly constant than any of the 'sub-features' of the various levels (particles, waves, organisms, etc.). But fundamentally the basic principle is: Nothing is permanent. Change is what is eternal. And each feature of the order of change is itself eternally changing. These changes of change in turn constitute new features, which lead on hierarchically to the totality of the process.

In its totality, this process is too vast and subtle to be encompassed by the measure of man. Man thus necessarily abstracts partial and relatively stable aspects. (But as has been indicated, these too are always changing.) It cannot be too strongly emphasized that, in the first instance, this abstraction is creative. That is to say, the process as a whole makes an 'impression' on the 'inward' side of man and this is somehow 'expressed'. By becoming aware of how the expression is related to new perceptual impressions, man begins to learn to abstract from the practically infinite complexity of the originally 'fortuitous' environment just those features which will ultimately prove to be 'relevant', 'essential', 'significant', 'basic'. Eventually, any particular line of abstraction tends to become fairly habitual and mechanical. Thus one tends to lose sight of the creative origin of our abstractions (both in early childhood of the individual and in the distant past of human society). But real perception in any field (whether scientific or artistic) demands the same kind of fresh creative abstraction that the young infant is carrying out when he is beginning to come into contact with the world.

▸ *Is nature more like an engineer or an artist ?* The whole question of the relationship of the field of function to what transcends function can be brought out very nicely by asking the question: Is nature more like an engineer or an artist ?

Of course, the whole notion of comparing nature in this way is only a metaphor.

Nevertheless, it is a useful one. For I submit that, tacitly, most scientists have been treating nature as if it were either an engineer or the work of an engineer.

This trend began to be taken very seriously when, in Newtonian times, the universe was compared with a vast clockwork, originally set in motion by God (who, according to some, also made occasional repairs). Nowadays, it tends to be the computer designer who replaces the clockmaker. But evidently the basic principle is not altered. In all sciences, including biology and psychology, inquiry is founded mainly on an analysis of functions. This is, in essence, the basic approach of the engineer who wants to design a system of functions, with a definite utilitarian end in view. One could perhaps say that in the field covered by biology, nature's utilitarian aim has been assumed to be the production of species of organisms that survive.

Now, if one considers the process of creation as I have been describing it, one sees that it is much more like that of the young child learning to draw and starting on the way to becoming a great artist than it is like that of the engineer who wants to design a system of useful functions. Each 'centre' of process can take in 'impressions' of the rest, and 'expresses' these in its 'environment'. These 'expressions' can now come in once again as new 'impressions', thus starting a cycle of process in which each 'centre' can 'assimilate' its 'environment' and at the same time transform this 'environment' to 'suit' or 'fit' what is 'within'. I suggest that all existence is of this general character, but that living, intelligent beings carry this process to much higher orders than is done in inanimate nature (e.g. the particle, as centre, 'expresses' itself in its environment through its 'field', which also carries into it 'impressions' of the whole universe).

If this comparison is at all valid, might we not also reasonably expect that nature's creations quite generally tend to move towards overall harmony, which is *felt* by us as beauty in our immediate perceptions and which is *analyzed* intellectually as coherent and rationally ordered function ? Thus, we are not surprised that almost anything to be found in nature exhibits some kind of beauty both in immediate perception and in intellectual analysis.

Of course, an underlying harmony, first sensed as beauty, tends to make for survival but, in general, does not guarantee such an outcome. It would clarify this issue if we considered nature to be largely neutral on the question of survival of any particular creation. (Indeed, when a real artist has finished a piece of work he doesn't care much what happens to it, as he has already started something else.) Anything that survives will of course have to have a certain harmony and will therefore be felt to have a corresponding beauty. But each type of harmony leads to survival only in limited conditions. Thus, a snowflake melts at room temperature, while a man dies at temperatures low enough to permit a long period of 'survival' for the snowflake.

Such a change of metaphysics becomes particularly significant when it enters the study of human beings. After all, it makes no sense to say that one lives merely in order to function, or even that one lives merely in order to survive, or to produce offspring that survive. A human being cannot do other than seek to live in harmony and beauty, without which survival has no value. Those who cannot do this will seek delusory substitutes in drugs and in exciting stimuli provided by various forms of violence. The same holds true for society. Any society which puts mere survival as the supreme value of life is already on the way to collective decay and the violent ending of life for many of its individual members. Survival makes sense only in a broader context, just as the engineer's organization of functions makes sense only when the ultimate aim of this function is beauty, harmony, and a creative life for all.

▸ *On the self-regulating hierarchy of process.* In the whole discussion, one theme that was common to almost everyone was the study of self-regulating systems in engineering processes, in computer structures, in rhythms of physico-chemical change at many levels of living organisms, in perception, in brain function, and in many other contexts. It therefore seemed worth while to extend the notion of hierarchic order to that of a self-regulating or self-governing process.

In this connection it is useful to recall that '-archy' means 'government'. So the notion of hierarchy is intertwined with that of government in a fundamental way. It is thus hardly accidental that each government tends to be organized hierarchically. In addition, we have to consider the relationships of different 'sovereign' governments which, of course, tend as a rule to approximate anarchy. To avoid the deleterious effects of such anarchy, there are efforts to organize governments of governments (e.g. Federation of the states of the USA, or perhaps ultimate world federation). Thus one obtains a series of levels of parallel governments, organized in super-hierarchies (i.e. hierarchies of hierarchies). Such a hierarchy of hierarchies might in principle extend onward and upward without limit.

All of this is not intended merely as interesting speculation. Rather, I refer to it to emphasize the vast order that is implicit in the full concept of hierarchy. It is something like this sort of order that is probably involved in a living being and even more in an intelligent being. Perhaps even the electrons and protons of an inanimate nature are also organized in some sort of very complex self-regulating hierarchy. The reason I suggest this is that in a metaphysics based on the notion of process we cannot take the continued existence (survival) of any particular aspect for granted. Because the basic order of process is eternal change of everything, we can no longer appeal to the mechanical notion that certain basic objects, entities, etc., 'simply exist' with constant and invariable properties. Rather, the survival of any particular thing, however 'basic' it may be thought to be, demands a complex process of *regulation*, which provides for the stability of this thing, in the face of the eternal change in all that serves to constitute what it is.

As an illustration of how the principle of regulation of hierarchies could operate let us begin by considering the hierarchy in a particular government. As shown in the diagram, a typical government operates on a number of levels, each of which contains various departments. Within a given level the main activity is, generally speaking, the carrying out of appropriate functions (which is done by 'functionaries' in the manner indicated earlier). But between the levels the main activity is that of *abstracting* information about the order of functioning. This abstraction proceeds in two directions. There is an upward movement in which the higher level officials are informed about what is 'essential', 'relevant', 'significant'.

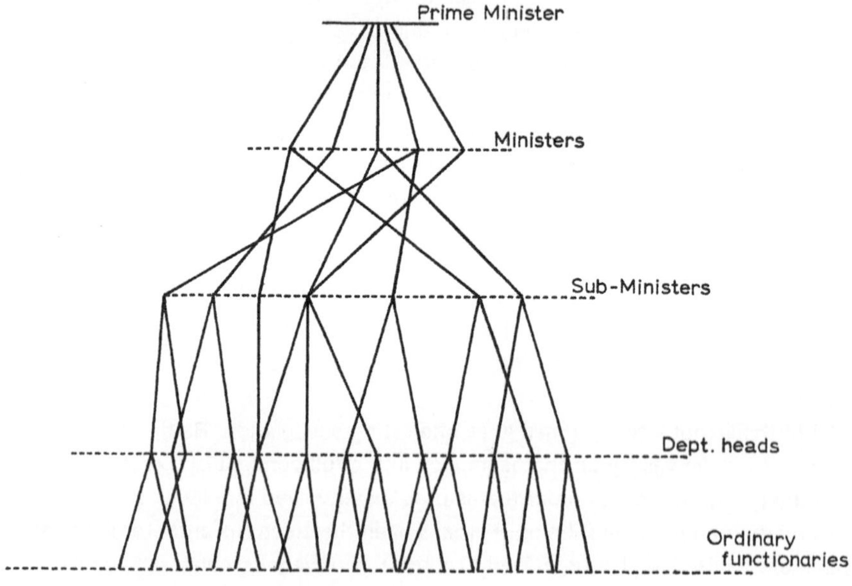

(Naturally, such a selection is necessary, as they would be overwhelmed with a flood of unnecessary data otherwise.) Then there is a downward movement in which the higher level officials inform those lower in the hierarchy how they are to order their actions in the light of the general aims of the government, and in the light of information of all sorts coming from other departments and levels.

Such directives are necessary, since without them the various departments and even various individuals would be likely to engage in conflicting actions. Indeed, even with such information, conflicts are likely to arise from time to time (e.g. the governing bodies of universities may be directed by the Minister of Education to expand, while the Treasury directs them to spend less money). In the upward flow of information, higher officials (in some cases even going up to the Prime Minister) should normally become aware of such conflicts and then give new directives that bring the activities of the departments concerned into a coherent and harmonious relationship.

The proper action of a government therefore depends on a circular process, in which information moves upward and downward all the time. In a properly functioning government there would be no need for any imposition of authority in this circular movement. Rather, each person would want only to do his job; that is, to help regulate the order of society in a harmonious way. The higher officials would always be eager to learn the real state of affairs from the lower officials, who in turn would be eager to consider directives based on the greater range of information available to the higher officials. Actual governments seldom come near to this mode of functioning, for people generally subordinate the real function of their jobs to the effort to extend and glorify the ego, or to make it secure. As long as human beings tend to operate in this way, neither governments nor any other organizations are likely to fulfil their proper functions. Therefore, if we are to use the notion of government as a model of a self-regulating hierarchy, we will have to refer not to actual governments of human beings as they are now, but rather to governments of beings (human or otherwise) who would be free of the ego. (Indeed, scientists find computers so interesting to work on just because of the tacit hope of constructing an artificial intelligence that would not be confused by an ego.)

The above considerations are particularly relevant to the ability of a government to adapt to contingencies. This is evidently dependent on the pattern of the hierarchical organization of the circular flow of information. To help define our terms more precisely, let us refer to this upward and downward flow of information as a 'vertical movement' while the various functions carried out on a particular level will be called a 'horizontal movement'. It is basically the vertical movement that 'regulates' the horizontal movement and thus avoids clashes and conflicts which can arise when unforeseen changes of the general situation cause previously coherent sets of functions to cease to be compatible with each other.

Generally speaking, a particular organization of the governmental hierarchy can meet only some limited range of contingencies. Beyond a certain point, conflicts may arise for which a given hierarchical structure has no means of resolution. At present such a situation tends to give rise to revolution, in which the order and structure of the hierarchy is violently altered. Such violence is generally so destructive that a vast range of new problems is introduced, which may well be even more difficult to solve than were the conflicts that originally gave rise to the revolution. But basically this violence arises because the people involved on all sides of the conflict generally subordinate their functional activities to the advancement of the aims of the ego, either the individual ego or the collective egos of a particular group as opposed to another. In order to use government as a model of a properly self-regulating hierarchy in this context it is therefore necessary to consider an ideal situation in which government officials were interested only in their proper functions but not in any sort of ego, whether individual or collective. When one or more governments are involved in a common crisis, all the people concerned would, regardless of individual or national background, cooperate towards working out a new hierarchical structure and then try to see how it actually functions. If it were still not adequate, they would try again and again until a solution was achieved.

In this kind of situation we would have a fairly clear illustration of the 'timeless' character of the vertical structure of the hierarchy. That is to say, the ordinary functional activities of the hierarchy involve the order of time in an essential way. But changes in the vertical order really do not thus involve the order of time in a correspondingly essential way. ( For example, it can be agreed by all that at such and such a moment one vertical order is to be replaced by another.)

Of course, it takes time to work out the consequences of a new vertical order. In the beginning, such a process would go through a complex period of readjustment, but if the new structure is an adequate solution, things would sooner or later 'settle down' to a quasi-equilibrium state that was relatively harmonious once again.

We are thus led to consider the 'timeless' ordering of the vertical structure and its role in the 'time' process of functioning on the horizontal level. In essence, it is the 'timeless' that regulates and dominates the time process. Anything really new and creative has to come in 'timelessly'. (Here one should consider that various parts of the hierarchy can also change their structures in this timeless way.) The order of time is then what carries the creative change forward to complete realization.

Probably the most immediate analogy is between this process and the operation of the human brain. In this connection one should advisedly think of a vast series of hierarchies of hierarchies. This serves to regulate the movements of the organism on all levels so that the latter is able to adapt and meet contingencies from moment to moment. The *functional systems* by which this regulation is achieved will include many kinds of non-linear cycles (having periods from microseconds and less to years and more), some of which were discussed in the Conference.

When an organism meets a new situation that is beyond the capacities of a given organization of cycles to handle, what is called for is a suitable reordering of the vertical structure. On the lower levels of living things this takes place genetically (at least to a considerable extent). That is, most of the individuals concerned die, but a few of them in whom the reordering has 'fortuitously' taken place in the genotype will survive and produce offspring. But at higher levels (including human beings) this reordering can often take place much more rapidly, far up in the hierarchy of the brain and nervous system of the individual. Such a change occurs when the individual is said to have had a 'flash of understanding'. In the point of view that is being suggested here, this kind of 'flash' is a reordering of the hierarchy, which results in a new mode of action of the organism. In essence, the flash takes no time, but it takes time to work out its implications and consequences in terms of new intentions, modes of thinking, and general orders of functioning.

Consider, for example, the set of cycles of body function described by Art Iberall, going from the shorter 'oxygen cycles' of a few minutes to longer 'kidney' cycles of many months. Observations have shown that conditions of stress, such as those commonly met by business executives, can disturb these cycles, to produce illnesses such as high blood pressure. Evidently such men have a pattern of thinking that leads them to engage in activities that are beyond the normal ability of the hierarchy of self-regulating cycles of functioning to cope with indefinitely. Sometimes when these men are told the facts about their situation there is a sudden flash of understanding. It is this flash that reorders the vertical hierarchy of the process of thought, and thus ultimately restructures the entire sets of hierarchies of function.

Of course, it takes time for the restructuring to take effect. First there is the development of a new intention (to live in a healthier way) along with a new way of thinking about these questions. And then, all the steps needed (e.g. exercises and diet) are taken to implement the new intention. After a year or so, such a man is usually visibly healthier than before. In a way, he has carried out a peaceful revolution in his whole way of life. This revolution started in the flash of understanding that altered the order of his thinking process, in which the crisis had its basic origins. And thus it also altered the organization of the whole of his being.

Is it not possible that analogous processes take place on the biological level ? Waddington has brought out how complex is the chain of process between genotype and phenotype. C. Longuet-Higgins has suggested (at least tacitly) that in this chain there is room for 'improvement of programmes', providing a kind of adaptation going beyond that determined merely by a change in the genotype. But is this 'improvement of programmes' not a particular kind of change in the 'timeless' order of the hierarchy ?

Now let us come to physics. I want to propose that electrons, protons, etc., are merely the names of aspects of a vast, self-regulating, hierarchical process, operating at the level of inanimate matter. Such a hierarchy would in a natural way show many of the properties that are now attributed to quantum mechanical systems. For example, consider a decision made on a given higher level of the government. As the effects of this decision percolate downwards, they spread out, somewhat like a wave. But then, from the lower levels, information is abstracted in an ever more 'concentrated' way, so that it may come back to another department of the original level at a given 'point'. If one looks only at that single level, it seems that the decision has 'moved through space' from one department to another. In such a view one could perhaps attribute all these decisions to a hypothetical 'little ball' that passed from one government building to another (this ball could perhaps be called the 'decidon'). But then, sociologists investigating the phenomena from this standpoint would be puzzled at the wave-like properties of the decidon. The analogy to the electron is quite evident. May it not be that the word electron is actually only the name of a certain complex and relatively stable mode of action of some universal hierarchy of material process ?

Such a notion gives a very good model of 'quantum transitions'. Thus each 'quantum state' would correspond to a quasi-equilibrium order of the hierarchy. When physical conditions were such that a given hierarchical order could no longer adapt its function to the actual situation, there would be a sudden 'timeless change of vertical order, resulting in a new quantum state. After a short period of readjustment, this would settle down to a new quasi-equilibrium mode of functioning. The change of order would be discontinuous while the function would be continuous. But because the complex details of function are not abstracted into the higher levels of information defining the structure of the hierarchy, it is impossible from a knowledge of this structure (i.e. the quantum state) to predict, in each individual case, exactly when the revolutionary change of structure will take place. At most, this can be done statistically, on the average, for an ensemble of systems with similar hierarchical structures.

This brings us to an interesting new concept, i.e. the *law* of the vertical order of the hierarchy. Now, in the horizontal mechanical order of functioning, the law of motion is known (in classical physics at least) to be some sort of second order differential equation. This can be expressed as a 'second order' discrete process, in which the basic law is the similarity of successive differences. That is, if A denotes the first state of the system, B the state after a short time interval later, and C the state after a corresponding interval still later, then we can say that B is to A as C is to B. This law then determines the whole time sequence (i.e. D is to C as C is to B, etc.).

Let us now, for the sake of exploration, suppose that in the movements thus far studied in physics the vertical order has a similar law. In other words, the differences of adjacent levels are similar. So we could write: $L_2$ is to $L_1$ as $L_3$ is to $L_2$, etc. (where $L_n$ is the $n^{th}$ level). This law would determine an ever-changing level structure in the timeless order of the hierarchy. If we introduce an order parameter $\tau$, this law would have some formal resemblance to the mechanical law determining the horizontal movements in terms of a time parameter t. Indeed, the timeless order parameter $\tau$ would even have some kind of vague and general statistical relationship to the time order of process, measured in terms of the time parameter t. (This is because it takes some time for information at one level of the timeless order to reach another level where its effects would be felt in the mechanical order.) Nevertheless, it would be of crucial importance to distinguish the two kinds of order, i.e. timeless and time.

I suggest that, in quantum mechanics, what has been called time (t) in Schrödinger's equation should actually refer to the timeless order parameter $\tau$ if one is to understand what it really means. In other words, Schrödinger's equation should not be regarded as determining the actual movement of things in time. Rather, it determines the 'timeless order' which is really the order of the hierarchical structure, and therefore the 'quantum state', in the sense defined earlier. This quantum state is then only statistically related to the mechanical time order of t. Real time involves a change of order, i.e. a change of quantum state, or a 'quantum jump'.

If these general notions are valid, then it follows that current quantum mechanics is incomplete, in the sense that it leaves out altogether (except in a crude and vague way) the treatment of the actual time process of function and change of state. This suggestion then points to a possible new direction of research, i.e. development of laws of the time order of physical process. I am now working on these with some success, but lack of space prevents me from giving further details on this point.

▸ *On the separation of the observer and the observed.* When we consider the notion that information is what is moving up and down in the vertical order of the hierarchy, we come to the yet deeper question of the observer and the observed (subject and object). It has generally been accepted in science that observer and observed are separate and distinct entities. Such an assumption (which is evidently a basic part of our generally accepted metaphysics) has led to a veritable hornets' nest of entangled and confused problems, growing out of the effort to understand how subject and object are related. But in the metaphysics of process, observer and observed cannot be taken as separate entities. Rather, they are only names of aspects of the total process. And, indeed, we see in the notion of the hierarchical structure that since information is moving upward and downward at all levels, there is no need for a separate 'subject' who would be 'doing the observing'. Rather, inanimate nature *is* both observer and observed.

The same holds true for living beings, which are also hierarchical structures, built out of higher order hierarchies of the hierarchies that constitute inanimate nature. Thus the vertical movement of information continues onward into living beings. In human beings this movement extends further into the hierarchies that constitute awareness, intelligent perception, and understanding. Therefore, at no stage is there a separate observer or subject. The observer is the totality of all that exists, and this is also the observed. Indeed, the movement of information is the dominant order in the whole process, from which the mechanical order is abstracted as 'subordinate'.

Of course, we always have to discuss some partial aspect, abstracted from this totality. In this aspect there are always the two streams—'factual' information coming upward and 'directive' information going downward. It is the upward stream going on through our own sense perception into consciousness that makes our own observation possible, while the downward stream makes possible our participation in the total process. Thus, as indicated earlier in the discussion of Piaget's observations on the young child who is learning to draw, perception and action are always two sides of one circular process (whether in the electronic level or in the brain). This is indeed the basic metaphysical position to which I would like to call your attention here.

Finally, I would like to consider the question raised at the Conference: Is the total hierarchy finite or at least denumerably infinite ? In my view, this is an inappropriate question based on an inadequate understanding of the general metaphysics that is under discussion here. In the metaphysics of process we have to start with the notion that the totality is vast and beyond the measure of man, not only quantitatively but in the potential and actual richness of its qualities. Indeed, we *always* begin with 'what is', which is the unknown. From this, mankind can, at any point in its history, abstract a certain knowledge, having partial, relative, and limited validity. This abstraction is basically creative, i.e. the outcome of a movement of 'outward' participation and an 'inward' perception of this participation. In this 'circular' movement, mankind creates its percepts, concepts, language, and so on. What has been created up to a given time provides the very terms of communication available at that time. These terms enable us to focus on some limited aspects of the vast field of what is to be perceived directly with the senses, and to treat these as 'basic', 'essential', 'significant', 'relevant', while other far more numerous and richer features are treated as 'fortuitous' and 'irrelevant'. Experience guided by these notions is then always showing us the limits of validity of the corresponding abstractions. These limits indicate the need for extending the 'circular' process to create newer abstractions leading to new percepts, concepts, language, etc.

Indeed, even the notions that I have discussed here are only a part of this unending process, in which nothing is fixed, not even the terms of discussion of the total metaphysical context. So it would not make sense to postulate a finite or denumerably infinite hierarchy, since the process metaphysics implies that, as our experience is extended, the whole notion of hierarchy will very probably have to change radically and fundamentally. Even the notion of process metaphysics itself may well change beyond recognition, for all that we can now know. But all that I am suggesting is that this metaphysics may be relevant in the present stage of development of human knowledge.

In this metaphysics, man is a part of the vast totality of all process. At the same time, it is man who has created the abstractions, with the aid of which he is able to recognize certain features of the process in perception, remember them, communicate them, etc. By means of these abstractions, man is able to assimilate the world within his consciousness while he is also participating creatively in the world, to help transform it, so that it will be more suitable to his needs (which in turn are changing in this process). These two movements (assimilation and creative participation) are two inseparable sides of one 'circular' process.

## Notes and References

1. These notions of function have been explained in an article: D. Bohm, *Proceedings of the International Conference on Elementary Particles*, Kyoto, 1965.
2. Is this not like the 'improvement' of computer programmes, suggested by C. Longuet-Higgins ?
3. It is the development of a quasifortuitous relationship between genotype and phenotype which permits the former to be nearly 'insulated' from the latter, thus making possible a certain reliability of hereditary transmission, as emphasized in Neo-Darwinism.

# Bohm's Metaphysics and Biology

**Marjorie Grene**
University of Texas

1 A philosophical observer at a scientific conference is a kind of ethologist (or epistemenologist ?) watching the conceptual behaviour of the other animals. The Second Serbelloni Symposium was outstanding not only because it brought together a group of extremely ingenious and well-trained performers, but also, and above all, because it produced a confrontation of two different conceptual patterns or, to borrow Kuhn's term, two paradigms, one orthodox and relatively restricted (and restricting) in its scope, the other heterodox and comprehensive. The result was not, as often happens in cases of deep-seated conceptual disagreement, simply the clash of two sub-groups. Rather, the first was literally comprehended–that is, described and explained–by the second, though its members were, with some exceptions, unaware that this is what had occurred. It was–I hope–a case of evolution in action: where the species doomed to extinction, innocently unconscious of its lack of 'fitness', continues happily to perform its traditional rites. The spectacle was instructive, but difficult to report, for two reasons. On the one hand, David Bohm in his original paper and in his 'Further Remarks' has himself indicated plainly how his 'metaphysic of process' assimilates and explains the truncated metaphysics of orthodox biology (and physics and computer science and psychology, etc.). Yet on the other hand, most of the contributors to and probably most of the readers of this volume, subscribing as they do to the still current orthodoxy, which as a matter of fact flourishes exceedingly at present, rather like the horns of the Irish elk, are unlikely to see the pertinence of Bohm's metaphysical remarks to their own methodology, and so are unlikelier still to see anything but *im*pertinence in my remarks on these remarks. The poor best I can offer in these circumstances is to try to put Bohm's speculations and, by implication, the metaphysics of the orthodox majority, into their historical context in terms of the major development of philosophical thought in the past three centuries or so.

2 It is otiose, yet necessary, to point out once more that the major trend of modern thought has been held captive by the brilliant success of the scientific revolution of the seventeenth century. The revolutions of the twentieth century have occasioned *some* fundamental rethinking of basic principles, and may yet–if Bohm's predictions are correct–have more far-reaching effects than they have,

explicitly, had so far. But the chief model of 'scientific method' is still that of the Galilean-Newtonian philosophy. And there was something deeply incoherent about this philosophy from the start. Bohm indicates the source of this incoherence when he points out that the acceptance of Newtonian mechanics depends in the last analysis on an act of faith. Why should our mathematicizing be true of nature ? There is no intrinsic reason, for example, why Newton's geometrical proof of Kepler's second law should demonstrate anything about what goes on in the sky. For an Augustinian, what we think, when we think clearly, and what there is in nature, both, if at different levels of perfection, express God's being. Descartes, with his sharp and simple dichotomy of cogitation (= mathematicizing) mind over against extended matter, has to invoke God *ex machina* to hold the two together. But he is still sufficiently an unquestioning Augustinian, so that, for him, the invocation works. With the secularization of thought (metaphysical as well as scientific: as Bohm quite correctly states, they are never wholly unrelated, since 'metaphysics' is just the most comprehensive range of anybody's thought, whether he knows it or not), the Cartesian dichotomy becomes unstable. Its uneasy synthesis in Kant, with nature reduced to phenomenon and mind to moral will, depends, still, on Kant's undoubting pietism for its ultimate support. That gone, it is only a short, inevitable step on the one hand to the *Nullpunktsexistenz* of the Sartrean for-itself, which frankly lives by contradiction, and on the other to the fruitless and equally self-contradictory objectivism of the contemporary philosophy of science (and of many scientists). (For the fruitlessness of the latter, see the outcome of Wittgenstein's *Tractatus*, for its self-contradiction, the argument of Russell's *Human Knowledge;* or see also the critique of E. Straus in *Vom Sinn der Sinne*, or of course Whitehead.) Of course each lingering remnant of the divided cosmology tries to account for the whole: Sartre's 'dialectical reason' serves up a caricature of nature; modern epiphenomenalism, a caricature of mind. Along these lines there is just nowhere further to go—and there never was; but it has taken us three hundred years, and indeed may take still longer, to find this out.

The incoherence lies not just in mind-body dualism, which has long since given way to a belief in matter-in-motion as the sole reality, but, as Bohm emphasizes, in the deep-lying divisiveness of our conceptual framework along a number of related lines. To cut off mind from nature is to cut off subject from object, so sharply that science itself (the product, after all, of subjects) becomes irrational and reality meaningless. Science becomes computation-for-the-sake-of-prediction-for-the-sake-of-computation-for-the-sake-of-prediction . . . , 'understanding' a merely subjective addendum, and 'truth' a dirty word, dropped in weak moments, like words with one less letter, but decently avoided for the most part in polite society. And the world so known ? It used to be, and, as Bohm points out, for many biologists still is, the seemingly solid one-level nature of Democritean atomism, where faith that God made and keeps united our thoughts and their objectives gives way to the equally, if not still more, irrational faith that more complex orders *must* be explained out of, and exhausted by, those that are simplest, and ultimately out of the one 'real' order of matter in motion. Taking subject and object together we have, in Whitehead's words, 'a mystic chant over an unintelligible universe'.

For what the subject-object dichotomy entails is a separation of order from the ordered, of meaning—which shrinks to a game with meaningless counters—from what is meant—which shrinks to an infinite aggregate of equally meaningless data. In philosophical jargon, it entails, as the literature (and indeed, literature) richly shows, a radical division between value and fact: in Bohm's terms, harmony becomes a little secret preference of our own, and beauty a private vagary, rather than, as it is, the criterion of our access to reality. In contrast, Bohm's linking of understanding, beauty, and the timeless orders that govern emergent process may herald, in my view, a comprehensive alternative to the self-denying ordinances of modern thought: self-denying in that it alleges itself to be only the compulsive outcome of its own neural processes. (Waddington is right, of course, in calling Bohm's view Whiteheadian, but it may prove more viable than Whitehead's own cosmology, since it can be developed, I believe, without recourse to a doctrine of eternal objects: a radical *in*coherence, I feel, in Whitehead's system.)

True, there are some other signs also of relaxation in the cramping cosmology that still governs most scientists' minds. On the 'subject' side, books like Hanson's *Patterns of Discovery* or Kuhn's *Nature of Scientific Revolutions* may help. The slowly growing influence of Polanyi's *Personal Knowledge* is a hopeful sign, although unfortunately many of those who profess to accept its conclusion have grasped only the most superficial theme of its complex argument. On the 'object' side, not only Bohm has argued (and Waddington in his summary accepts this) that physics itself has come to the end of its Democritean chapter, and so a new and richer synthesis may be in sight. But for one thing Bohm's position on physics still appears to most professionals as extreme heterodoxy, and on the other hand authoritative biologists, especially in molecular biology—both Crick and Watson, for example—still argue that a complete one-level, particulate ontology is, if not already here, just around the corner. And often, I believe, even when they seem to moderate their position by pointing to the richer complexities of modern physics, biologists are just concealing, to themselves and others, their real reductivism behind a screen of Gibbs statistics, Volterra-Lotka equations and the like. There is still a long way to go before we are out of the woods.

3 Meantime there is a further fundamental disability in the Cartesian-Newtonian world view which was exhibited most beautifully at Bellagio, both in its representatives and in Bohm's manner of transcending it. This is a disability also stressed by Whitehead: the incapacity to develop an adequate conception of life.

Not only Descartes' *bête-machine* but the principal thrust of Kant's transcendental analytic make it plain that in terms of the chief modern tradition this must be so. At the very start of the first Critique Kant distinguishes between acts which I perform and so have as things in themselves, but cannot know, and intuitions (*Anschauungen*) which are passively present to me, and which alone can supply content for my knowledge. The knowable, in other words, is passive; agency cannot be known. But as Whitehead argued, as Suzanne Langer argues, as Bohm argues, what we know as alive we know precisely in and through its *activity*. Bohm adduces here Piaget's account of child development; one could cite also von Weiszäcker's *Gestaltkreis* or Goldstein's concept of preferred behaviour. This is also, I believe, the fundamental (if sometimes concealed) import of Waddington's stress on epigenesis: organisms *are* not simply, they *act*. And acts are, in Bohm's terms, *creative* processes: they bring into being an order that was not. But thought in the Cartesian tradition has restricted itself, in viewing the object of knowledge, to its passivity: to what Bohm calls equilibrium and dynamic processes, excluding creative process. In the beginning was God's creation, thereafter a 'clockwork' universe, but no creative creature. This is again a consequence, if you will, of the subject-object dichotomy, of which the logical outcome on the subjective side is the Sartrean for-itself, and on the object side Crickian reductivism. Bohm argues, of course, that even in physics a concept of creative process has been found indispensable; my point here is simply that without a concept of act, that is, of creative change of order, there can be no adequate concept of life.

But, it will be objected, that is just what the conference did move to develop: we had brilliant applications of automata theory, statistical dynamics, chemical engineering, and so on, to biological problems. Of course there is no limit to such applications, and in the appropriate context they are all to the good. The question remains: what *is* the appropriate context?

The context within which the most articulate participants at the Symposium approached their problem was thoroughly functional. Almost everyone used, constantly and as self-evident, the term 'biological' as synonymous with 'functional', 'adaptive', 'conducive to survival'—strictly, in evolutionary terms, conducive to leaving descendants—or 'produced by Natural Selection'. The best, indeed the perfect, specimens of this breed of thought were Longuet-Higgins, Gregory, and Maynard Smith; even Waddington, though not *quite* orthodox, appears in his summarizing notes as an interesting mutant of the same species.

The fundamental principles of this reigning form of biological thought are two: first, uniformitarianism extrapolated to the faith that ultimately all explanation is one-levelled in terms of least particulars. (It's all, after all, physics and chemistry—see Arbib's note on L.-H., p. 336) Secondly, that the only allowed supplementation of such a monolithic materialism is the reference to adaptation personified in the concept of Natural Selection, that is, of the 'mechanism' by which the less adapted are eliminated in favour of the better adapted. Now naturally, I hasten to say, I too cross myself when speaking of Natural Selection. Yet for all my efforts I am still unable to understand quite what it really means, especially when I hear from Dobzhansky about 'selection *sub specie aeternitatis*' or from Simpson that 'whatever we see, selection sees much more', and so on. Let me try once again.

Look at Longuet-Higgins' summarizing statement: 'The secret of life is the ability of living creatures to improve their programs.' 'Improve', when challenged, he altered to 'adapt'. Now this is plainly the neo-Darwinian *credo* using the contemporary tools of population genetics and information theory. Whatever computer techniques it embodies, what it adds to the Democritean platform—the ultimate reduction of all process to its simplest level—is simply the axiom of adaptivity: the thesis that living things are adaptation machines. This thesis is then supposedly sufficient to generate evolution. What does it mean and how does it work?

Medawar recently stated the core of contemporary evolutionary theory as consisting in the two propositions: (i) that the terrestrial populations existing at a given future time will differ statistically in some degree from those existing today; and (ii) that the genetic constitution of those populations will have some connection with their changed phenotypic characters. We may add to these Waddington's emphasis on the role of the phenotype in the selective process controlling (i) and (ii). These are all perfectly harmless statements which no one, philosopher or otherwise, would want to challenge. But they tell us nothing at all about the epistemic relation between discourse about least particulars: gene pools or populations of gene pools, and discourse about cells or organ systems or organisms, nor about the ontological import of such discourse, all of which, in Longuet-Higgins' 'Of course in the end it's all physics and chemistry', they presuppose. Nor do they tell us how from a time when there were no living cells or organ systems or organisms there came to be such entities. When we try to go further here, we find a peculiar muddle, a muddle which needs to be disentangled before an adequate theory of evolution, emergent or otherwise, can be formulated.

I have tried to analyse this peculiar muddle a number of times elsewhere and so, of course, have numerous other people. Langer in her new book has some good arguments; and indeed Bohm's argument on the necessity of the transfunctional seems to me absolutely conclusive. But perhaps in my less imaginative way I may briefly make one more attempt.

Behind Medawar's harmless statement, the principal presupposition of modern evolutionary theory is the thesis formulated in 1932 by R.A. Fisher: 'Evolution *is* progressive adaptation, and consists in nothing else', the thesis echoed in Longuet-Higgins' 'self-adapting programs'. Yet adaptation on its own, or progressive adaptation on its own, as neo-Darwinism takes it, is by no means self-explanatory, or indeed explicable. For one thing, the concept of evolution as identified with progressive adaptation is basically ambiguous. On the one hand, such adaptations are supposed to be mechanically self-generating—through mutation, natural selection, recombination and isolation—and so to entail no teleological reference. Yet on the other hand, adaptation, like its Victorian twin, 'utility', is itself a teleological concept: it is adjustment of something to something for some end. To interpret organisms as adaptation machines, as neo-Darwinism does, therefore, is to interpret them as complexes of means for ends, and therefore teleologically. Secondly, if the means-end reference is admitted, one must ask further: means to what end ? But the 'end', for Darwinism, dare not be some 'higher' form of life, the next 'level' to which evolution aspires: that would be to reintroduce a forbidden version of 'unscientific' teleology. It must then, and is usually said to, be *survival* that adaptation is 'for'. But in that case the whole 'theory' becomes a tautology, a complicated way of saying simply that what survives survives. (And Medawar's two propositions, which were stated in response to the accusation of tautology, simply open out the tautology a little way, only to let it form again when we try to interpret those propositions in their full theoretical context.) Finally, and fundamentally, the trouble is, really, that the concept of adaptation is essentially a relative one—relative to the existence of two things: the organism to be adapted and the *environment* to which it is to be adapted. Helmuth Plessner makes a similar point in an early work, *Die Einheit der Sinne,* in the context of an analysis of sense perception. Referring to the physiological investigation of perception, he points out that perception cannot be explained wholly in terms of adaptation, since there must be *something there already to become adapted. Adaptability,* which is a potential relation between organ, medium, and object of perception, must precede adaptation. Thus the adaptable entity, the organism which *can* achieve adaptation, must be assessed in its own right, by its appropriate norm, before the detailed conditions of its adaptation can be specified. The same is true in the context of evolutionary theory, or indeed wherever the attempt is made to rely on utility or adaptation as a principle of ultimate explanation. Adaptation is a crypto-teleological concept, but teleology even when explicit is itself dependent on the prior evaluation of the ends evoked. And in the case of evolutionary biology, the end is not simply survival but the survival of—a type, a mode of living, an order of orders, in Bohm's language, adjudged as significant in itself. That is the only judgment that can fill in the tautology of survival-for-survival-for-survival.

4 Such a judgment, however, you are all, or nearly all, unwilling to make, or to admit that you make, and that despite the undeniable force of Bohm's argument on function/transfunction. Why not?

There are temperamental reasons: the passion for model making, and social reasons: the prestige and power granted to machine makers in our society. But the fundamental reason lies, I believe, in the basic ambiguity of the concept of 'mechanism' itself. The world of the seventeenth century's 'new mechanical philosophy' was, apart from its Infinite Designer, a one-level universe, whose laws would ultimately be specifiable in terms of its 'hard, impenetrable' least parts. Its laws are, in Bohm's phrase, automorphic, expressing the simplest order of matter-in-motion. Take away the Designer and you have a self-regulating system, just what, three hundred years later, automata theorists have triumphantly learned to produce. But a machine, as Polanyi has demonstrated (both in *Personal Knowledge* and recently, August 1967, in *Chemical and Engineering News*) is, essentially, not a one-level, but a hierarchical system. It demands operating principles ordinally complementary to, i.e. depending on, but controlling, the laws of physics and chemistry which govern the behaviour of its parts. Because, however, we have put it together out of discrete parts which we control and can specify, we can easily neglect this two-level structure and hold that physics and chemistry alone 'produce' and 'explain' the machine. The clockwork seems to make the clock. Now of course the clock needs its clockwork. But you cannot in terms of physics and chemistry alone say anything about telling time. You cannot in terms of physics and chemistry alone distinguish any message, whether the time of day or the hereditary program of an organism, from a noise.

Why is *this* message so hard to put across ? Because of the compulsion of Democritean thinking times the self-deception of 'utility'. Machine thinking, as I have argued for the case of evolutionary theory, is crypto-teleological. But when you make the end explicit, in engineering terms it is still a means; and you can keep running, like Yellow-Dog Dingo, without ever stopping to face the fact that *some* intrinsic value, some harmony, in Bohm's language, some timeless order, is the controlling principle which all the while governs your unending course. Or if you're pushed, it is still the minimal order, the maximally meaningless order, survival, or motion-for-the-sake-of-motion, that turns out, allegedly, to be your goal.

Bohm asks: Is nature more like an engineer or an artist ? Admittedly, the two have much in common. The artist must control his material craftsmanly: he is also an engineer. And the great engineer also achieves beauty. Moreover, the emphasis on engineering concepts as against *mere* physics and chemistry (see for example Gregory's *The Brain as an Engineering Problem*) is in fact an advance from a one- to a two-level ontology, and that is all to the good. But the difference in ends, and therefore in the logical structure (the order of orders) of the two enterprises is what matters here. The engineer makes artefacts, without intrinsic significance or intrinsic reality. They are in essence means to the perpetuation of what is. The artist makes new realities, richer orders that never were on land or sea, dependent of course on conditions specifiable in terms of lower levels, but neither predictable nor explicable in terms of them. Such harmonies, such emergent orders, have to be apprehended, not through manipulation of means for means for going on going, but through understanding: understanding, again in Bohm's terms, as the union of observer and observed in the presence of beauty.

To the physics or engineering minded, such formulations may well appear absurd, wildly metaphysical, 'subjective', 'irrational'. Of course, so did Galileo to the good Aristotelians. I can only point, in conclusion, to convergences in other contemporary writers with Bohm's cosmology. The three principal orders he specifies in his paper are identical with those distinguished by Merleau-Ponty in *The Structure of Behaviour*, and Merleau-Ponty like Bohm points to artistic creation as the paradigm on which we ought to lean if we would understand our own way of being (the order of intelligence). And of course the analogy of nature and art (not artefaction, or invention, but creation, artistic discovery) forms the central theme of Suzanne Langer's recent work. There is no doubt, in my view, that we must acknowledge and implement philosophically Peirce's insight: that while logic is subordinate to ethics, in the sense of practical or engineering knowhow, ethics *sive* engineering is subordinate to aesthetics; our sense of beauty, of the intrinsically meaningful, dominates, whether we will or no, our grasp of what is real, of what is worth making real or allowing to perish. We all seek, in our own way, as Plato saw, to achieve immortality, to find a timeless order, through begetting on the beautiful. And only he who has seen the beautiful itself 'can breed true virtue, since he alone is in contact not with illusion but with truth'.

## Comments by C.H.Waddington

I want to make two brief comments on metaphysical points—but certainly not to attempt any full discussion of all the metaphysical issues raised.

1 A remark in passing to Marjorie Grene. I know that the fashion of the time is very much against notions which have such transcendental-sounding names as Whitehead's 'eternal objects'; but I shall require convincing that David Bohm's 'timeless orders' are either very different, or much more likely to prove acceptable to the philosophical establishment.

2 A more serious comment, to David Bohm. I agree with the argument, on page 44, that function implies the trans-functional, i.e. must ultimately be referred to something beyond the field of function. The question is, how is this trans-functional to be conceived of ? There is, I think, a danger that people may think that it must be not only trans-functional but also transcendental. David seems to me to come near to implying that in his discussion of 'vicious circles' ('they eat to live and live to eat'). Now I do not think all such 'circles' are vicious; and I doubt if we can find, and am reluctant to believe that we need, anything beyond such circular and therefore self-sufficient systems. In fact the title of a chapter in one of my books (*The Ethical Animal*, 1960) was 'The Shape of Biological Thought, or the Virtues of Vicious Circles'. What is unsatisfactory about David's 'Eat to live' example is not the circularity, but that the circle is too small, eat is inadequate to act as the complement to live. David seems to get nearer to my outlook when he writes (p. 45) that the reference 'beyond the field of function'

may be to a 'natural necessity', that is, to the fact that the world just is structured in that particular way.

I think the extent of my agreement with Bohm, and the directions in which I should like further clarification from him, may be best expressed if I quote two paragraphs from my discussion of 'function' in *The Ethical Animal*.

'When we assign a function to something, we in fact assert two propositions about it. First, that it forms part of a causal network; and secondly, that the results of the causal network, when observed over the range in which they are expressed, exhibit some general property. Another way of expressing the latter point is to say that the causal network is organized. The concept of function is in fact very closely connected with that of organization, and can be regarded as a derivative of that more general notion. Is organization, then, an illegitimate concept? It probably is so in terms of a crudely mechanical materialist picture of the world i.e. a picture in which we consider that all existing things can without loss be reduced to the movements and interactions of some ultimate constituent particles. But such a picture has never been more than a theoretical aspiration in biology, and is at present out of date even in the physical sciences. There are now, I think, few scientists who would consider it illegitimate to conclude that groups of elementary constituents may, by entering into close relationships with one another, build up complex entities which then enter into further causal interactions with one another as units. It is this fact, of the integration of groups of constituents into complexes, which in certain respects operate as units, which is spoken of as organization. In so far as it occurs, the concept of function is a legitimate one. If we have some complex entity A which acts as a unit, we can regard it as exhibiting organization of its constituent elements. Suppose that within A we can discern certain sub-units, P, Q, R, then the function of P within the organized system A is the contribution which P makes towards those types of behaviour in which the unitary character of A is exhibited' (p. 62).

The question whether the trans-functional, Bohm's 'timeless order', is also transcendental is of course of particular importance and has a particular urgency in the ethical connection with which I was concerned. I quote again: 'Adopting the usual terminology of biology, we can say that the function of ethicizing is to make possible human evolution in the socio-genetic mode. Now, once we have assigned the function to a general type of activity we have a rational criterion against which to judge any particular example of that activity' (p. 29).

'It is as well to consider what is implied by such a mode of approach by taking as an example some aspect of human activity which is less emotionally loaded than ethical beliefs. Consider for example the activity of eating. The human newborn infant has first to develop into the sort of creature that goes in for eating. In this development innate factors probably play a much greater role, and extrinsic factors a lesser one, than they do in the development of the infant into an ethicizing being, but this alteration in the relative importance of the two types of factors is of minor consequence in the present context. Next the child will acquire certain specific food habits, becoming accustomed to and accepting a particular diet. This is a process analogous to the development of specific formulated ethical beliefs. In order to find a basis for criticizing these food habits we have first to enquire what is the function of eating. We find that it is to make possible the growth of the body. Inspecting the growth of human beings on a wide basis, we discover that it manifests a general character which we describe as health. We can then ask of any particular food habit or diet how effective it is in bringing about healthy growth. The criterion we are applying here is one of general accordance with the nature of the world as we observe it. If any individual approaches a nutritionist and says that he prefers to grow in an abnormal and unhealthy manner, the nutritionist can do no more than tell him that if he does so he will be out of step with nature. The criteria, of biological wisdom in the case of ethics, or healthy growth in the case of eating, which can be derived in this way, are immanent in nature as we find it, not superimposed on it from outside. However, even if one considers that there is some overriding supernatural being from whom our ethical standards are ultimately derived, it is surely blasphemous to suppose that the nature he has created is such as to deceive us as to his true wishes. Thus, even an immanent criterion, if we have discerned it aright, would not contradict, though it might of course fall short of, a supernatural one' (p. 30).

# The Practical Consequences of Metaphysical Beliefs on a Biologist's Work: an Autobiographical Note

C. H. Waddington
University of Edinburgh

Several of the more 'hard-headed' characters at the Second Symposium expressed from time to time, at cocktails or after dinner, a suspicion that metaphysical considerations of the kind introduced by David Bohm have ultimately no real impact on the directions in which science advances. They suggested that they were merely part of the froth churned up while the theoretical physicists flounder and thrash about trying to find a firm footing in the deep and dangerous waters of quantum theory, sub-nuclear particles, and the like; and that when this footing has been found the froth will settle down and disappear. It is software, they suggested, so soft as to be deliquescent and ultimately evanescent.

I do not agree with this. I should like to argue that a scientist's metaphysical beliefs are not mere epiphenomena, but have a definite and ascertainable influence on the work he produces, by reminiscing for a moment about my own career. I am quite sure that many of the two hundred or so experimental papers I produced have been definitely affected by consciously held metaphysical beliefs, both in the types of problems I set myself and the manner in which I tried to solve them. I do not want to argue now—though I'll do so later if anybody wants me to—that these were really the most interesting problems, and that I set about them in the right way. Maybe my metaphysics was leading me up the garden path (though I don't think so); but the point I want to make now is that it was leading me somewhere and was therefore something more than a set of decorative flourishes on the proscenium arch, giving on to the stage in which the real action takes place.

As David Bohm points out, metaphysics is a sort of poetry; and we are all talking poetry all the time—if you doubt it, look again at what Richard Gregory has to say about such a basic activity as perception. As poetry, metaphysics can be absorbed through communication-channels other than extended rational exposition. So, to begin this metaphysical-experimentalist's autobiography, I will mention two (or perhaps three) notions which infiltrated into my thinking at a very early stage, without much benefit of academic dignity, and which have remained there ever since.

When I was a schoolboy there was a peculiar period after you had taken your entrance or scholarship examinations to the university in December of year n, and before you actually went to the university in October of year n + 1. This was the time when your schoolmasters could really have fun with you. I had been in the classical sixth form, learning mainly Latin and Greek and a certain amount of mathematics; but when it dawned on me that all my friends were leaving and I had better leave too and go to the university, I decided I had better try to get in on the basis of chemistry. Fortunately we had a chemistry master, E. J. Holmyard, who was something of a genius of a teacher. During one summer holiday and autumn term, he taught me the whole of chemistry, at least enough to push me into the lowest grade of assisted entry to the university, an Exhibition. After that he could really break loose and teach me what he was interested in. His passionate interest happened to be the functions of the Alexandrian Gnostics, and the Arabic Alchemists derived from them, in transmitting both the philosophical ideas and the technical knowhow from the Greek civilization which expired around AD 200, to the European one which began to come alive at the Quattrocento. So he made me learn a smattering of Arabic and look at a large number of very odd late Hellenistic documents. Two ideas stuck:

- *The world egg* 'Things' are essentially eggs–pregnant with God-knows-what. You look at them and they appear simple enough, with a bland definite shape, rather impenetrable. You glance away for a bit and when you look back what you find is that they have turned into a fluffy yellow chick, actively running about and all set to get imprinted on you if you will give it half a chance. Unsettling, even perhaps a bit sinister. But one strand of Gnostic thought asserted that *everything* is like that.
- *The Ouroboros*, the snake eating its tail. This famous symbol, which is as well known in ancient China as in Alexandria, expressed the whole gist of feedback control almost two millenia before Norbert Wiener started 'creating' about the subject at MIT and invented the term 'cybernetics'. Here is a drawing of an ouroboros which I made for an essay which I wrote while I was still at school, presumably around 1923. I reproduce it because you will see that inscribed within the ouroboros is a third subsidiary notion; the slogan 'εν το παν,' 'hen to pan', 'the one, the all', a phrase which implies (in a cybernetic context, be it remembered) that any one entity incorporates into itself in some sense all other entities in the universe (Fig. 1).

Before these highly poetic metaphysics had any practical influence on my

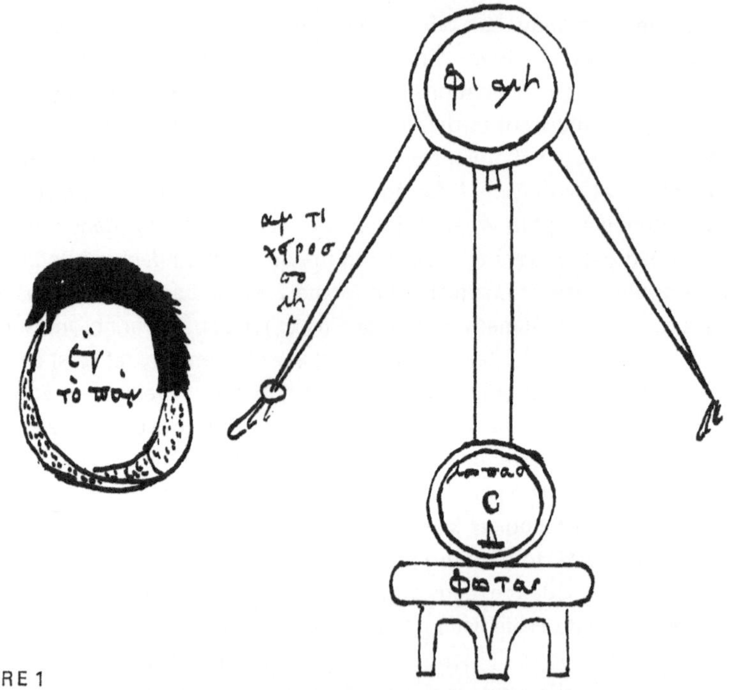

FIGURE 1
The Ouroboros, together with an alembic (distillation vessel) redrawn from an alchemical document known as the Chrysopeus (i.e. gold-maker) of Cleopatra.

scientific work, there was added to them a large body of much more explicitly rationalized thinking; in the first place that of Whitehead, to whose writings I paid much more attention during the last two years of my undergraduate career than I did to the textbooks in the subjects on which I was going to take my exams. Later this was joined by some infusions of thought which claimed to be materialist–either 'fancy' (dialectical), which preceded Whitehead and seemed to me to be in the main left behind by him; or 'crude', the prime example being Morgan and his school, who insisted that the gene is not just a logical construct from Mendelian ratios [cf. Woodger's definition

$$\mathrm{mend} =_{\mathrm{Df}} \mathrm{Aeq}'\hat{\hat{x}}\hat{y}\,(x, y\varepsilon\,\mathrm{whz.}\quad \mathrm{Apr}'\mathrm{Zyg}'x \rceil \mathrm{K} \lceil \mathrm{Apr}'\mathrm{Zyg}'y\varepsilon 1 \rightarrow 1)],$$

but is just a simple lump of stuff. But one was anyway surrounded by materialists, and the whole of science was dominated by essentially Newtonian conceptions of billiard-ball atoms existing at durationless instants in an otherwise empty three-dimensional space. It was, for me, Whitehead who suggested new lines of thought.

What was this Whiteheadian metaphysics ? I will sketch very briefly what were the salient features in my eyes.

1. The raw materials from which we start to do science—or with which we finish the scientific testing of a theory—are 'occasions of experience'.

2. An occasion of experience has a duration in time (cf. David Bohm, 'there are no things, only processes').

3. An occasion of experience is essentially a unity. Any attempt to analyse it into component parts 'injures' it in some way. Yet we cannot do anything with it unless we do analyse it. Our first step towards an analysis is to dissect the unity into an experiencing subject and an experienced object. The dividing line between these two is both arbitrary and artificial. It can be drawn through various positions, and wherever it is drawn it is never anything more than a convenience.

4. The content of any occasion of experience is essentially infinite and undenumerable. Moreover, wherever we draw the line between experiencing subject and experienced object, the latter will always remain undenumerable. If this were not so, we should merely have to describe the totality of the content of the experience, and the experience itself would be created.

5. The experience, which Whitehead refers to as an 'event', has, however, some definite characteristics. Whitehead refers to these as 'objects'—the word which is usually used, and has indeed been used above, in a quite different sense. Definiteness of the Whiteheadian objects in an event implies that, although the event has some relation to everything else past or present in the universe, these relations are brought together and tied up with one another in some particular and specific way characteristic of that event. (Whitehead was writing *before* quantum mechanics became a dominant influence in our thought, but compare these notions with the idea that a particle must also be thought of as a wave function extending throughout the whole of space time.) For this tying-together of universal references into knots with individual character, Whitehead used various different phrases at different periods in the development of his thought. For present purposes I am content to stay with the least far-reaching of these, when he spoke of the coming together of the constituent factors in an event as a 'concrescence'. Later he described the way in which an event here and now incorporates into itself some reference to everything else in the universe as a 'prehension' of these relations by the event in accordance with its own 'subjective

feeling'. This is a metaphysics very close to that advocated by David Bohm when he speaks of creativity, and argues that nature is more like an artist than an engineer. Privately my own thought runs along similar lines; and I think they may be extremely important to the way in which one behaves in one's whole personal life; but I do not see that they have had any direct influence on the way in which I have conducted experimental work, which is the subject which we are discussing here. As far as scientific practice is concerned, the lessons to be learned from Whitehead were not so much derived from his discussions of experiences, but rather from his replacement of 'things' by processes which have an individual character which depends on the 'concrescence' into a unity of very many relations with other processes.

So, without going into further metaphysical sophistication, let's get down to brass tacks. What did I actually do as a practising biologist, and how was this influenced by this metaphysical background ?

I began work as a palaeontologist, studying the evolution of certain groups of fossils. And I chose, as my main interest, a group which forces on one's attention the Whiteheadian point that the organisms undergoing the process of evolution are themselves processes. The Ammonites were cephalopods, related to squids and the Nautilus, which laid down spiral shells. The animal occupied only the latest-formed part of the shell and from time to time moved forward a little, leaving behind it the part it had previously inhabited. Thus following the whorls of the spiral shell outwards, from the centre to the periphery, one has a record of the whole life-history of the animal; it never appears just as an adult whose juvenile stages have vanished. The whole developmental process is preserved so that one cannot avoid examining it. And the process is, of course, complex, with many facets. On the surface of the shell there are 'ornaments'—ribs, knobs, tubercles, etc.—which change as development proceeds; the cross-sectional shape of the tube may change, and so may the closeness with which it is coiled; and behind each living-chamber a partition is laid down which meets the outer shell in a complex 'suture' which is one of the most characteristic features of a species (Fig. 2).

In most types of animal, in which the adult form of the individual replaces the younger stages, the only way to study the developmental history is to collect a large population containing juvenile as well as adult stages. Age can usually not be determined directly, but one can take some measure of size as an indication of it. Fig. 3 shows a graph I made, around 1927, of the variation in ratio of breadth to length in a collection of fossil shells (a Brachiopod, *Terebratula*).

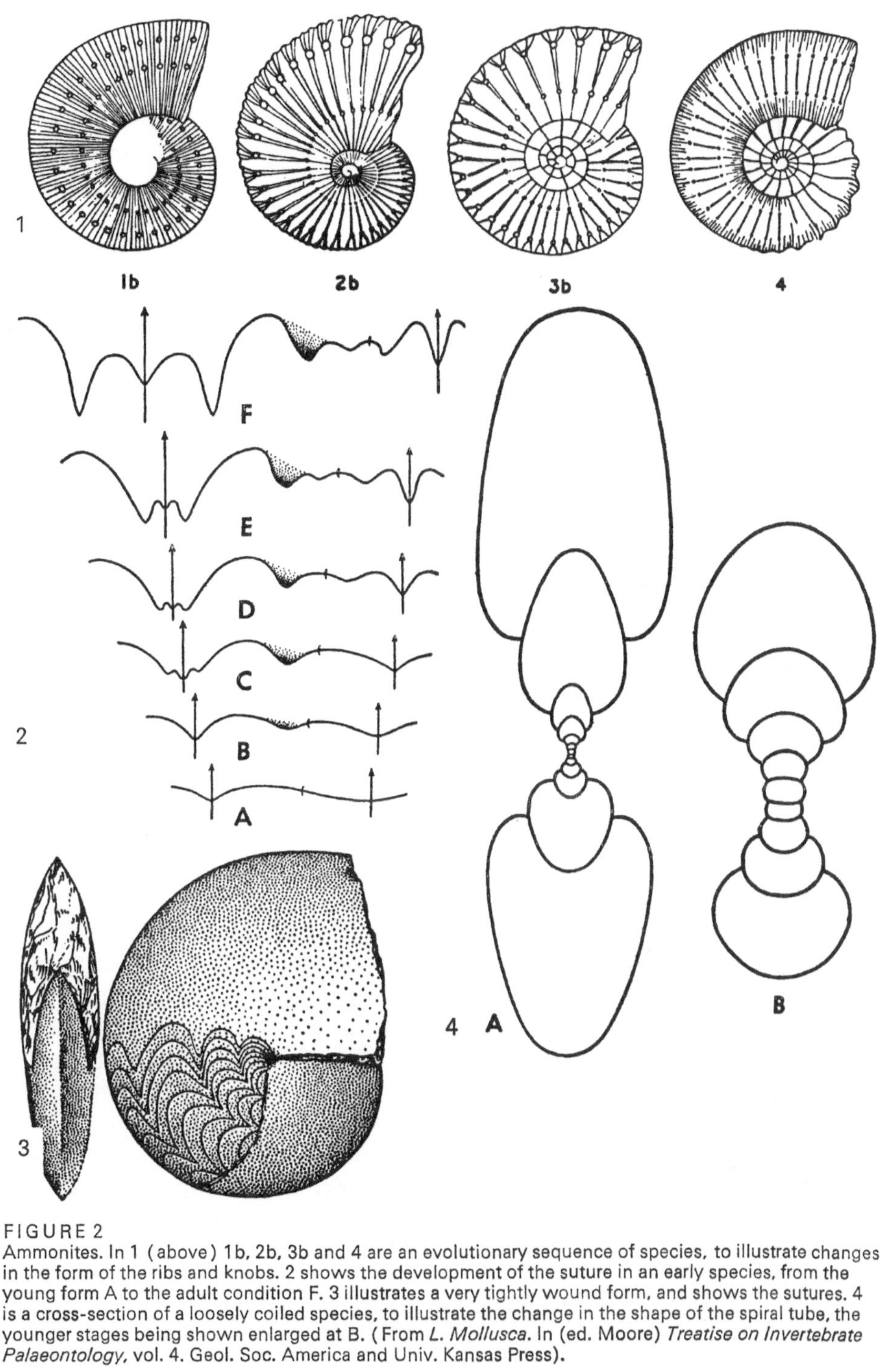


FIGURE 2
Ammonites. In 1 (above) 1b, 2b, 3b and 4 are an evolutionary sequence of species, to illustrate changes in the form of the ribs and knobs. 2 shows the development of the suture in an early species, from the young form A to the adult condition F. 3 illustrates a very tightly wound form, and shows the sutures. 4 is a cross-section of a loosely coiled species, to illustrate the change in the shape of the spiral tube, the younger stages being shown enlarged at B. (From *L. Mollusca*. In (ed. Moore) *Treatise on Invertebrate Palaeontology*, vol. 4. Geol. Soc. America and Univ. Kansas Press).

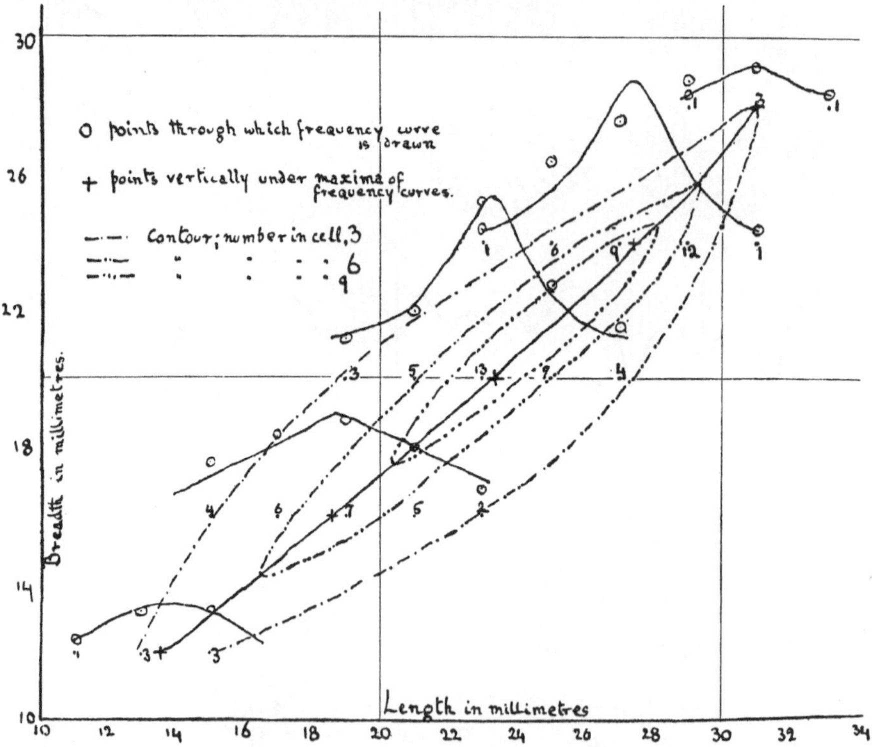


FIGURE 3
Frequency distribution of length and breadth in a population of fossil *Terebratula*, shown as a contour map with superposed frequency curves at five intervals. (From an unpublished exercise of my own, used in 1927 by my friend Reuben Heffer to try his hand as a printer in preparation for joining the family book shop.)

These early exercises left me with a deeply ingrained conviction that the evolution of organisms must really be regarded as the evolution of developmental systems–which was the title of one of the first articles I wrote about evolution when I returned to the subject some years later. It is of course related to such old ideas as Haeckel's 'biogenetic law'–phylogeny repeats ontogeny–but I took it also as a guiding principle in population genetics. I still think that when modern population geneticists express the variation in a population by means of a timeless frequency curve, which deals only with the adults, this is a simplification which needs justification–which of course it may often at least partially possess, for instance when one is dealing with the imagos of an insect like *Drosophila*, which metamorphoses suddenly into an adult form which thereafter does not change. But when I started doing experiments on *Drosophila* evolution, in the forties and fifties, I treated even that insect as a developmental system, and by manipulating the environment in which it developed was able to uncover the rather novel process of genetic assimilation. Thus my particular slant on evolution —a most unfashionable emphasis on the importance of the developing phenotype—is a fairly direct derivative from Whiteheadian-type metaphysics.

In my early career there was a considerable period in which I was concerned directly with developmental systems themselves, rather than with their evolution. My approach to experimental epigenetics was again strongly influenced by Whiteheadian metaphysics, but also by genetics. In fact, when I decided that I wanted to do something more experimental than is possible in palaeontology, I first tried to become a geneticist. My first two published papers were, one in plant genetics, and another a collaboration with J. B. S. Haldane along classical neo-Darwinist lines, in which we studied the effect of inbreeding on the segregation of linked genes, writing down, and trying rather unsuccessfully to deal with, great series—up to 27 members, I think—of finite difference equations. So I had my taste of the thin gruel of mathematical formalism, as well as of the strong nourishing soup of time-extended populations of fossilized developmental systems. But my attempt to become a geneticist was a failure, because at that time in Britain there simply was no way in which one could earn a living at the subject. So I became an experimental embryologist. For some years my most immediate interest was in solving the technical problems of carrying out meaningful experiments on types of embryo which no one else had tackled successfully, such as birds and mammals, or of doing biochemical work on the very small tissue-fragments isolated from various regions of an egg.

However, the theoretical structure of the subject also needed a good deal of attention, and this is where the Whiteheadian approach came in. In the twenties, T. H. Morgan had argued that epigenesis should be considered in terms of the activities of genes, but he had little effect on the workers in this field. Embryological theories involved such notions as the 'segregation' or 'differential dichotomy' of 'potencies' to develop into organs such as the nervous system, kidneys, gut, etc.; or, at a more advanced but still very imprecise level, notions such as 'the organizer', 'induction', and the like. I wanted to return to Morgan's idea that the only 'potencies' it is meaningful to talk about are the potential activities of genes. So did several people who were primarily geneticists, but who had become interested in development without having actually worked on it very much.

## Practical consequences of metaphysical beliefs

Two rather radically different lines of approach were followed. The one which was—and in fact still is—favoured by most geneticists depended on what I think may be called an 'atomistic' metaphysics. It set out from the assumption of the existence of single genes, and it asked, at first, what does A do, and, later, what controls whether gene A is active or not ? There is only space to mention, just to remind you, some of the key figures. Goldschmidt concluded that genes control rates of processes; Muller that they manufacture more, or less, or none of some substance, or sometimes an anti-substance, or even a quite new substance. Garrod (on human metabolism), followed by Haldane (on flower pigments), led on to the identification by Beadle and Ephrussi, in the early 1940s, of these substances as enzymes. Again in the forties, Hadorn in *Drosophila* and Grüneberg in mice studied the manifold developmental consequences which may follow an alteration in a single gene. Finally, about 1960, Jacob and Monod produced their story of the mechanism which controls the activity of single genes (or small operon groups) in prokaryotic organisms in which the chromosome normally lacks protein.

Clearly, this line of approach has paid off very well. But to my mind it does not really deal with the questions which the epigeneticist faces. For one thing, there is the difficulty that the Jacob-Monod control system could scarcely work, without considerable modification, in cells with proteinaceous chromosomes. But the problem is deeper. In cells of higher organisms we are not usually, if ever, confronted by the switching on or off of single genes. What we find is a whole complex cell becoming either a nerve or a kidney or a muscle cell In the late thirties I began developing the Whiteheadian notion that the process of becoming (say) a nerve cell should be regarded as the result of the activities of large numbers of genes, which interact together to form a unified 'concrescence'. This line of thought had several ramifications. For instance, just before the Beadle-Ephrussi era I showed in detail how the development of the wing of *Drosophila* is affected by the activities of some forty different genes. Again a few years earlier it had become apparent that the 'gene-concrescence' itself undergoes processes of change; at one embryonic period a given concrescence is in a phase of 'competence' and may be switched into one or other of a small number of alternative pathways of further change—but the competence later disappears and if you've missed the bus the switch won't work. My main preoccupation, however, was with the nature of the switches, which was the subject of the experimental biochemical work I was doing, with Needham and others, on 'embryonic induction'. Influenced—probably over-influenced—by genetics. I insisted that the switch must have sufficient specificity to recognize particular genes. We showed that, in these terms, the specificity resides inside the cells which react to induction—we called it 'the masked evocator'. This is very similar to the situation discovered by Jacob and Monod many years later in bacteria, where again the specific repressor molecules are internal to the cells which react to enzyme-inducing substances. If I had been more consistently Whiteheadian, I would probably have realized that the 'specificity' involved does not need to lie in the switch at all, but may be a property of the 'concrescence' and the ways in which it can change. Because of course what I have been calling by the Whiteheadian term 'concrescence' is what I have later called a *chreod*, a notion which René Thom has explicated; and the switches are Thom's *catastrophes*. The specificity *need* not be in what precipitates the catastrophe, but could reside only in the possible stable regimes (limit cycles in the simplest case) into which the system could be flipped.

Let us return to the beginning. The whole of this, I am afraid rather long-winded, exposition has not been aimed at showing that my line of thought, which I have derived from Whitehead, is the correct or best one. I should in fact admit that it has not so obviously paid off as the 'what does a single gene do ?' line; although I also continue to feel that it tackles deeper, because more embracing, problems. However, the point was to illustrate the fact that metaphysical presuppositions may have a definite influence on the way in which scientific research proceeds. And that point I have, I think, established, even if you feel that my metaphysics has led me up the garden path. And, after all, I am a biologist; it is plants and animals that I'm interested in, not clever exercises in algebra or even chemistry. The garden path has its attractions for the likes of us, and all of us who want to understand living systems in their more complex and richer forms are fated to look like suckers to our colleagues who are content to make a quick (scientific) buck wherever they can build up a dead-sure pay-off.

Since I am an unagressive character, and was living in an agressively anti-metaphysical period, I chose not to expound publicly these philosophical views. An essay I wrote around 1928 on 'The Vitalist-Mechanism Controversy and the Process of Abstraction' was never published. Instead I tried to put the Whiteheadian outlook to actual use in particular experimental situations. So biologists uninterested in metaphysics do not notice what lies behind—though they usually react as though they feel obscurely uneasy—and philosophers like Marjorie Grene may get so far as to conclude (p. 65) that I am not a wholly orthodox mechanical materialist.

# The Status of Neo-Darwinism

J. Maynard Smith
University of Sussex

*Introduction.* By Darwinism is meant the idea that evolution is the result of natural selection. Neo-Darwinism adds to this idea a theory of heredity. In its most general form, the theory of heredity is Weismannism, i.e. it is the theory that changes in the hereditary material are in some sense independent of changes in the body or 'soma'. In particular, the theory of heredity is Mendelian, i.e. it assumes that heredity is atomic, and obeys either Mendel's laws or some modification of them explicable in terms of the behaviour of chromosomes (e.g. linkage, polyploidy).

There are two reasons for discussing neo-Darwinism at this conference. The first is that only in the study of evolution is there a body of biological theory in any way comparable to the theories of physics; a conference on theoretical biology can hardly refrain from discussing it. The second is that the theory in at least some formulations is tautological. 'The survival of the fittest' appears to mean merely that survivors survive. There seems little point in trying to explain evolution by a tautology.

In this article, therefore, I shall attempt first to formulate the theory of a non-tautological form. I shall then discuss what types of observation might refute it, because this is the best way of seeing whether the attempt at a non-tautological formulation has been a success. Finally, at a less philosophical level, I shall discuss what problems the theory can cope with, what problems are at present unsolved because of a lack of data or of adequate mathematical tools, and what problems seem at present inaccessible to solution without introducing new concepts.

▸ *The formulation of neo-Darwinism.* The main task of any theory of evolution is to explain adaptive complexity, i.e. to explain the same set of facts which Paley used as evidence of a Creator. Thus if we look at an organism, we find that it is composed of organs which are at the same time of great complexity and of a kind which ensures the survival and/or reproduction of their possessor. Evolution theory must explain the origin of such adaptations.

At the outset we are faced with a difficulty: we have no way of measuring the degree of complexity of a structure. Thus although most of us would readily agree that the organs of a man are more complex than those of an amoeba, and those of an amoeba more complex than those of a bacterium, we have no agreed criteria on which to base this decision, and no way of deciding by how much one organism is more complex than another.

It may therefore seem odd to start formulating a theory of evolution by introducing a term which cannot be fully defined. However, I see no escape from doing so. If organisms were not both complicated and adapted, living matter would not differ from dead matter, and evolution theory would have nothing to explain.

Evolution is explained in terms of three properties, multiplication, heredity and variation, which organisms can be observed to possess. They will be considered in turn.

(i) *Multiplication.* All living organisms are capable of increasing in numbers in at least some environment. Multiplication is necessary because without it natural selection is impossible; you cannot cull a herd which can only just maintain its numbers. It is a corollary of this condition that life must consist of individuals and not of a continuum if it is to evolve.

(ii) *Heredity.* Briefly, like must beget like. More precisely, before we can say that entities have heredity, a number of different kinds of entities, A, B, C, etc., must exist, and each must tend to produce offspring like itself. Thus fire, if supplied with fuel, will multiply; but it does not have heredity, because the nature of a fire is determined by the fuel it is burning, and not by the nature of the fire from which it was lit.

(iii) *Variation.* If heredity were perfect, evolution would be impossible. Occasionally an offspring must differ from its parent. Viewed in this light, variation is merely the unreliability of heredity, and as Pattee [1] has emphasized, the problem is to explain why the reliability of replication is so high, not why mistakes are sometimes made.

However, this is not the whole story. If variation is to lead to evolution, then some variations must alter 'fitness', and at least some of these must increase fitness. By fitness is simply meant the probability of survival and reproduction. A melanistic moth is by definition fitter if it is more likely to survive, and a myopic man may be fitter if his myopia enables him to escape the draft. [Much confusion has arisen because 'fit' is not used in this sense in the phrase 'the survival of the fittest'. If it were so used, the phrase would indeed be tautological. A more precise though less elegant (and hence less 'fit') phrase would be 'the survival of the adaptively complex', i.e. organisms are adaptively complex or, as Bohm might say, 'harmonious', because such organisms survive better than less harmonious ones.] It follows from this definition that fitnesses can only be compared in a specified environment or range of environments.

Given entities with those properties, variants of higher fitness will replace their less fit ancestors: according to neo-Darwinism, this replacement constitutes evolution. Very early in evolution there arose a distinction between 'genotype' and 'phenotype', because those genotypes which gave rise to a phenotype were fitter than those which did not. By 'genotype' I mean that part of an organism which is replicated; by 'phenotype' I mean a structure or sequence of structures developing under the instructions of the genotype, and whose function it is to ensure the replication of the genotype. (These are not quite the accepted meanings of the words in genetics, but I would rather misuse words than invent new ones.) To paraphrase Butler's unexpectedly perceptive remark that the chicken is the egg's means of ensuring the production of another egg, the phenotype is the genotype's way of ensuring the production of another genotype. Once there is a distinction between phenotype and genotype, there is a process of epigenesis, and a process of decoding whereby the instructions in the genotype are translated into the structure of the phenotype.

Something must now be said about the origin of new variations, i.e. of mutation. It has been said that mutation is 'random'. Apart from the difficulty of defining the word, the statement is in one sense untrue, because different mutagenic agents produce different kinds of change in the genetic material.

Observation suggests that two things are in fact true about mutation: (i) most mutations lower fitness. If this were not so, evolution would proceed without natural selection; (ii) if a variant phenotype arises because development occurs in a changed environment, this will not produce corresponding changes in the genotype, such as to give rise in the next generation to the variant phenotype. This is the Weismannist assumption, expressed colloquially by saying that acquired characters are not inherited. Note that it does *not* say that changes in the phenotype cannot cause mutations, because of course they can. The apparent randomness of mutation arises because genotype and phenotype are connected by an arbitrary code.

The Weismannist assumption is expressed in molecular terms in the 'central dogma', which states that information can pass from DNA to protein, but not from protein to DNA; more precisely, if a new kind of protein is introduced into a cell, this cannot direct the synthesis of a new DNA molecule able to direct the synthesis of more of the new protein.

So far I have been describing a set of properties of organisms or, more precisely, a set of properties which neo-Darwinism assumes all organisms to have. This is not by itself a theory of evolution. The theory of neo-Darwinism states that these properties are necessary and sufficient to account for the evolution of life on this planet to date.

The limitations of time and place are important: of time, because in future we shall doubtless control our own evolution and that of our domestic animals and plants by the direct biochemical manipulation of DNA; of place, because we have as yet no grounds for asserting that if evolution has occurred elsewhere in the universe, it has done so by neo-Darwinist processes, although I would be willing to conjecture that it has.

It may help to clarify my position if I say that I accept Bohm's (this symposium) argument that to understand biological function we must appeal to what is beyond function, and also his statement that 'current metaphysics' (= neo-Darwinism) appeals to survival value as the trans-functional feature. Where I think he goes wrong is in regarding this procedure as tautological. He has been misled by the phrase 'the survival of the fittest'. Of course Darwinism contains tautological features: any scientific theory containing two lines of algebra does so. That it is not tautological *in toto* is best demonstrated by showing that it can be falsified, as I will now try to do.

▸ *Possibilities of refuting neo-Darwinism.* If this formulation of neo-Darwinism is not tautological, it must be possible to suggest observations which would refute it. Such observations could take two forms: (i) it could be shown that the assumptions made by neo-Darwinism are not in fact true of all organisms; (ii) patterns of evolution may occur which are inexplicable on the neo-Darwinist assumptions.

The possibilities will be considered in turn. It seems unlikely that we can show that organisms do not multiply or do not vary. However, the assumptions about heredity and about the origin of new variation could readily be disproved if they are false. Thus it should be possible to demonstrate Lamarckist effects if they occur, or 'inertial' effects whereby if one mutation in a given direction has occurred the next mutation is more likely to be in the same direction, or even 'teleological' effects whereby a succession of mutations occur which are individually non-adaptive but which together adapt the organism to a new environment.

By and large, such types of mutational event seem not to happen, and it is difficult to see in molecular terms how they could happen. It is impossible to *prove* that they do not, just as it is impossible to prove that heat never flows from a cold body to a hot one. All one can do is to assume they don't until someone demonstrates that they do.

## The status of neo-Darwinism

I will turn now to the possibility that patterns of evolution occur which cannot be explained by neo-Darwinism. The first possibility is that evolutionary changes occur more rapidly than can be explained by neo-Darwinism. This would be quite easy to demonstrate if it occurred on a small scale in the laboratory. Thus suppose for example a population of fruit flies were kept at an unusually high temperature. By measuring the genetic variance of temperature tolerance in the population before starting, it would be possible to predict the maximum rate at which temperature tolerance would increase. If in fact it increased faster than this, then the population would have evolved by a mechanism other than neo-Darwinism.

However, most critics of neo-Darwinism accept that the theory works at the level of laboratory experimentation. They suggest instead that there are large-scale features of evolution which call for additional types of explanation. Here we are up against the difficulty that we do not understand epigenesis, and we therefore do not know how many mutations would be necessary, for example, in the genotype of a small dinosaur to turn it into a bird. Therefore we do not know how many generations of selection would be needed to produce the change. This difficulty, combined with the imperfections of the fossil record, mean that we are unlikely to be able to disprove neo-Darwinism by showing from an examination of the fossil record that evolution has proceeded too rapidly. All one can say is that where we do have a reasonably continuous record, the observed rates of change are many orders of magnitude slower than those which can be produced in the laboratory.

If, however, neo-Darwinism were false, one would expect to be able to demonstrate its falsity by examining the end-products, i.e. existing organisms. Thus it follows from neo-Darwinism that if we find an adaptively complex organ, then the organ will contribute to the survival or reproduction of its possessor. One apparent exception arises in cases such as the worker bee, which have organs favouring the survival of their close relatives; but since their close relatives share many of their genes, this is explicable on the grounds that the phenotype of the worker bee ensures the multiplication of its own genotype.

If one invents counter-examples, they seem absurd. Thus if someone discovers a deep-sea fish with varying numbers of luminous dots on its tail, the number at any one time having the property of being always a prime number, I should regard this as rather strong evidence against neo-Darwinism. And if the dots took up in turn the exact configuration of the various heavenly constellations, I should regard it as an adequate disproof. The apparent absurdity of these examples only shows that what we know about existing organisms is consistent with neo-Darwinism. It is of course true that there are complex organs whose function is not known. But if it were not the case that most organs can readily be understood as contributing to survival or reproduction, Darwinism would never have been accepted by biologists in the first place.

Thus there are conceivable observations in the fields of genetics, of evolutionary changes in the laboratory, and of physiology, which could disprove neo-Darwinism. In palaeontology, although there is perhaps no possibility of a formal disproof because in our present state of ignorance about epigenesis it would always be possible to argue that a sudden evolutionary change was due to a single mutation, there are in practice many conceivable observations which would throw grave doubts on the theory. It therefore seems to me absurd to argue that the theory is tautological, though I readily admit that it is often formulated tautologically.

At present there are in my opinion no adequate observational grounds for abandoning the theory. This is of course no reason for not seeking for such grounds—I am all for people looking for Lamarckian effects, or for exceptions to the central dogma. But in the meanwhile, the theory explains so much that it is impossible to operate in biology without accepting it, just as it is impossible to operate in physics without accepting Newtonian mechanics, or some other theory which subsumes Newtonian mechanics as a special case.

▶ *The successes and failures of neo-Darwinism.* I have suggested that neo-Darwinism has not as yet been refuted. But is it of any interest ? Does it tell us anything not immediately obvious ? Does it solve problems ? Since the ability to solve problems seems to me one of the essential characteristics of a scientific theory, these questions are important. Perhaps the easiest way of answering them is to list some of the problems which can be thought about within the context of neo-Darwinism, and which would be unanswerable in any other context. This is not to say that all these problems have been solved—I would say that the first four have been largely solved, the fifth only partly solved, that we lack essential data for the solution of the sixth, that there are accepted but partially erroneous solutions to the seventh and eighth, and that both conceptual difficulties and lack of information prevent the solution of the last:

(i) How rapidly will gene frequencies change under selection ?

(ii) How can one predict the effects of selection for a continuously varying character?

(iii) What processes are responsible for the genetic variability of sexually reproducing species ?

(iv) How many selective deaths are needed to replace one gene in a population by another ?

(v) Will selection bring genes affecting the same character on to the same chromosome ?

(vi) Can selection be responsible for the evolution of characters favourable for the species but not to the individual ?

(vii) Can one species divide into two without being separated by a barrier to migration ?

(viii) In what circumstances will sexual reproduction accelerate evolutionary change ?

(ix) Has there been time since the pre-Cambrian for selection to program the length of DNA known to exist in man ?

These problems—except for the last—illustrate the field within which neo-Darwinism has been successful. Even when problems are unsolved, this is because of a lack of data or of mathematical technique rather than of concepts.

The failures of neo-Darwinism arise because of the absence of theories in the adjacent fields of epigenesis and of ecology. Lacking a theory of epigenesis, we cannot say how many gene substitutions are required to convert a fin into a leg, or a monkey's brain into a human one. Consequently we cannot say how many generations of selection of what intensity were needed to produce those changes. There is one exception to this statement of ignorance. Knowing the genetic code, we also know how many mutational steps are needed to convert one protein into another. Consequently we can speak with more precision about the evolution of proteins than about the evolution of legs or brains.

The difficulties which arise from our ignorance of ecology can best be illustrated by discussing the related problem of whether neo-Darwinism can explain the evolution of increasing complexity. Neo-Darwinism predicts that *in the short term* individuals will change in such a way as to increase their fitness in the environment or range of environments existing at the time. This may lead to an increase or a decrease in complexity. Sometimes, as in the evolution of tapeworms or viruses, it has led in the direction of decreasing complexity, albeit in an increasingly complex environment.

Thus there is nothing in neo-Darwinism which enables us to predict a long-term increase in complexity. All one can say is that since the first living organisms were presumably very simple, then if any large change in complexity has occurred in any evolutionary lineage, it must have been in the direction of increasing complexity; as Thomas Hood might have said, 'Nowhere to go but up'. But why should there have been any striking change in complexity ? It is conceivable that the first living thing, although simple, was more complex than was strictly necessary to survive in the primitive soup, and that evolution of greater fitness meant the evolution of still simpler forms.

Intuitively, one feels that the answer to this is that life soon became differentiated into various forms, living in different ways, and that within such a complex ecosystem there would always be *some* way of life open which called for a more complex phenotype. This would be a self-perpetuating process. With the evolution of new species, further ecological niches would open up, and the complexity of the most complex species would increase. But this is intuition, not reason. It is equally easy to imagine that the first living organism promptly consumed all the available food and then became extinct.

What we need therefore is first a theory of ecological permanence, and then a theory of evolutionary ecology. The former would tell us what must be the relationships between the species composing an ecosystem if it is to be 'permanent', i.e. if all species are to survive, either in a static equilibrium or a limit cycle. In such a theory, the effects of each species on its own reproduction and on that of other species would be represented by a constant or constants. We want to know what criteria these constants must satisfy if the system is to be permanent. A start on this problem has been made by Kerner [2] and Leigh [3].

In evolutionary ecology these constants become variables, but with a relaxation time large compared to the ecological time scale. Each species would evolve so as to maximize the fitness of its members. If so, a permanent ecosystem might evolve into an impermanent one. For example, a predator-prey system might be permanent because the prey could burrow and so escape total extinction. But if the predator evolved the capacity to burrow too, the ecosystem would become impermanent.

What then are the criteria to be satisfied if an ecosystem is not only to be permanent, but is to give rise by evolution to permanent ecosystems of greater species diversity ? We have no idea. But the first living organism, with its food supply, had to comprise such an ecosystem if evolution was to lead to increasing complexity.

## References

1. Pattee, H.H. *Towards a Theoretical Biology 1* (Edinburgh University Press 1968).
2. Kerner, E.H. *Bull. Math. Biophys. 19* (1957) 121–46.
3. Leigh, E.G. *Proc. Nat. Acad. Sci. 53* (1965) 777–83.

## Addendum on Order and neo-Darwinism by David Bohm

After finishing my remarks on order for the Bellagio Conference I remembered a very interesting conversation with Maynard Smith, the significance of which suddenly struck me. So it seemed appropriate to add a brief discussion of what we talked about, because it so nicely brings out what I want to say about the effects of tacit metaphysical assumptions on the whole course of one's thinking.

In this discussion I shall *underline* metaphysical words such as *all, only, always, never, basic, relevant, significant*, etc.

The conversation in question began when I recalled an article that I had read about population limitation by birds. In this article it was stated that birds (and other animals) had somehow evolved methods of controlling population so as not to outrun their food supply. In particular it was noted that not all the birds mated, and that when the flock was large, the fraction that mated was correspondingly reduced. To explain this behaviour it was proposed that it was related to the swarming of the entire flock, which occurred before mating. The suggested explanation of population limitation was, then, that when the birds swarmed, each one could note how large the flock was. The perception would somehow influence the propensity of each bird to mate, in such a way that when the flock was large the fraction would decrease.

This explanation was very severely criticised by Maynard Smith on the basis of neo-Darwinist conceptions of the evolutionary process. In essence, the argument was that those birds that tended not to mate in given circumstances (e.g. the perception of a large swarm) would not pass on this characteristic to succeeding generations. On the other hand, those birds that did tend to mate would pass on the tendency to mate. So in the long run all birds would tend to mate, no matter how large the swarm. Therefore the explanation cannot be accepted as a possible one.

In saying this, Maynard Smith emphasized very strongly that hereditary characteristics are passed on *only* in the individual organism, and *never* by any collective entity that could be called the 'species' or the 'group'. Indeed, the notion that any given characteristic has survival value for the species or the group as a whole was implied to have no *direct relevance* for giving an account of how the genetic process actually takes place, because there is no way *at all* in this process for the state of the species or the group to play a causative role. After all, one would hardly want to assume, for example, that each individual bird has some 'altruistic' tendency to put the welfare of the flock first, when it comes to determining whether he will mate or not. Such behaviour might conceivably occur in individual human beings, who can understand abstractions such as 'the community as a whole', but almost certainly, individual birds cannot formulate such abstractions.

At the time, I found Maynard Smith's arguments very convincing. But now I can see that they are based on a number of tacit metaphysical assumptions which could perhaps usefully be questioned. In essence, these assumptions are: (1) Changes in the genotype are *always* related to the changing experiences of the phenotype in a *completely fortuitous* way; (2) Systematic changes in the behaviour of the phenotype are *always* the necessary results of changes in the genotype, and *never* occur without such changes.

In explanation of assumption (2) it should be stated that, of course, no one would do something so absurd as to suppose that *every* change in the phenotype has to be caused by a corresponding change in the genotype. What is assumed in neo-Darwinism is, as far as I have been able to tell, that changes in the phenotype not thus caused by changes in the genotype have a *completely fortuitous* relationship to the transmission of hereditary characteristics.

Thus, if different birds had different tendencies to refrain from mating when the flock is perceived by them to be large, these tendencies would either have to correspond *only* to genetic differences between the individual birds, or else they would have to be *totally fortuitous* in the way in which it entered into the genetic process as a whole. If the tendency is determined *only* by the genotype of the individual bird, then, as has been seen, that genotype cannot survive, and if it is *entirely fortuitous*, then by hypothesis, there is no means by which it could be transmitted.

Let us now see whether we cannot get outside of these rather limited metaphysical possibilities, in a way that does not violate our feelings as to what seems reasonable or possible, in the light of all that has been learned over the past few hundred years. To do this, let us begin by supposing, for the sake of argument, that the genetic constitution of the birds is such that the tendency to mate in each individual is not completely determined by the genetic structure alone. That is to say, we allow room for variations or fluctuations in this tendency, even among birds that all have the same genes. However, instead of supposing that the variations or fluctuations in the behaviour of individuals is subject to an *absolute* randomness (disorder, lawlessness, etc.) we propose that they are *only* 'relatively' fortuitous. That is to say, they are fortuitous in relationship to the genetic constitution of the individual birds, but not in relationship to the actual situation of these birds in their overall environment.

Let us assume, for example, that the propensity for mating of a given bird depends not only on the size of the flock when swarming took place, but also on his position in the flock, as well as on myriads of other factors, such as what food he has eaten, what is the state of the wind as it ruffles his feathers, etc. So even in birds of the same genetic constitution, there would inevitably be 'scatter' in mating behaviour, which would however vary systematically in such a way that a large flock would tend to reduce its rate of mating.

Of course, one could use the theory of probability, with its 'normal distributions' to *describe* such a 'scatter'. But at present we conventionally refer also to *laws* of probability, implying thereby a kind of 'ironbound' necessity for *complete randomness, disorder,* and *lawlessness* in the variations and fluctuations of behaviour of individual birds that make up the 'scatter'. Once we realize that probability theory is *only* a *conventional* (as well as approximate) *description* and not a *law,* we are free to consider limits in the degree and kinds of fortuitousness of this behaviour, which lead to radical changes in the meaning of the whole picture.

One of these changes is that hereditary characteristics can now belong to the group or species without being analytically deducible from properties of the individual considered in isolation from the group. For the genetic constitution of the individual may determine *only* a range of possible behaviour. The actual behaviour of each individual is fortuitously dependent on the *total* environment, and this latter can include the *actual* group. For example, the tendency of birds to refrain from mating may prevent the group from exhausting its food supply. Because *all* the individuals have, in this regard, the same genetic constitution, those that mate do not thereby propagate to their offspring any more tendency to mate than is possessed by those birds that refrained from mating.

In thus noting the possibility of going outside current metaphysical assumptions I do not wish to blame or criticize those scientists who hold particular views, such as neo-Darwinism. Indeed, such views have evidently been very fruitful, and have helped lead people to many useful and interesting discoveries. But now it may well be necessary to give attention to various tacit restrictions that could be getting in the way of further progress.

In this regard, the situation in biology is not really basically different from that prevailing in other sciences, such as physics. For example, quantum theory has certainly been a gigantic step forward. Nevertheless, it contains the tacit metaphysical assumption that when the quantum state (i.e. the wave function) is determined, then the fluctuation in behaviour of individual atoms is *completely random*, and must remain so, no matter what question the physicist may come to inquire into, nor what conditions may come to be established for these atoms. So, in effect, there is an assumption of a *law of lawlessness* (which is evidently in some ways an inherently self-contradictory notion). Would it not be more reasonable to suppose instead that the behaviour of individual atoms is fortuitously related to the quantum state, and that it might therefore be non-fortuitously related to other things (some of which may perhaps be thought of only in the future)? Then, both for the birds and for the atoms, we could consider that some of the factors responsible for fortuitous variations could change systematically in certain (generally as yet unknown) contexts. It is clear that it will be useful to remain alert to these extended possibilities rather than to close our minds with tacit assumptions of whose metaphysical character we are in general not even aware.

## Comment on Bohm's Addendum by J. Maynard Smith

The example which Bohm discusses serves very well to bring out the difference between us. The theory of population regulation which Bohm mentions is due to Wynne-Edwards [1]. My own interest in the theory was aroused because on the one hand it appeared, for the reasons given by Bohm, to contradict neo-Darwinism, but on the other hand it provided an explanation for certain phenomena not otherwise easily explicable. I am enough of a Popperian to think that a proponent of neo-Darwinism should pay particular attention to observations or ideas which appear to contradict his theory.

In thinking about Wynne-Edwards' ideas I was led to make almost precisely the assumption about the effects of genotype and environment on breeding behaviour suggested by Bohm. I considered [2] the evolution of a species divided into reproductively isolated groups, each group consisting of genetically identical individuals. Some groups consisted of individuals which would refrain from breeding in certain circumstances; thus when a group grew too large, some members would refrain from breeding, but which members refrained would depend, not on genetic constitution which was the same for all members of the group, but on such things as age, position in the flock, etc. It was supposed that groups consisting of such individuals were at a selective advantage, because less likely to outrun their food supply, over groups consisting of genetically different individuals which continued to breed irrespective of circumstances.

## The status of neo-Darwinism

Thus although I considered the possibility mentioned by Bohm, I also took into account the possibility that there would be genetic as well as environmental differences in readiness to breed. I think Bohm makes the tacit metaphysical assumption that all individuals are genetically identical. The justification for taking genetic differences into account is that I know of no phenotypic character for which there is not a genetic component of the variance. The crucial question is therefore this: granted that there are genetic differences (e.g. in the density which must be reached before an individual in given circumstances ceases to breed), how will the species evolve ? The answer depends, among other things, on the size of the groups and the amount of migration and interbreeding between them. As it happens, we have little information about populations from this point of view, and it was for this reason that in my article on 'the status of neo-Darwinism' (written before I saw Bohm's addendum) I listed the problem of the evolution of 'altruistic' traits as one for whose solution we lack factual information.

Now there is nothing in all this which runs contrary to neo-Darwinism. It may well be that the particular model I chose for the evolution of an altruistic trait was the wrong one. What is important is that it is possible within the framework of neo-Darwinism to analyse the evolution of population regulation, of altruistic traits, and of the related phenomenon of sexual reproduction, and to make fairly precise statements about the conditions which must be satisfied if particular characteristics are to evolve. But if the Weismannist assumption were relaxed, it would not be possible to say anything of comparable precision. It follows that if experimental evidence were to oblige us to abandon Weismannism, our theory of evolution would lose most of its power to make precise and testable statements.

Thus while I agree with Bohm that we should make our assumptions explicit, that we should question them, and that if the evidence requires it we should alter or abandon them, I cannot see why he should think that 'it may well be necessary to give attention to various tacit restrictions, that could be getting in the way of further progress'. I think science progresses by making assumptions more restrictive, not less so. It would be as easy to abandon Weismannism in genetics as to abandon the law of conservation of momentum, but the result in both cases would be a loss and not a gain in explanatory power.

### References

1. Wynne-Edwards, V. C. *Animal Dispersion in relation to Social Behaviour* (Oliver and Boyd 1962).

2. Maynard Smith, J. *Nature, 201* (1964) 1145.

## Note on Bohm's Addendum by C. H. Waddington

If the conversation between Bohm and Maynard Smith really took the form reported here, it seems to me to have been rather unsatisfactory. It is made to appear that Maynard Smith adopted the most narrowly orthodox neo-Darwinian line, implying that selection acts directly on genotypes, so that the selection against birds which fail to mate in dense populations is bound to be effective. Bohm rightly makes the point, which I have been insisting on in several places in these essays, that selection acts on phenotypes, and that there are many steps between the phenotype and the genotype. This point is actually often admitted in some neo-Darwinian genetics which can be considered relatively orthodox, though more frequently in the context of artificial selection; it is less often that its full implications for the theory of evolution in nature are considered (see my article on The Paradigm for an Evolutionary Process).

One does not, however, have to be very unorthodox to admit that selection may sometimes be ineffective, so that the failure of certain birds in a population to have offspring may have negligible or even no genetic consequences. The subject of 'selection limits' is of great importance to practical breeders concerned, for instance, with increasing the numbers of eggs laid by hens, where there would at first sight seem to be obvious selective advantage under natural conditions, let alone artificial ones. It is well known that natural selection does not in practice lead to a continual increase in the number of eggs produced by an individual; and poultry breeders find that, practising artificial selection, they can push up the number quite a way, but then come to a standstill. Geneticists concerned with the fundamental mechanisms of animal breeding have made considerable studies on the problem, e.g. from my own laboratory [1–4].

There are a number of mechanisms which can be appealed to: (i) The individuals in the population may be genetically identical as far as the selected character is concerned. (More strictly, the additive genetic variance for this character may be exhausted.) This is the explanation put forward by David Bohm. It is likely to occur only after the selection has been effective over a considerable number of generations. (ii) The phenotypes selected may be determined by heterozygous genotypes. In the simplest case, if, in a population consisting of *AA*, *Aa* and *aa* individuals, the *Aa* ones breed less (or more) than both the *AA* and *aa*, the situation settles to an equilibrium, after which no effect is produced on the frequency of the genes in the next generation; (iii) The phenotype selected for by one criterion may, for physiological or other reasons which are very difficult to circumvent, be selected against by some other criterion, which will probably operate at some other point in the life history. This seems to be usually the most powerful determinant of a selection limit; (iv) It is possible that there are inherent physiological limits, which random mutation has not succeeded in finding a way round. For instance, the number of eggs laid by a bird may be limited by the energy-expenditure required to digest enough food to have a surplus to lay down as stores in eggs. But it is doubtful whether one can find any cases of this which cannot more usefully be considered as examples of (iii).

Finally, there is a mechanism of a quite different order which has been invoked. Suppose that there are no selection limits of the kind just described which restrain the progressive increase in birds which will mate however high the population-density rises; it is suggested that a flock which has gone far along this line may, in a bad season, find it eats itself out of a living, and thus leaves an empty space into which there expands some nearby flock in which the deterioration of the control of population numbers had not gone so far. The reality and power of such 'inter-population' or 'inter-deme' selection is at present a highly controversial issue. The true-blue orthodox neo-Mendelians won't have it at any price, maintaining either that it can't or doesn't actually occur in nature, or that, if it did, it would do no more than slow up the processes that would occur without it; the more unorthodox argue that it might slow them so much as to deprive them of any practical validity; or, going further, that it would lead to the formation of gene-pools in which the control system was built in, for instance by producing populations in which the available genes fix a selection limit by one of the mechanisms (i) to (iii) above.

The only point of listing all these technicalities of selection-genetics is to show that we know what the form is sufficiently to put them on one side and come down to what, I take it, David was really getting at.

He used what he understood Maynard Smith to say (whether he really did so or not) to make the following main points:

1 To argue that, if selection favours individuals with a certain character, there will be a genotype for that character, and the frequency of that genotype will increase, is to indulge in metaphysics. And, moreover, in a metaphysics of a 'thing' kind rather than a 'process' kind. A genotype, i.e. a hereditary memory-store, is a static thing, a phenotype—the results of epigenesis—is a process. This is a point that I also have been making, *ad instantum* rather than *in abstracto*, throughout these meetings. The point to be emphasized is that much orthodox conventional scientific theorizing is full of unacknowledged metaphysical assumptions, which it is as well to have out into daylight.

## Notes by Marjorie Grene

2 He wants to point out that if an entity has characteristics which are strictly determined in contexts in which you have earlier been interested (organisms with a specified genotype, atoms with a determined quantum state), they may exhibit properties in other contexts, and these properties need not be *completely random* or *lawless*, but may be related in very interesting ways to other variables in the situation which have previously not attracted attention but which may repay study. I think no biologist would deny that the phenotype variations which can be produced among a group of genotypically identical individuals are not completely random or lawless, but are related, in interesting ways, to the environmental circumstances in which epigenesis occurs (cf. the studies on identical twins reared apart). Here the main point would seem to be that, even when we have 'scientific' answers, we need to realize that they are partial answers to only *some* of the questions worth asking.

### References


1. Falconer, D.S. and King, J.W.B. *J. Genet.*, *51* (1953) 561.
2. Falconer, D.S. *Introduction to Quantitative Genetics* (Oliver and Boyd, Edinburgh and London, 1960).
3. Roberts, R.C. *Genet. Res.*, *8* (1966) 347.
4. Robertson, A. *Proc. Roy. Soc. B*, 153, 234.


## Notes on Maynard Smith's 'Status of neo-Darwinism' by Marjorie Grene

Darwin demonstrated that given multiplication, heredity, and variation, natural selection follows. The first three were taken as given facts which together entail the fourth. (So powerful was his argument that it now seems natural selection is what we have, and multiplication what we infer from it: 'multiplication is necessary because without it natural selection is impossible'!) But Darwin was aware, at least sometimes, of the limitations of his theory in a way in which many of his twentieth century successors are not. First, he recognized that natural selection controls only adaptive characters and that some of the complexities of biological phenomena may not be adaptations. (They cannot survive if they are maladaptive, but natural selection may leave room for viable alternatives.) True, he was inclined to think of organisms primarily in Paleyan terms, that is, as mechanisms for self-maintenance and survival, but he did admit the possibility that not all characters might be explicable in these terms. And secondly, he recognized, at least occasionally, that natural selection theory does not offer any explanation of increasing complexity. This point is made whimsically in Kingsley's legend of a land where animals 'progress' backwards—from man to amoeba, as it were.

## The status of neo-Darwinism

Maynard Smith's statement seems to me to illustrate admirably the confusion of contemporary theorists with respect to these two limitations. Of course, as long as natural selection theory remains a theory for studying natural selection it provides a powerful framework for experimentation on such questions as Maynard Smith enumerates. But what happens when he tries to show that this framework is adequate to explain evolution as a whole ? He wants to show that the theory is not tautological : that it is not just a device for measuring the survival of what survives. To do this he has to distinguish the 'fittest' as the 'harmonious' or the 'adaptively complex' from the genetically 'fit', i.e. those most likely to leave descendants. But either, as he sometimes argues, such 'complexities' are again to be assessed as superior in adaptivity, that is, in genetical fitness, and so the theory again collapses into tautology. Or 'complexity' is not reducible to survival value, and then, as he also admits, Darwinism or neo-Darwinism has no theory of increasing complexity at all. But increasing complexity is an essential feature of evolution. So neo-Darwinism provides an excellent and ingenious set of analytical instruments for measuring selective phenomena, plus either a vacuous theory of evolution or none at all.

Further insights may come, Maynard Smith suggests, from theories of epigenesis and of ecology. Bohm suggests that they may come from a study of creative process in non-living as well as living nature. Such theories would by no means undermine natural selection theory in so far as it limits itself to its proper range ; but they would supplement it by explaining evolution in terms of laws and processes for which selection provides the necessary, but not the sufficient, conditions. Sufficient conditions it seems to provide only when it (1) extrapolates adaptivity to all biological phenomena, thus pretending to make the functional self-sufficient in a manner in which it cannot be so (see Bohm's notes and my comments on them) and (2) surreptitiously assimilates the concept of complexity, or of an increasing order of order, to that of adaptation, so extrapolated.

## Some comments on Maynard Smith's contributions by David Bohm

Maynard Smith in his *Status of neo-Darwinism* has made what I regard as a very useful contribution to the clarification of the metaphysical assumptions underlying this subject. As indicated in my earlier articles, the main point about anyone's metaphysics is not to criticize it (though this can often be an appropriate thing to do). Rather is it to bring out what the metaphysics is, to make clear what is fact, and what assumption. I think that, in general, Maynard Smith has done just this with regard to neo-Darwinism. However, it seems to me that in a number of significant respects he is still overlooking metaphysical assumptions by tacitly accepting as fact certain notions that are in reality either suppositions or conventional definitions of terms.

The question at issue concerns Maynard Smith's reply to the frequently-voiced criticism that the basic conceptions of neo-Darwinism are tautological. He answers this on two levels, which I shall discuss in succession.

First, he says that there is a clear and unambiguous (i.e. tautology-free) definition of fitness, which is simply 'the probability of survival and reproduction'. (Adding also that this notion is more precise and elegant than that of 'the survival of the adaptively complex' or of my own notion of 'harmony'.) Now, in my view, the words 'probability of survival and reproduction' have little or no meaning, except in certain very narrow and strictly limited contexts, which could well fail to exist, for the most part in typical natural environments.

It should, however, be emphasized here that I am questioning not merely the metaphysics of biology, but also that involved in the use of probability theory in physics, and, generally, in all applications of statistics. The basic point, the significance of which has, in my opinion, been almost universally overlooked, is that probability theory has meaning only when a situation of randomness prevails. As indicated in my earlier articles, the problem of randomness has never been clearly defined, and is indeed full of sources of confusion and contradiction. Nevertheless, one sees that mathematicians and scientists do have at least certain (more or less tacit) notions in mind when they apply such ideas. Consider, for example, a series of coin throws. Not only does the average frequency of heads or tails tend to come near to a half in a long series of throws. Equally important, if we select arbitrary sub-sequences in which any particular results have already been obtained (e.g. ten heads in a row), then the frequency of heads still tends to come near to a half in a long series of throws. This is (at least in part) what is meant by randomness of the sequence. If this requirement is not satisfied, then the application of probability theory can lead to confused and erroneous results.

Now, evidently, the above condition for randomness is satisfied in coin throws. It is also satisfied in many applications in physics, chemistry, biology, and social statistics. But it is not clear that wherever there is a 'scatter' or variations of results, it is always possible to define a meaningful probability of occurrence of a certain kind of event. For this 'scatter' may, in certain interesting and significant ways, fail to be random.

Of course, statisticians are well aware of this problem. To try to deal with it they have introduced the notion of probabilities of correlation of various classes of events (e.g. in a non-random sequence, the probability that B follows A is different from the probability of B in the whole sequence). But this depends on the (usually tacit) assumption of a yet higher order of randomness which applies to the distribution of correlated sets of events. In other words, the definition of probabilities of correlation does not remove the problem of establishing randomness; it only pushes it into the more obscure problem of what happens to higher order combinations of correlated events. For reasons similar to those given in connection with simple sequences of events, it follows, therefore, that the failure of randomness cannot always be comprehended in terms of probabilities of correlation of various classes of events.

Consider, for example, the Wynne-Edwards theory of limitation of bird populations, discussed in my 'Addendum' and in Maynard Smith's 'A Comment on Bohm's Addendum'. In the 'Addendum' I proposed that a given set of birds may have a common genetic constitution, with, however, a variable propensity for mating, that depends on various factors in the environment. (Of course, Maynard Smith is right to point out in his 'Comment' that the consideration of a fixed genetic constitution is also an assumption. However, it was my intention to propose this merely as a convenient basis for argument, and not as a fixed and generally valid metaphysics.) Now, these environmental factors are in general changing in a complex and non-repeating way. Not only is this happening in the external environment. Even more, it is being proposed that the flock or group of birds is itself a key part of the environment of each bird, in the sense that the propensity for mating of such a bird depends on the size of the flock that he perceived in swarming; where he was, in the flock, when he perceived it; and on many other factors, most of them not precisely or clearly specifiable.

Of course, one may, if one wishes, assume that all these factors, specifiable and unspecifiable, fluctuate at random in a way that has no *significant** relationship to the survival of the individual or the group. However, the theory of Wynne-Edwards specifically proposed that (among other factors) the propensity for mating depends on the size of the flock. This is therefore not a 'randomly' fluctuating variable. On the contrary, it is (in this theory) a key causal factor

* In this article I am using the convention of italicizing metaphysical words.

on which depends the whole process of reproduction and survival of the flock.

Now, consider the situation of a biologist who is trying to discover what it means to assume a 'probability of survival and reproduction'. Of course, in the relatively strictly-controlled conditions prevailing either in a laboratory or in the practical work of animal or plant breeders, the *relevant* factors may for the most part either be fixed or else fluctuate 'at random'. In these cases, such probabilities can be meaningfully defined. But then, in nature, there is a vast range of processes that vary in a way that is neither known nor controlled. Under natural conditions these processes may sometimes be random, but not *always*. It is pure metaphysics to suppose that assumptions of randomness that are valid under artificial or carefully selected natural conditions will continue to hold for *all* or even for *typical* natural conditions. The theory of Wynne-Edwards is only one example of how, under certain natural conditions, new variables (the size of the flock) may appear which are significant for the whole process of reproduction, and which are neither fixed nor 'randomly fluctuating' in relationship to the process. But as far as anybody can tell *a priori*, there is a potential infinity of such as yet unknown variables. Whenever one assumes that there exists a 'probability of survival and reproduction', one is assuming tacitly that the *relevant* effects of *all* that is unknown are *always* either fixed or else fluctuate at 'random' in relationship to the process of survival and reproduction.

Even if one assumes a 'probability of correlation of the process of survival and reproduction to this potential infinity of unknown variables', one does not solve this problem. For, as has been already indicated, such a probability has meaning only if the corresponding yet higher order combinations of variables are either fixed or fluctuate 'at random'. But after all, any particular natural process of evolution covers a limited period of time. If there is an essentially unlimited number of fluctuating 'variables' of very many kinds, and only a limited number of individuals and of generations of individuals, it is not at all unreasonable to suppose that the total set of events is not in general large enough to satisfy the necessary conditions for the meaningful application of the concept of 'probability of survival and reproduction'.

Unless the conditions for randomness described above are actually satisfied, the neo-Darwinist conception of 'probability of survival and reproduction' becomes confused. At best, it can be regarded as a tautology and, at worst, as a definitely self-contradictory notion. If one says that 'probability of survival and reproduction' means the actual relative frequency of creatures of a certain kind that did in fact survive in a specified historical process, then it is evidently a tautology. If it is taken to mean more than this, in the sense that one uses the probability concept to draw further inferences, that process is not logically justifiable, and may in fact lead to erroneous and contradictory results.

In my view, René Thom and Waddington have indicated a possible way out of this difficulty. Instead of thinking of the random fluctuations implied by the concept of probability, one thinks of a 'chreod', describing a vast range of possible ordered developments having certain relatively stable features. In this 'chreodic' process, it is essential to emphasize that selection operates on the phenotype rather than on the genotype. The main role of the genotype is to contribute to the conditions of selection. Thus, as Waddington has often emphasized, in general, a particular gene does not determine either survival or probability of survival (except possibly under certain limited conditions). Rather, like any other factor in the overall environment, it makes a certain contribution that is relevant in the total process of life that may lead either to survival or to non-survival. Whether there is survival or not depends in fact on the degree and kind of harmony that exists between phenotype and environment. Changes of the genotype will in general change this harmony in certain ways (either favourable or unfavourable). Changes of the environment will likewise also alter the harmony. But it is all one process in which effects of genotype and environment on the phenotype are interwoven in an inseparable way.

The tendency of neo-Darwinism to treat genotype and phenotype as two quasi-independent kinds of processes is based on tacit assumptions of randomness that can be expected to have in general only a limited applicability. So while neo-Darwinism may well lead to correct results in certain fields of study, it seems necessary to note that the extension of this theory to the *whole* of natural evolution is a kind of metaphysical assumption that there is good reason to question.

Now I come to Maynard Smith's second level of answer to the criticism that neo-Darwinism is tautological. This is that one can show the non-tautological character of neo-Darwinism by suggesting observations that would refute it.

The difficulty with such a response is that if the basic concepts of neo-Darwinism should (as seems more likely than not) turn out to be inapplicable or confused in their meanings outside of certain rather limited kinds of contexts that are not typical of natural evolution, it is hard to see why the refutability of the theory in its limited context of applicability would answer the criticism of tautology in broader contexts. Here it is surely significant that Maynard Smith attributes what he calls the 'failures of neo-Darwinism' to the absence of theories in the adjacent fields of epigenesis and ecology. As I have already indicated, it is just in these domains that one is led to question seriously whether concepts like 'probability of survival' have any real meaning (because conditions for randomness are not satisfied). Indeed, it may well be that the efforts of scientists to think in terms of such concepts is a principal barrier to progress in these fields. Rather, concepts like 'chreod' and 'harmony' may be much more relevant here.

One sees here a characteristic difficulty in finding evidence that could really refute a general and basic set of metaphysical assumptions. It is generally admitted that each theory has its 'failures'. But only very rarely does one regard these failures as evidence tending to refute the metaphysics. Rather, one places the fault in particular failures to solve certain 'problems' that arise within the framework of the theory. But if one is not fully aware of the metaphysical assumptions, he may not see that these problems could actually be inherently and basically insoluble, because his metaphysics makes assumptions whose meanings are confused when extended out of certain limited areas, where they may work fairly well.

It is here that it is relevant to consider Maynard Smith's 'Comment' on my 'Addendum'. Towards the end he says that science progresses by making assumptions more restrictive, not less so. I wonder whether this is *always* true, or whether it is only sometimes true. If we accept it as *always* true, we have made ourselves a particular kind of metaphysics concerning the nature of scientific research. Of course, modern science did in fact grow up, at least in part, in a struggle against the vague, speculative, and untestable features of mediaeval metaphysics. But does it not now tend to overemphasize sharpness of definition of concepts and testability (or refutability), thus jumping to the error of the opposite extreme ?

I think that Popper's emphasis on refutability has been wrongly understood, in many ways. It seems to me almost as if many scientists would give the highest *value* to any theory, merely because it happens to be easily refutable, while vague notions of potentially very great scope are frequently brushed aside, as having little or no *significance.* Here I would like to stress that the mere restrictiveness and refutability of a theory is not *always* the main consideration determining its *value.* Consider, for example, Ohm's law in physics, which is wonderfully precisely defined, restrictive, and refutable, but which is not generally regarded as having a deep and broad *significance.* On the other hand, early in this century there were vague notions, such as that light shining on metals will liberate electricity. It was very hard at that time to make precise and restrictive assumptions about this phenomenon. Yet, in time, the study of it led to the photo-electric effect, and helped lead to the quantum theory. So I would like to emphasize, not the immediate restrictiveness and refutability of a theory, but rather that these are features which we can reasonably aim to achieve ultimately. It does not, therefore, follow that a theory that is immediately more restrictive than another is *always* a better one for the progress of science.

Perhaps one could here usefully recall the notion that Nature is more like an artist than an engineer (which was discussed in my 'Further Remarks on Order' paper). Therefore it requires a basically artistic attitude to understand it. Now, when an artist wants to learn to make a picture, for example of a man, he begins by indicating the general form first in a vague way and then step by step learns with the aid of careful observation how to articulate this picture in more detail. The attempt to draw an extremely detailed picture at too early a stage inevitably leads to confusion, because until one has drawn the general outlines one generally does not know what is the next level of articulation of structure that it is appropriate to look at. Similarly I would propose here that in science it is generally necessary to begin with vague general notions at an early stage, and to articulate these notions later as the subject develops. The metaphysical assumption that a more restrictive and more easily refutable theory is *always* a better one is therefore liable to lead at best to superficiality (as the accidentally defined details of the theory are given basic *significance*) or to confusion (as its concepts are extended too far beyond the limited experimental domain in which they have been demonstrated to work).

With regard to biology, I want to say that in my view, at least, this subject is now in an extremely early stage of development. Therefore somewhat vague notions like chreod, harmony, etc., may well have a key part to play. Rather than *base* everything on precisely defined ideas like 'genotype determines probability of survival and reproduction' (which could very easily have no meaning in typical areas of the domain of natural evolutionary process), it seems to me that it is more *relevant* to consider both genotype and external environment as different but closely interwoven aspects of what may be called the 'total environment' or 'total set of conditions' of the phenotype.

These considerations on falsifiability and on the question of vagueness *v.* restrictively precise definiteness of a theory help, I think, to explain a certain misunderstanding between Maynard Smith and myself that comes out in the 'Addendum'. When I received Maynard Smith's 'Comments' on this 'Addendum', I was surprised to learn that as far as I could see he had no very strong objections to the Wynne-Edwards theory of regulation of bird populations, and that indeed he had himself tried to work out a model of such a theory. How, I asked, was it possible for such a misunderstanding to develop ? The answer that occurred to me is that it was due to our deep metaphysical differences. To me, nothing that Maynard Smith wrote seems to constitute any real evidence against the theory of Wynne-Edwards. But as far as I can see, the fact that this theory has not yet been formulated in a precise, restricted, and easily refutable way seems to mean to Maynard Smith that it is not conducive to the progress of science. Since I did not, at the time, understand his point of view very well, I tacitly assumed that his severe criticism of the theory implied that he had much stronger arguments against the theory than those that he actually wanted to make (i.e. that it was not restrictive enough).

# Paradigm for an evolutionary Process

C. H. Waddington
University of Edinburgh

The theory of evolution has passed through two main phases. In the first ('Darwinian'), the essence of the process was held to be natural selection operating on a hereditary system characterized by 'blending' inheritance, in which new hereditary variation was brought into being by some unknown mechanism. In the second ('Post-Darwinian'), the hereditary system on which natural selection acts had the Mendelian properties of dependence on discrete alternative states of the hereditary factors and the production of new variation by mutation. The enormous advances made in our understanding of evolution by calling on the resources of Mendelian genetics need no emphasis.

The formulation of the logical structure of the typical, or paradigm, process of evolution, assuming Mendelian heredity, was initially carried out with perhaps greater attention to refuting certain loudly expressed objections to the new outlook than to developing fully its own inherent character. Of the great triumvirate who laid the first foundations for a fully logical, i.e. mathematical, formulation of Mendelian evolution, Haldane and Fisher were English—and England had just seen one of the most ferocious (and silly) of all academic battles, in which the anti-Mendelians, led by Pearson and Wheldon, had gone down to defeat at the hands of the believers in Mendelism and Discontinuous Variation, whose champion was Bateson [1]. In the heat of the fight there had been some corrosion, on both sides, of trust in Darwinian mechanisms of evolution. Bateson, emphasizing discontinuity not only in heredity but also in phenotypic variation, toyed with ideas of Mutationism allied to those of de Vries, while believers in various types of over-riding evolutionary forces, such as orthogenesis, found that the vagueness of the material basis for the hereditary system postulated by Wheldon and Pearson offered scope for their own equally nebulous ideas. Thus the two English progenitors of what came to be called 'Neo-Darwinism' considered that one of their main tasks was to establish, as clearly as possible, that Darwinian natural selection would, after all, 'work' in Mendelian populations. The other main pioneer, the American Sewall Wright, was less affected by these predominantly British squabbles. His mathematical formulation is far less drastically

This essay is dedicated, with respect and affection, to my good friend Theodosius Dobzhansky, in celebration of his seventieth birthday.

simplified for polemical purposes; but its richer intellectual content calls for subtler and more difficult mathematics, and until recently it has been less influential than those of Haldane and Fisher.

Before attempting to re-formulate the essential logical features of an evolutionary process, it will be as well to remind ourselves of the formulations which were given by the pioneers. The first and simplest of them was Haldane, beginning in a series of papers in the obscurity of the Proceedings of the Cambridge *Philosophical Society* from 1924 onwards. This work was summarized in his book *Causes of Evolution*, 1932. An indication of the atmosphere in which it was written can be found from the quotation he chose as the motto for the Introduction: 'Darwinism is dead.—Any sermon'. On page 20 we find the statement, surprisingly apologetic to modern eyes: 'But I propose to anticipate my future argument to the extent of stating my belief that, in spite of the above criticisms, which are all perfectly valid, natural selection is an important cause of evolution.' The argument, when it comes, is related to systems of which the paradigm case is described as follows (pp. 180, 181): 'In a random-mating group a population composed of the three genotypes in the ratio $u^2AA:2uAa:1aa$ is stable in the absence of selection, and any group whatever reaches this stable equilibrium after a single generation of random mating. . . . Now after selection the population $u_n{}^2AA:2u_nAa:1aa$ is reduced to $u_n{}^2AA:2u_nAa:(1-k)aa$.'

Haldane developed this paradigm mainly by studying the rates of change of gene frequency, for genes of various kinds, in populations with different types of mating system.

The most drastic simplifications involved in this paradigm are:

(i) The system essentially implies an equilibrium, in which the frequency of a gene selected against is reduced to zero, or to the level at which it is maintained by recurrent mutation; but the paradigm assumes that the initial conditions are not at equilibrium. Nothing is explicitly stated about why this should be so; it might be, for instance, because a change in environmental conditions has altered selective values, or because a totally new gene has occurred by mutation.

(ii) There is no explicit mention of the phenotype, and certainly no hint that phenotypes can be affected by environments as well as by genotypes.

(iii) There is no mention of the fact that the effect of a given gene is influenced by the rest of the genotype. In some of the later developments Haldane does discuss specific interactive effects between two or more genes, but he leaves on one side the pervasive effects of 'the genotypic milieu', to use the terminology of that time.

Fisher's paradigm avoids the last of these implications, but otherwise differs more in mathematical technique than in logical structure. Instead of measuring selection coefficients by linear coefficents, such as Haldane's *k*, he uses 'Malthusian parameters' expressed in exponential terms, which state the numbers of offspring produced by organisms of the genotype in question. In his paradigm, the difference in the Malthusian parameters for two alleles enters an expression which also involves the frequencies of the alleles and a parameter which expresses the phenotypic difference produced by altering one allele to the other in the actual population, taking into account its breeding structure (e.g. amount of inbreeding or assortative mating) and all the other genes present. He develops his paradigm initially into 'The Fundamental Theorem of Natural Selection', in the form: *'The rate of increase in fitness of any organism at any time is equal to its genetic variance in fitness at that time.'*

This is a statement which has proved extremely difficult to interpret. In the first place, it is clear that the word 'organism' must be shorthand for 'population of organisms'; but though this makes it easy to attach a meaning to 'genetic variance in fitness', it does nothing to elucidate what may be meant by an 'increase in fitness of an organism ( = population)'. It is usually held to imply, if not to be synonymous with, an increase in the numbers of the population; and since genetic variance is essentially a positive quantity, this would lead to the conclusion that all animal populations must always increase in numbers, which they do not. Some way has to be found to get around this difficulty. We will return to it later in connection with Maynard Smith's remarks. Here I would only point out that Fisher's paradigm still involves the first two simplifications characteristic of Haldane's.

It is not so easy to disentangle and describe any particular situation as the paradigm adopted by Sewall Wright. He was not concerned to demonstrate the point on which Haldane felt he had to make a bold assertion: 'that natural selection is an important cause of evolution'. He took this for granted, and was more interested in the circumstances in which the operations of natural selection are mitigated or even overcome by other factors. However, his basic picture of an evolving population has a further element of inclusiveness and flexibility over and above that introduced by Fisher. Wright deals in the selective values of whole genotypes, considered as combinations of alleles at large numbers of loci. These values are envisaged in terms of a hyper-surface in a space in which fitness provides one dimension, while the others express the vast number of possible gene combinations. In his earlier papers at least, Wright was mainly concerned with an initial situation whose non-equilibrium character he carefully defines. He conceives of the fitness hyper-surface as comparable to a rough piece of country, with many hills and valleys; and he sets out to consider the mechanisms by which a population, which for some contingent reasons finds itself at the top of one hill, may travel across a valley and thus reach the top of some other, possibly higher, hill in the neighbourhood. Much of his work is therefore concerned with what might be called quantization processes in evolution; and the mathematical tools he uses deal largely with changes in the frequency distribution of gene frequencies in populations, particularly with processes which lead to certain alleles becoming 'fixed' at frequencies of 0 or 100%.

Wright's treatment began by avoiding simplification (i), i.e. failure to specify an initial state of disequilibrium, but only by invoking a rather special case, one in which a population has, by chance, got into a metastable position. His formulation can without too great difficulty be modified to deal with other types of initial non-equilibrium conditions, such as heterogeneity of the environment in space or time, but it remains true that these are not explicitly incorporated into any general paradigm. Further, he makes little more open reference to phenotypes than do Haldane or Fisher, and he does not incorporate into his scheme any suggestion that the phenotypes on which selection acts are affected by environments. He therefore employs simplification (ii) just as the others do.

Now, on the face of it, the two great problems of the Theory of Evolution—once we have granted that natural selection is an effective agent—would seem to involve just those points omitted by simplifications (i) and (ii). One problem is adaptation, and the focus of the long-continuing debate about Lamarckism is precisely that organisms so often exhibit adaptations which *look as though* they were responses to the environment, but which turn out not to be so in any direct way. Any paradigm which omits the effects of environments in altering phenotypes would seem to make it difficult, if not impossible, to deal with this (leaving it to 'random mutation' is not dealing with it). It was only by taking this factor into account that a solution could be found, in the form of genetic assimilation. Again, the second main problem is that of speciation. Here again everything leads us to the conclusion that diversity of the environment in space and/or time is of the essence, and that a paradigm which implies that the fitness of a genotype is single-valued is likely to prove inadequate

▶ *Maynard Smith's defence of neo-Darwinism.* The comments made above do not in any way imply that we should abandon neo-Darwinism*; they only

* Hereafter I shall contract this to neo-D which I shall also treat as an adjective.

suggest that some of the simplifications on which the mathematical theory has been based have outlived their usefulness and should be revised. The nature of the problem involved may be better appreciated if one looks at the article by Maynard Smith in this volume, in which he tries to demonstrate that neo-D is not a mere tautology by enquiring how one might refute it. I shall argue that none of his suggested 'refutations' would really require us to abandon the theory. (As a side-issue, I should like to remark, as I have done earlier [*Ethical Animal*, p. 151], that I have never been convinced by Popper's argument that, while hypotheses cannot be proved, they can be disproved; in practice they can always be suitably amended to deal with the objections raised. Popper encourages a fashionable current of thought in the philosophy of science, which states that the thing to do with a hypothesis is to try to refute it. This is the treatment Maynard applies here, and in my opinion the result suggests that it is not a very useful line of approach—searching for an improved paradigm, as suggested for instance by Kuhn, may prove more rewarding.)

Let us consider, in turn, the various possible 'refutations' which Maynard Smith describes.

The first, which he dismisses undiscussed, would arise if 'we can show that organisms do not multiply', and multiplication he has defined as 'increasing in numbers in at least some environment'. Now it is perhaps a quibble, but it might be pointed out that a once-numerous species which had lost the power of increasing in numbers in any environment might still undergo neo-D evolution for some period while its numbers were declining; it is not unlikely that the Sequoias, for instance, are in this situation. But in general I agree that no serious line of attack is likely to emerge in this connection. I also agree that it would be fatal to neo-D if we could show that organisms do not vary; but it would be fatal to the idea of evolution in general, not only to the neo-D version of it.

Maynard Smith then speaks of 'the assumptions about heredity and the origin of variation', and he has stated that a refutation would occur if 'it could be shown that the assumptions made by neo-D are not in fact true of all organisms'. Now, of course, many different sorts of non-Mendelian heredity have been demonstrated in a variety of organisms—episomes in bacteria; chloroplastal, mitochondrial or more general types of non-chromosomal genes, such as Sager's; organelles in the cortex of Ciliates; and so on. Some types of bacterial transformation, or episomal heredity, could even be interpreted as examples of Lamarckian phenomena. And there are certain rather weak, 'inertial' effects of the kinds he mentions; for instance the occurrence of a duplication of a locus makes possible the evolution of a protein dimer with two related polypeptide chains. If these were really refutations of neo-D it would have been refuted already. But in fact, of course, they are regarded as mere details and special cases. The main body of the theory is not noticeably weakened, though it has to give up the claim—which its more enthusiastic protagonists sometimes announce—to be the sole and sufficient explanation of all evolutionary phenomena.

All these attacks on neo-D are, in fact, attacks simply on Mendelism, not on anything which neo-D has added to Mendelism. To refute neo-D in this manner requires no more, and no less, than a refutation of Mendelism.

Maynard Smith then turns to some suggested refutations based on rates of change, but points out, justly in my view, that we cannot make any quantitative predictions in this field, and that therefore no refutations are possible.

Finally he turns to 'examining the end-products—the existing organisms'. He tries to invent animals whose organs exhibit an order which is clear enough to be undeniable but which it is implausible to attribute to any form of adaptation for reproductive efficiency. I think he fails to realize that he has come into a region where there are also epigenetic rules and types of organization to be considered. For instance, his first example is: 'If someone discovers a deep-sea fish with varying numbers of luminous dots on its tail, the number at any time having the property of being always a prime number, I should regard this as rather strong evidence against neo-D.' I should not draw such a conclusion so quickly. Which prime numbers ? If they were 1, 3, 5, 7, 11, 13, 17, 19, I should suppose that the spot-producing mechanisms worked with some threshold-type action, so that at a low level it could produce 1 spot, and at higher levels added more spots two at a time until it got to 7, then the next effective jump gave an extra 4. We might explain 13 by saying that going from 7 through 9 to 11 put us well above the relevant threshold, and therefore the next jump goes back to being only a 2-jump; and after that, of course, we would get back to a 4-jump and reach 17 (Fig. 1). Then we would deal with 19 as we did with 13, and 23 as with 17. Then we'd go up another notch to a 6-jump, and get 29. After that I'd be willing to pass it back to Maynard Smith and remind him of his article in the 'Prolegomena' about epigenetic mechanisms for counting large numbers.

He goes on: 'And if the dots took up in turn the exact configuration of the various heavenly constellations, I should regard it as an adequate disproof.' If we are to take 'exact' quite literally, this might carry some weight; but if we allowed some latitude in the configurations, the observation, if made, would in my opinion refute not neo-D, but some as yet unformulated theory of epigenetic

Paradigm for an evolutionary process

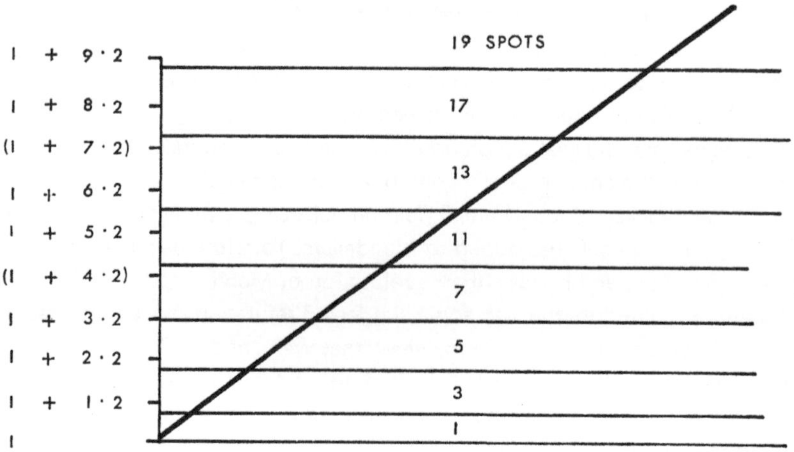


FIGURE 1
Hypothesis for explaining prime numbers of spots. One spot appears in any case. Further spots are added, two at a time, according to the concentration of a substance produced under the control of the genotype, which varies within the population. Each phenotype has a certain range of stability with regard to this substance, these ranges (shown on the graph) gradually increasing at higher concentrations of the substance. The adjustment of these stability-ranges to get the first eight prime numbers is not wildly implausible.

mechanisms. All neo-D can say about the configuration of spots is that they are useful; there is no *a priori* way of telling whether the fact that they look like something else is relevant or not; if they look like something that seems unlikely to be relevant, the first hypothesis is that the epigenetic system imposes limitations which make it impossible to obtain the useful effects without this associated surprising side-effect.

I should like to give two concrete examples in this connection. There is a family of Lepidoptera known popularly as 'Lantern Bugs'. Many of the species have large prolongations on the front of the head. In the species *Laternaria lucifera* this prolongation bears a number of spots and patches which give it an extraordinary resemblance to a crocodile's head. Now the prolongation is only a few millimetres long; it is difficult to believe that any animal which has been predated by crocodiles sufficiently to have evolved avoiding mechanisms will ever mistake the moth's head for a real reptile. What adaptive value can have led to its evolution? Is not this almost as odd as Maynard Smith's 'absurd example' of a fish with spots on its tail arranged like the stars of the Great Bear? I cannot explain it, but I do not propose to abandon neo-D on that account.

My other example is, please believe me, not wholly frivolous. When those neo-D stalwarts J. B. S. Haldane and Julian Huxley were young men, they co-operated in writing an extremely good elementary book on *Animal Biology.* For the frontispiece they went, like Maynard Smith, to the fauna of the deep sea. And they picked out illustrations of two species which no one could fail to see as very recognizable caricatures of the two authors (Fig. 2). Now Haldane and Huxley could certainly be formidable adversaries to run across unexpectedly in the darkness of some deep argument; but no, that cannot have been the neo-D reason for the evolution of these fish. The resemblances are as unexplained as would be the constellation-spots. One can say little more than that, if enough

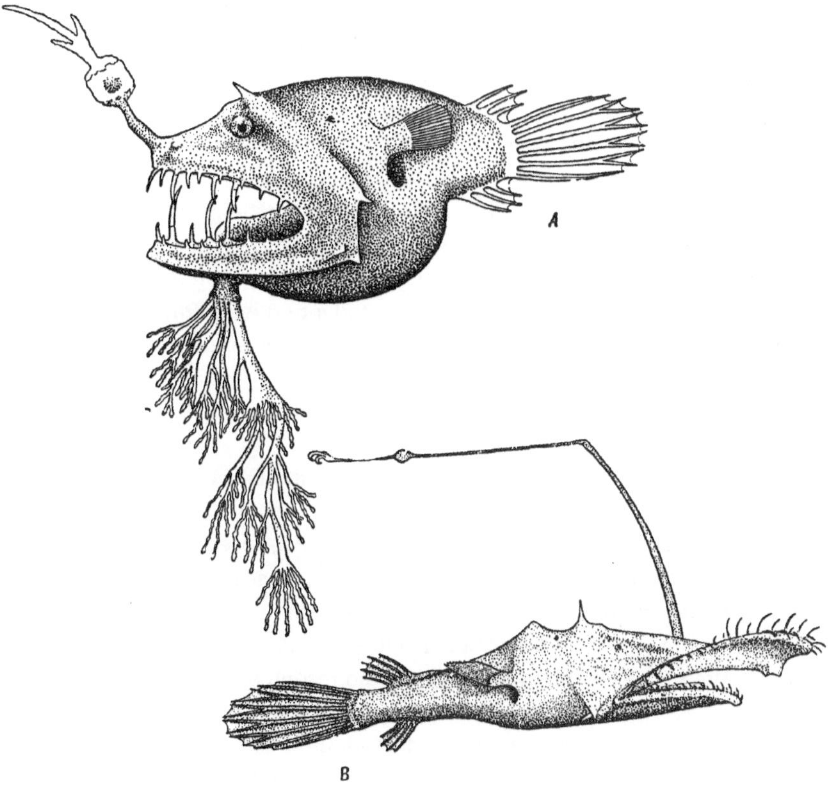


FIGURE 2
Two deep-sea angler-fishes, to show their extraordinary structure and their adaptations to their mode of life.

species go on evolving long enough, a lot of funny things are likely to turn up.

The point that is non-frivolous about this example is that the deep-sea fauna is one of those which presents in dramatic form one of the great problems of evolution. How did it come about that any animals ever went into such an uninviting and difficult environment ? The standard neo-D paradigm, particularly in the Fisher-Haldane version, has altogether too little to say about the colonization of new habitats, especially of habitats which are poorer than the original home of the species, as the abysses are poorer than other parts of the sea.

▶ *Towards a post-neo-Darwinian paradigm* As a matter of fact, Maynard Smith actually gave away the whole story about 'refuting' neo-D before he started to show how he thought this might be done. On page 83 he admits that the phrase 'the survival of the fittest' *is* a tautology (and therefore irrefutable), if the word fitness is used in its neo-D sense. He claims that it should be used, in that phrase, in some other sense: he suggests 'adaptive complexity', but does not discuss this much further, except for the statement that complexity cannot be precisely defined. One way of describing the aim of a search for a new paradigm would be to say that it is an attempt to define more fully what this concept of 'fitness' should be. It is, after all, very generally accepted that fitness is a difficult and obscure concept. The simple Haldane-Fisher fitness has to be modified in situations which are technically more complicated, e.g. with non-random mating, or selection intensities dependent on gene frequencies [2]. More radical modifications have been made by people studying heterogeneous environments or more complex genetic systems [3].

A few authors have tried to formulate a concept of general parameter which will always change in one direction during evolution, as entropy always increases in physical systems, and as Fisher seems to have thought that his fitness would always increase.

If such a parameter could be defined, one could deduce from it the nature of the 'evolutionary force', which keeps evolutionary processes on the move in the face of the many factors which tend to bring it to a halt at some position of equilibrium (see page 115). Among possible parameters, MacArthur [4] has suggested that one measuring efficiency of using limited resources would always increase in evolution; Slobotkin [5] proposes that homeostasis is always increased. But both these authors, in my opinion, began their arguments from too close to the conventional neo-D paradigm, and did not pay enough attention either to the epigenetic effects of the environment or to changing distributions of organisms throughout a heterogeneous environment. I have myself [6] suggested that the parameter which is continually increased in evolution would have to express ability of members of the system to find some way or other of keeping alive and leaving offspring. This is perhaps not very far from MacArthur's suggestion, but involves the possibility that the organisms may not simply become more efficient at using available resources, but may begin exploiting new resources. The way to give more precision or more penetration to such ideas is, I suggest, a new investigation of the logical structure of the evolutionary process.

▸ *The post-neo-Darwinian paradigm* 1. Suppose you have a material structure $P$ with a characteristic $Q$ such that the presence of $P$ with $Q$ produces $Q$ in a range of materials $P_i$ under circumstances $E_j$. Then you could have 'natural selection' to increase the range $P_i$ and $E_j$. (For instance, a crystal dislocation in material $P$, where $Q$ would be replicated in a variety of materials $P_i$ in several environments $E_j$.) But the whole system would not qualify as 'life' because, in my original phrase, it is not interesting enough: in rather more objective language, because this set of postulates makes no provision that $Q$ has an effect on $E$. But $Q$ can only affect $E$ if it is not merely a memory-store to be replicated, but is also an operator. To say that $Q$ becomes an operator is the same as saying that $Q$ becomes a phenotype. This might, logically, involve no, or very little, or very much, translation of the memory-store $Q$ into the effective operator phenotype $Q^*$. But in practice—and perhaps because of a profound law of action-reaction—it is difficult (impossible?) to find a $Q$ which is stable enough to be an efficient store and at the same time reactive enough to be an effective operator. Thus a considerable translation from $Q$ to $Q^*$ is characteristic, and may be necessary, for all systems which can be accorded the name of 'living'.

2. Having now got a system in which natural selection will evolve something effective (or interesting) enough to be called 'life', we need to specify conditions under which evolution will continue. If the range of conditions $E_j$ is single-valued, evolution will produce a single optimum phenotype $Q^*$ (probably, but not necessarily, depending on a single optimum genotype $Q$) and then stop. Similarly, if $E_j$ has only a finite number of values, evolution will eventually reach a stable end-state, with a number of different $Q^*$s, each optimum for one value of $E_j$. The paradigm situation therefore demands two further conditions: (i) that $E_j$ is an infinite-numbered set; and (ii) that there are sufficient $Q$s to provide $Q^*$s suitable for an infinite sub-set of $E_j$s.

The infinity of $E_j$s is ensured by the fact that $Q^*$s are components of $E_j$s. Thus any evolution of a particular $Q^*$ into $Q'^*$ automatically changes a number of $E_j$s in which this $Q^*$ is a component. In more biological language, the environment which exerts selection on one organism is influenced by the presence of other organisms; and as the other organisms change in evolution, so the environment of the first organism is altered, and it must evolve too. In a still cruder formulation, we can say that the evolutionary appearance of a new species automatically creates potential new environmental niches ready for exploitation by some further new form. ('Big fleas have lesser fleas, upon their back to bite 'em, and lesser fleas have smaller fleas, and so *ad infinitum.*') [7]

The second requirement, that the available genotypes must be capable of producing phenotypes which can exploit the new environments, requires some special provision of a means of creating new genetic variation. It might at first sight seem simplest to ensure this by a mechanism in which the new environment would itself stimulate the production of new appropriate hereditary variation. This is the Lamarckian hypothesis. It is indeed a fact that new environments can often produce new appropriate phenotypes ($E'_j$ produces $Q'^*$). But except in very special circumstances $E'_j$ does *not* produce $Q'$. Instead, what we normally encounter are systems in which $Q$ is continually giving rise to a range of altered forms ($Q'_1, Q'_2 \ldots$) by a process which is 'random' in the sense that it is unrelated to any selectively-effective $E_j$ (though it must certainly be controlled by rules of occurrence not related to some $E_j$).

The relative importance of these two 'evolutionary forces'–new environments, new genes–needs some consideration. The classical neo-D position is that all gene mutations occur with a specifiable frequency, so that there is no question of genuinely novel, first-time-ever mutated genes; and that mutation frequencies are so low ($10^{-5}$ or less) that they can almost never overcome the selection pressures. Such considerations gave rise to a dogma of the ineffectiveness of mutation as an evolutionary force; it merely provides some raw material on which other more effective forces might act. But this dogma grew up in the pre-molecular period of genetics, when one was thinking of, e.g. the mutation rate of the plus allele of the 'white eye locus' in Drosophila to an allele which gave a white eye phenotype. We have to rethink the situation now that we realize that the protein produced by the *w* locus (or group of proteins, from a group of loci [8]–but that is an irrelevant complication from the present point of view) can be reduced to complete ineffectiveness by changes in any one of a large number of different peptides. If one characterizes mutations, not by such crude criteria as 'producing white-eyed rather than red-eyed phenotypes', but in molecular terms as substituting amino-acid $P$ for $Q$ at position $W$ in a polypeptide, then the rate of particular mutational steps is reduced, probably by several orders of magnitude. This means that they are even less able to overcome the effects of selection by sheer frequency of occurrence; but it also raises the possibility that some mutation rates may be so low that in comparison with the time-scale of the relevant evolutionary events, it may not be justifiable to consider them in terms of continuous rates at all. We may have to bring back into our thinking the possibility of radically new, never-seen-before, mutations–considerations which were in the forefront for the earliest Mendelian evolutionists, such as Bateson and de Vries in the first decade or so of this century.

Before one can evaluate the relative importance in evolution of (i) changes in environment, as against (ii) mutations-as-alterations-of-molecules or mutations-as-determinants-of-epigenetically-complex-phenotypes, one needs to realize the range of spectrum of levels of organization in which evolution may occur. At one end of the spectrum is:

(a) the purely macromolecular. Is it (selection-wise) a 'good thing' to substitute valine for glutamic acid at position 6 from one end of the chain of haemoglobin (producing the sickle-cell character)? In higher organisms there probably is a gene for doing this floating around in the population anyway–or you could find some other way of dealing with the environmental selective pressure. The question is more important in relation to the very earliest phases of evolution, when the living systems comprised little more than a bare minimum of molecules which could just keep going. How did one improve the very first, barely effective, DNA polymerases, oxidative enzymes, or chlorophylls? The course followed by evolution may, at such stages, really have been influenced by the nature of the polypeptide substitutions that mutation threw up.

(b) From this, there is a complete continuous spectrum of increasing levels of complexity–through the 'hypermolecular', which is what classical neo-D concerns itself with–to the 'ecosystem' evolution involved in such a question as: how did the London sparrow cope with the success of the petrol engine in driving off the streets all those horses, with their offerings of dung full of delicious seeds? How did the rabbit of the gentle English fields evolve a method of flourishing in the harsh Australian outback, with not a single species of its normal food plants, or the red deer of the Scottish heather-covered hills find its evolutionary way into a set-up in which, ecologically married up with the Australian wombat, it dominates much of the wilder country of New Zealand and, incidentally, grows to about twice its normal weight in its home country? [9]. It is extremely unlikely that any of these evolutionary episodes had to wait for any new polypeptide sequences. All the *Q*s necessary to produce the appropriate

$Q^*$s were, one imagines, already present in the population, and the effective 'evolutionary force' was the occurrence of the new environments ($E_i$s).

The relative importance of the two 'evolutionary forces' arising from new random mutations and new environments must therefore change as we pass from the macromolecular end of the evolutionary spectrum through the hypermolecular to the ecosystem end, but wherever we are within this range, each of these 'forces' has *some* importance.

It is important to emphasize that the new genetic variation must not only be novel, but must include variations which make possible the exploitation of environments which the population previously did not utilize. In Longuet-Higgins' terminology, it is not sufficient to produce new mutations which merely insert new parameters into existing programmes; they must actually be able to rewrite the programme. Essentially the same point was made by Pattee. In a letter (12 February 1968) discussing tactic copolymer growth as an elementary genetic system, which he raised again in the *Prolegomena*, he writes: 'The fundamental reason that evolution is limited in such single tactic copolymers is that the rule for monomer addition or the code is intimately dependent on some fixed length of the description. If evolution is to lead to unlimited complexity, then the code must accept a description of indefinite length and the description itself must be able to grow indefinitely without changing the code.'

The predominance of 'random mutation' over Lamarckian mechanisms in tellurian ('this-earthly') biological systems does not arise because the latter could not, in theory, support evolutionary processes. It is presumably related to the problem described at the end of paragraph 1, how to combine a store which is unreactive enough to be reliable, with something which interacts with the environment sufficiently actively to be 'interesting'. To ensure reliability the store must be rather unreactive; the solution has been generally adopted by tellurian systems to provide the necessary variability by a random process not depending on reaction with the environment rather by endowing the storage material with some residual reactivity [10].

3. We need also to provide that the variant $Q$'s actually come in contact with the variety of new $E_j$s. This may be brought about simply by arranging for very wide but quite indiscriminate geographical dispersion of newly fertilized zygotes; and this is the mechanism on which the whole plant kingdom relies. The world of non-sessile, motile animals has a more economical method of achieving this result by employing behavioural mechanisms, which lead animals either to explore situations more or less at random (which is little better than plants

can do) or, more typically, to choose those environments in which they can most easily earn their daily bread and escape their enemies. This is an elaboration, of a rather Lamarckian character in that it involves a reaction to the immediate circumstances, which is over and above the absolute necessities of the evolutionary paradigm; but one which is very important in animal evolution.

4. In order to accommodate the numbers of altered or mutated $Q$'s required to evolve into the new environments $E'_j$, the paradigm situation must be one describable only in terms of populations, not of individuals.

5. The new environments $E'_j$ are, as we have seen, essentially complex entities, being functions of a number of variables, among which are, for instance, a number of phenotypes $Q^*$s of various species, as well as various physical quantities, etc. The evolution of an optimum phenotype for a particular new environment, $Q^*_{E'_j}$ will therefore in many cases involve combination of a number of different variant genotypes, $Q'_p$, $Q'_q$, $Q'_r$, . . . , etc. It is therefore not surprising to find that most evolving systems have developed some method of encouraging the production of appropriate combinations of Qs. Provision for this is not an absolutely necessary constituent of the evolutionary paradigm, since some organisms get away with nothing more radical than very large numbers of short life-cycles, though there are few, even of the viruses and bacteria, who cannot do anything better than this. Nearly all organisms have evolved mechanisms which facilitate recombination of variant $Q$s, usually by some sexual or parasexual process.

6. In so far as the necessary heterogeneity of the environment comes into existence as a consequence of interactions between existing organisms (i.e. the $E'_j$s are functions of $Q$'s), two consequences will follow: (i) There will be a continual increase in the number of species which are optimum in some or other of the ever-increasing number of $E'_j$. Thus the diversity of the organic world will continually expand; (ii) Since a new, attainable $E_j$ may be a function of an indefinite number of existing phenotypes, there will be a tendency for the evolution of even more complex phenotypes, capable of operating optimally in environments of increasing complexity. When there are both water-snails and water-visiting mammals, evolution can produce parasites with life-cycles involving interactions with both hosts; when there are night-flying insects, then, but not before, evolution can produce larger night-flying predators, such as bats, which need a sonar system to prevent them running into obstacles or to locate their prey.

7. The necessary heterogeneity of environment has another consequence, which cannot be omitted from the paradigm situation. Selection operates on phenotypes, and phenotypes are affected by environments as well as by genotypes. Further,

since there is a necessity for mobility (whether passive or active) to ensure that new environments are explored, it will not in general be true that the environmental influences which contribute to the formation of the phenotype are identical with those which exert the most important selection on it. (Consider, for instance, an annual plant; the phenotype of a plant and its seeds will be influenced by the weather, etc., in the year *n*, while the main selection may be exerted by the weather in year $n + 1$, when the seeds germinate.) Thus the paradigm situation must incorporate two theoretically separable effects of any environment $E_j$: on the development of a $Q$ into a $Q^*_{E_j}$, and on the selective values of the variety of $Q^*_x$ which arrive within or pass through it.

The complete paradigm must therefore include the following items: A genetic system whose items (Qs) are not mere information, but are algorithms or programs which produce phenotypes (Q*s). There must be a mechanism for producing an indefinite variety of new Q'*s, some of which must act in a radical way which can be described as 'rewriting the program'. There must also be an indefinite number of environments, and this is assured by the fact that the evolving phenotypes are components of environments for their own or other species. Further, some at least of the species in the evolving biosystem must have means of dispersal, passive or active, which will bring them into contact with the new environments (under these circumstances, other species may have the new environments brought to them). These environments will not only exert selective pressure on the phenotypes, but will also act as items in programs, modifying the epigenetic processes with which the Qs become worked out into *Q*'s.

From the standpoint of this paradigm, how should we envisage the classical evolutionary concepts of adaptation and fitness? I confess I have not sufficient mastery of the sophisticated mathematics which would be necessary for a rigorous exposition. Instead of attempting that, I will suggest, for adaptation, an allegorical illustration which I originally applied to man in his world (motto to Chap. 4 of *Behind Appearances*), but which I think applies almost equally well to any organism:

*Man in the world is like a caterpillar weaving its cocoon. The cocoon is made of threads extruded by the caterpillar itself, and is woven to a shape in which the caterpillar fits comfortably. But it also has to be fitted to the thorny twigs—the external world—which supports it. A puppy going to sleep on a stony beach—a 'joggle-fit', the puppy wriggles some stones out of the way, and curves himself in between those too heavy to shift—that is the operational method of science (and of the evolution of biological systems).*

As regards fitness, we have to define that concept in terms which allow for the existence of heterogeneous and evolving environments, and of organisms capable of active or passive dispersal through a range of environments which are acting both as agents of selection and as subsidiary programs affecting the development of the phenotypes which are selected. The fitness even of a single phenotype cannot, therefore, be represented by a single-valued coefficient, but only by a matrix, or a continuous distribution of values, which specifies also the variety of environments, in which selection may occur. If we wish to attach 'fitness' to a single, multigenic *genotype*, we should have to increase the dimensionality of the matrix so that it could take account also of the epigenetic-programming aspects of the environments; and if we wished to emulate the classical neo-Darwinists and speak of the fitness of single genes, we should have to increase the matrix again to incorporate all the various genetic combinations in which it might occur in the population in question (it seems highly dubious to me that any process of averaging over all these combinations, as advocated, e.g. by Fisher, has any biological validity).

In contrast to the difficulties of conceiving of 'the fitness of a gene' from this point of view, the concept of the fitness of a population comes close to giving one an opportunity to use the adjective 'perspicuous' in the precise sense given to it by that pedant Fowler (*Modern English Usage*)–'means (the being) easy to get a clear idea of'. The fitness of a population is the degree to which its gene pool gives it the ability to find some way or other of leaving offspring in the temporarily and spatially heterogeneous range of environments which its dispersion mechanisms offer to it.

But, even if not too difficult to formulate, this concept has a joker in it–which is what makes it interesting and challenging. How to know what the 'temporarily heterogeneous range of environments' may bring? A new Ice Age, a new virus, a new predator? The gene pool can, of course, preserve for some time genes which turned out to be useful in relation to critical situations in the more or less recent past: Lewontin [11] has discussed these possibilities in a very stimulating way, and shown that, unless the environment behaves itself–is not capricious–they are rather limited. From a more general point of view, one can envisage a number of possible strategies. As (i) to canalize your development, build up well-buffered chreods, and insist on developing into some good all-purpose almost invariant form in spite of whatever environmental effects get thrown into the epigenetic programs–as mice concede little more than a fractional elongation of the tail to the difference between growing up in a hothouse rather than a cold-storage depot [12]; or (ii) acquiring a gene pool which allows an extreme flexibility in the end-results of development–good examples are some small crustacea, such as *Daphnia* and *Artemia,* in which the minute physico-chemical variations between every pond of water are reflected in the shape of the adults, or plants such as the 'water arrowroot' *Sagittaria sagittifolia,* in which the leaves of one and the same plant have radically different forms when they are growing wholly under water, on the surface, or in the air. To make a success of this gambit, it is of course necessary to ensure that the plasticity of epigenesis is in general, or at least in really crucial situations, such that the environmentally-modified forms are selectively useful in the environments that produced them. It is no use allowing your muscular development to be influenced by the environmental demands on the use of muscles if the dependence takes the form that using muscles causes them to be consumed and to wither away; (iii) another ploy is to develop some general defence mechanisms which do not have to know in advance what they will have to defend against. 'Random' mutation provides this facility to some extent, but only over periods of many generations of selection. If the population can afford to wait that long, as bacteria can, to add a general mechanism for rapidly spreading a good defence, once it has been acquired, by such systems as Infective Resistance Factors, is obviously a useful second stage. Perhaps the best example of a generalized defence is the vertebrate antibody-production system, which seems able, even within one lifetime, to protect against an enormous range of invading foreign substances.

The systematic exploration of the evolutionary strategies in facing an unknown, but usually not wholly unforecastable, future would take us into a realm of thought which is the most challenging and very characteristic of the basic problems of biology. The main issue in evolution is how populations deal with unknown futures; is this problem so different from that described by Gregory, when he says that 'perception involves the continual solution of a series of puzzles'? In epigenesis we find systems which will develop into perfectly good lenses or livers, even when there is something non-standard in the conditions which normally guide the cell's synthetic machinery into those paths. In all these cases we are forced to consider the nature of mechanisms which can operate effectively on the basis of inadequate information. This seems to be one of the central general problems of Theoretical Biology. Life might be defined as the art of getting away with it; and Theoretical Biology as the attempt painstakingly to explicate just how it is done.

C. H. Waddington

## Notes and References

1. Wheldon's polemical journal, *Questions of the Day and of the Fray*, did not cease publication till 1924 (?); and the anti-Mendelian influence was strong enough to ensure that Fisher's first major genetical paper, arguing amongst other things that continuous variation is explicable on Mendelian principles, was rejected by the Royal Society of London, and published by the less prestigeful Royal Society of Edinburgh, in 1918.

2. Wright, S. Stochastic processes in evolution, in *Stochastic models in Biology and Medicine* (ed. Garland). (Univ. Wisconsin Press 1964) p. 199.

3. Levins, R. *Amer. Nat. 96* (1962) 361; *Amer. Nat. 97* (1963) 75; *J. Theoret. Biol. 7* (1964) 224; *Evolution 18* (1965) 635; *Genetics 52* (1965) 891.

4. MacArthur, H. H. *Proc. Nat. Acad. Sci. Wash. 48* (1962) 1893.

5. Slobotkin, L. B. *Am. Sci. 52* (1964) 343.

6. Waddington, C. H. *The Nature of Life* (Allen and Unwin 1961) p. 109, *Prolegomena* (1967) p. 22.

7. Several ecologists argue that there will be, for ecological reasons, a selection pressure towards the development of more complex ecosystems, involving an ever-increasing number of species. The fundamental selection is towards increasing the stability of the ecosystem; when there are only a small number of interacting species, there is a tendency for violent fluctuations in numbers, which may lead to the extinction of some species and the complete collapse of the whole system. See, for example, Hutchinson, G. E. *Amer. Nat. 93* (1959) 145; MacArthur, R. H. *Ecology 36* (1955) 533; Dunbar, M. J. *Ecological Development in Polar Regions* (Prentice Hall 1968).

8. Green, M. M. *Proc. 11th Int. Cong. Genet. 2* (1965) 37

9. For this and other examples see *The Genetics of Colonising Species*, I U B S Symposium (ed. Baker and Stebbins) (Acad. Press 1965).

10. In connection with the nature of the mutational events which may play a part in evolution I would like to advance some ideas, which are, I freely admit, so speculative that I have relegated them to a note rather than the body of the text; but many biologists will consider them so outrageously heterodox that they may refuse to consider them at all—and these I should beg to think again.

One of the major problems of evolution theory is to understand how the sharp discontinuities between major taxonomic groups—Phyla, Families, Species-Groups, and so on—have come into being. A simple-minded empirical inspection of the facts would suggest, as it did for instance to Goldschmidt when he wrote his *The Material Basis of Evolution* 1940 and *Theoretical Genetics* 1955, that it might be profitable to contemplate the possibility of the very occasional occurrence of what Goldschmidt called 'systemic mutations', which result in a complete restructuring of the genome, achieved either in a single step, or at least in rather few generations. When Goldschmidt wrote, no clear-cut example could be given in which the occurrence of such a process had been observed. The orderly minded orthodox biological world closed its ranks against this suggestion that revolutionary processes may happen. It became accepted that the only respectable doctrine is that evolution never involves anything but step-by-step Fabian gradualism, plodding along a weary way similar to that by which the annual milk yield of dairy cows or egg yield of hens is slowly improved—the occurrence of a little allopolypoidy or rearrangement of chromosomes by two or three breaks could be admitted, but would only push the basic philosophy from Bourgeois-Liberal to right-wing Social Democrat.

It is still impossible—so far as I know—to quote a compelling instance in which a systemic mutational event has been observed in an evolving multicellular organism. But events which appear to be essential of this kind are becoming well known in the field of cell culture. It is a common experience that cells isolated from vertebrate tissues usually grow in culture for a fairly restricted number of cell generations—a hundred or two—and then die out, *unless* they undergo some sort of change which brings into being cells capable of forming an 'established line', which can then be sub-cultured in perpetuity. The nature of the change from a 'strain' to an 'established line' is highly obscure, but it often involves what looks like a complete restructuring of the genome; there may be a considerable reduction in number of chromosomes, accompanied in some cases by considerable changes in chromosome morphology. (For a recent review see *The mammalian cell as a differentiated microorganism* by Howard Green and George J. Todaro, *Ann. Rev. Microbiol. 21* (1967) pp. 574–600.)

The fact that cells in culture can throw up, within at most a few cell generations, new types of cells capable of giving rise to 'established lines', and that the change may involve a very drastic reshuffling of the genome (usually, in the cases observed, with a loss rather than a gain of chromosomal material), is evidence that something like a 'genetic revolution' or 'systemic mutation' can occur. It is, of course, more difficult to see how such an event in an evolving population could be propagated so as to affect the future, but if such events are not ruled out of court by the nature of genetic processes it seems silly to close one's mind to the possibility that evolution has found some way of making use of them.

An example of an evolutionary phenomenon which suggests a very radical reorganization of the genome between nearly related species is the astonishing difference found by Forbes Robertson (in press, *Genet. Res.*) in the D N A sequences of *Drosophila melanogaster* and *simulans*, as tested by molecular hybridization. R N A manufactured *in vitro* on a *melanogaster* D N A template hybridizes only one-third as well with *simulans* D N A as does the R N A made on the *simulans* template; the results of a reciprocal experiment are very similar. Although it seems certain that the hybridizations only occur between substances related to the highly reiterated stretches of D N A, this is still strongly suggestive that the differences between these two species involve much more radical and pervasive alterations of base sequences than were contemplated a few years ago, when it seemed that all that was involved was a few inversions, and translocations of large sections of the chromosomes.

11. Lewontin, R. C. *Bioscience 16* (1966) 25.

12. Barnet, S. A. *Biol. Rev. 40* (1965) 5.

## Some comments on Waddington's paradigm by J. Maynard Smith

*1. Neo-Darwinism and Mendelism.* Wad argues that 'to refute neo-D requires no more, and no less, than a refutation of Mendelism'. This of course depends on how you define neo-D. In my 'defence' of neo-D I made Weismannism rather than Mendelism the central assumption. Otherwise the refutation of neo-D is trivial: bacteria do not Mendelize but they do evolve. But the Weismannist assumption (roughly, if the phenotype of an individual is altered by an altered environment, this will not cause that individual to produce offspring with the new phenotype) is not disproved by the types of non-Mendelian heredity mentioned by Wad.

Nevertheless, I agree with the main point Wad is making—the most direct way (but not the only way—see below) of refuting neo-D is to show that its genetic assumptions are wrong.

*2. Refutation by the end-products of evolution.* I argued that complex structures which did not contribute to the survival of their possessors would refute neo-D. Wad argues that since development is based on algorithms, it can lead to inexplicably complicated (=funny) results. I think that I am right, but I agree that no single example could be decisive. My point is that when biologists are confronted by a structure, they analyse its function in terms of its contribution to survival, and this method of analysis usually works. If it didn't—i.e. if it often turned out that an organ when studied in detail could not be interpreted as contributing to survival—then biologists would have abandoned Darwinism long ago. But it is of course true that there are at any one time plenty of structures, usually ones which have been little studied, which cannot be interpreted adaptively.

*3. 'The survival of the fittest' and Waddington's paradigm.* It is perhaps a pity that this phrase was ever introduced into biology, because it is a standing invitation to philosophers to argue that Darwinism is a tautology. There seem to be three ways of treating the phrase:

(a) Assume 'fittest' means 'most likely to survive', and you have a boring tautology.

(b) Replace 'fittest' by some more sophisticated definition of survival capacity, and you may have an interesting tautology. Thus a conclusion may follow necessarily from certain assumptions, but still be interesting, e.g. the conclusion that planetary orbits are elliptical, given Newton's laws of motion and gravitation. This is particularly likely to be the case if the conclusions are more easily tested than the assumptions, as is the case in the Newtonian example.

Thus one might try to deduce from the laws of heredity some property which will be maximized. This is what Fisher's 'fundamental theorem' does, and what Wad does in his article. So far the approach seems to me to have been unfruitful, mainly because the assumptions (i.e. of genetics) are easier to test than the conclusions (i.e. the course of evolution).

If Fisher's 'fundamental theorem' is interpreted to mean that evolution must lead to an increase in the rate of growth in numbers (i.e. 'fitness') of a population, then the theorem is simply false. If, alternatively, one interprets fitness as a mathematical function of the frequencies of genotypes in a population and of their *relative* probabilities of survival, then the theorem is, with certain qualifications, true, but, as Wad points out, difficult to apply.

The snag with Wad's less mathematical attempt to find something which increases in evolution is that it leads to a false conclusion. Thus he concludes 'the fitness of a population is the degree to which its gene pool gives it the ability to find some way or other of leaving offspring in the temporarily or spatially heterogeneous range of environments which its dispersion mechanisms offer to it'. Now fitness in this sense is not necessarily maximized. For example, plant species commonly and animal species occasionally lose the capacity for sexual reproduction. Such a change usually leads to extinction. It is a lowering of fitness in Wad's sense, yet I see no reason to doubt that the change occurs by natural selection.

(c) The third approach, which I adopted in my 'defence', is to reformulate the phrase 'the survival of the fittest' in a non-tautological way, by taking fitness to refer not to some function, sophisticated or otherwise, of survival capacity, but to the properties of 'adaptive complexity' or 'harmoniousness' or what have you – i.e. to those properties of living organisms, and sometimes of their artifacts, which distinguish them from inanimate matter, and which call for an explanation.

I do not think approaches (b) and (c) are mutually exclusive (although obviously confusion arises if 'fitness' is used in two senses). The difference between Wad and myself lies not so much in our views about the mechanism of evolution, which are rather similar, but in what we were trying to do. In my article on 'the status of neo-Darwinism' I was trying to defend the present orthodoxy from criticisms of a philosophical and fundamentally Lamarckist type. Wad, perhaps rightly, regards this argument as no longer very interesting, or as something for molecular biologists to worry about, and has therefore been trying to say something new about evolution.

## Reply by C. H. Waddington

To my way of thinking, John's comments introduce some confusions into this discussion, which it may be well to try to clear up. They are largely terminological, and connected with the fact that when I offer a criticism of neo-D, John tends to reply with an impassioned defence of Darwinism or Weismannism or some other well-accepted historical precursor of the views I was discussing. Let us first, then, agree on what we mean, at least roughly, by the doctrines attached to these various names.

## Reply by C. H. Waddington

By *Darwinism*, I mean the theory that organisms come into existence by a process which involves material heredity from their progenitors under the control of natural selection. This is certainly *not* a tautologous statement, since there is an alternative to it—which was in fact generally accepted before Darwin, namely that organisms are brought into being by 'Special Creation', or something of the kind. What John refers to as 'refutation by the end-products of evolution' would be refutation of Darwinism itself, not merely of neo-D.

By *Weismannism*, I mean the same as John does, according to the statement in his first paragraph. This again certainly is *not* tautologous: It is a necessary but not a sufficient condition that evolution should be of the neo-D type. I am not attacking it. Perhaps John is right in thinking that it still needs defending in any company in which philosophers are present, but I don't think we need waste much time on rehashing the old arguments in this meeting.

By *Mendelism* I mean the theory that heredity is transmitted in the form of discrete factors which can segregate and recombine. I think it is confusing to introduce the verb 'to Mendelise', a piece of lab jargon dating from the days of Bateson and Punnett, when it had the meaning 'to exhibit the phenomenon of segregation into classes with one or other of the classical 'Mendelian ratios', i.e., 3 : 1, 9 : 3 : 3 : 1, etc.' Many types of organisms (e.g. polyploids) do not Mendelise in that sense, but no one would deny that they exhibit Mendelian heredity. I think the same is true even of bacteria, and do not accept John's contention that their behaviour refutes Mendelism.

By *neo-D* I mean the view that Weismann's doctrine—that there is no influence of the phenotype on the genotype—can be transferred from the individual level to the population level, and that an adequate theory of evolution can be formulated in which 'fitnesses' are attributed to genotypes. John slides altogether too easily between the Weismannist point that the environment of an individual does not affect the heredity he transmits, and the quite different argument that the environment of a population does not affect what they transmit. I maintain that a population's environment does influence, quantitatively, what they transmit, because natural selection acts on phenotypes which are partially environment-dependent. John also gives away too much in his para. 3(a), where he suggests that if we define fitness as 'most likely to survive' we have only a boring tautology, and that we need to define it in a more sophisticated way to get a tautology as interesting as Newtonian mechanics. I have never wished to deny that the results obtained by classical neo-D were as interesting and valuable in their field as Newton's conclusion that planetary orbits are elliptical—and just as unavailable to commonsense unassisted by algebra. The point I am making is comparable to the criticism which might be offered against Newtonian mechanics—that it deals with point-masses, frictionless surfaces, an unresisting medium. It is good mathematics, but as science it is good in outer space, but poor in the sticky conditions on earth. Now, as soon as we have phenotypes we are in a realm whose correlate in the physical world would be that of friction, turbulence, bodies occupying volumes, and all the other complexities which have to be added to the Newtonian picture before it can be actually used. This is what we now need to do to neo-D.

But we do not, in my opinion, need to give up the basic point that fitness is essentially 'survival capacity' (i.e. capacity to leave offspring). I was surprised to find John, in his para 3(c) ready to allow this position to be overrun by the enemy. But it is the fundamental strategic strong-point of the whole of Darwinism. Once you concede that the thing that survives, i.e. that contributes most to evolution, is something other than the thing which leaves most offspring, then you might as well go straight back to Special Creation and have done with it. The point is not to compromise on the issue that the only way to contribute to evolution is to leave offspring, but to ask more sophisticated questions about just which organisms *do* leave more offspring.

Finally, I do not follow the last paragraph in John's section 3(b). Of course some plant and animal species sometimes have lost the capacity for certain types of reproductive performance, including sexual reproduction in general; but I see no reason to doubt that this loss is produced by natural selection of those individuals with the greatest capacity to leave offspring in the environments immediately available. Of course, again, selection, which operates on the differences in fitness between contemporary individuals, may push a population into an evolutionary situation in which it is unable to cope with changes in its environment, and so becomes extinct. But to state that a certain property is being maximised within a population does not imply that it is always getting greater. Natural selection will pull up fitness as far as it can, but that may still not be far enough to ensure the survival of the population. The point which John is making here against my concept of 'fitness' is the same as that which he advanced to David Bohm against Wynne-Edwards concept (p. 90); and I feel justified in defending myself even more strongly than I defended Wynne-Edwards on p. 95.

# Gibbs Ensemble and Biological Ensemble

*Reprinted from the Annals of the New York Academy of Sciences*
*Volume 96, Article 4, pages 975–84, 2 March 1962*

Edward H. Kerner
University of Buffalo

With apologies to practising biologists, especially ecologists, who labour first-hand with perhaps the hardest observational material in all science, I am going to trespass a bit in difficult and unfamiliar terrain and venture some ideas at second-hand.

My proposal is to consider seriously as a first step in a general theory of ecology the scheme of population dynamics advanced by a great mathematician, Vito Volterra (1931), almost 30 years ago. The dynamics can be written down in one line:

$$\frac{dN_r}{dt} = \epsilon_r N_r + \frac{1}{\beta_r}\sum_s \alpha_{sr} N_s N_r$$

($N_r$: population of $r$th species).

The meaning may be simply stated: beyond an intrinsic growth or decay rate leading to the Malthusian $N_r(t) = N_r(0)e^{\epsilon_r t}$ ($\epsilon$ of either sign) if each species were isolated from all others, the growth rate of each is conditioned further only by binary $(N_r, N_s)$ interactions between species ($\alpha_{sr}$ measuring the strength of interaction) in a purely reciprocal or predator-prey manner ($\alpha_{sr} = -\alpha_{rs}$).

It would seem rather odd to suggest so simplistic a scheme as a guide to understanding what occurs in the real ecological world. Objections abound. Since birth, death, and growth are complicated stochastic processes and population numbers are discrete, not continuous variables, how can deterministic differential equations be useful? What of the age distributions of different species? Where is account taken of the complex of breeding habits and peculiarities of the life cycles of the different species? Where is any description given of populations in space as well as time, of immigration currents and herding instincts, or of season, climate, geography, and topography? What about symbiotic and other kinds of interactions besides the antagonistic ones?

I believe that these and other questions are answerable only by considering what one may expect of a theory or of an initial theoretical attempt. In ecology we are faced with a stunning, even paralysing, array of variables and factual data. For theoretical purposes is it not a matter of trying to sift out of the array some few of the most basic elements, the grossest realities, and of trying to see what organization amongst them alone may first be found? In short, should we not try to build that first rudimentary model that captures some of the distilled, if distorted essence of the reality?

Let us imagine that our knowledge of atomic phenomena had preceded that of celestial mechanics and that our empirical data in physics were wholly at what we now call the quantum level, our objects of study being atoms, molecules, and quanta. Things would be difficult. Such theories as might be attempted would likely deal directly in probabilities, the data being rather lawless and experiments being imperfectly reproducible. It would be at some stage, we may guess, a far-fetched speculation to suggest, for instance, that in a collision of two large molecules we forget the empirically visible complicated internal characteristics of each, forget the myriad electrons and nuclei and the massive data about them ever before our eyes, and treat the collision as a *two*-body collision, inventing a—to us—gross fiction of some kind of 'billiard ball' representation of the molecules. Nevertheless, let us trust that the invention would be seen to be useful for explaining *some* things, although surely not everything. The further perception by a latter-day Newton of some simple limiting kind of deterministic 'laws' that could *for certain purposes* be used to help interpret the discretenesses, discontinuities, and probabilities of our daily experimental fare would not be without merit. To be sure, we should not be fooled by such laws, knowing full well that reality was not so simple.

I suggest that the Volterra generalized predator-prey scheme (1) grasps a root matter of macroecology, (2) has the necessary generality and simplicity to allow a wide assortment of observationally testable results to be made, that is, has real theoretical viability, and (3) admits appreciable elaboration, both in respect of the primitiveness of the predator-prey scheme itself and of comprehension of population variations in space as well as time, and also, in principle, of effects of physical factors such as temperature and radiation intensity. Its crudity is not its weakness but its strength. At the least I think it illustrates the *kind* of thing a general theory can possibly do in ecology, although the theory itself may be wrong or incomplete. By its generality is meant its capacity to view many species in association, without recourse to the abstracting of one or a few species from the rest of the biological world.

The root matter referred to is simply this: cells eat cells and thereby beget cells, but some classes of cells, instead, live off nonvital matter and energy (the interconnection of the nonvital and vital elements may at first be set to one side). This has a peculiarly biological character, apart from the specific meaning of 'cell', without very much counterpart elsewhere in the observable world (an exception may be found in certain classes of biochemical reactions *in vivo* where chemical concentrations may be coupled by Volterra-type equations [Hinshelwood, 1951; see also Lotka, 1920]; more generally, the equations of chemical kinetics are not unrelated to those of Volterra). It is universal (or trite, if you like, but surely not trivial); it is perhaps the single overmastering phenomenon in the ecosphere, without which any comprehensive view of the large biological world is bound to be vacuous. Volterra's dynamics is but a mathematical paraphrase of it.

Let us not forget that at least in *some* microecological cases (figure 1) the Volterra scheme speaks the essential truth, and that on the macroscopic scale it answers, as none other does, to possibly the grossest fact of general observation: that of the unremitting fluctuations in populations beyond the incidents of season and migration.

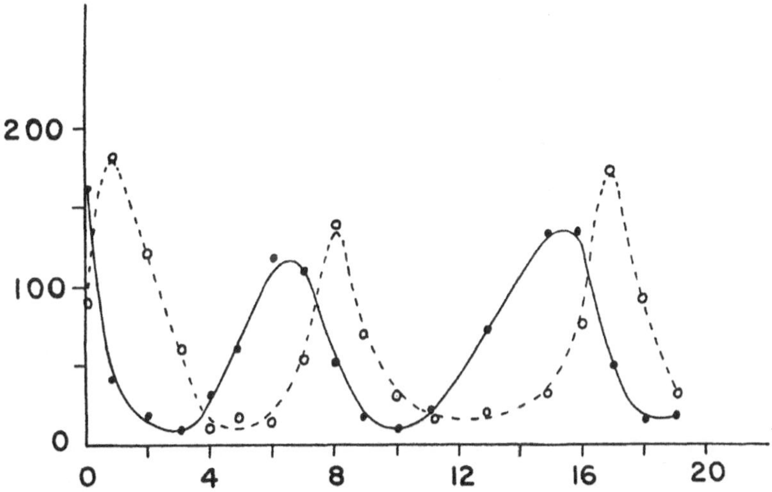


FIGURE 1
Population variations of *Paramecium aurelia* (*broken curve*) and *Saccharomyces exiguus* (*full curve*). Ordinate shows the number of individuals in 0·2 gm samples; abscissa is time in days. Adapted from Gause in D'Ancona (1954)

In thus advocating the blurred rather than the sharp view of the biological scene, toward the end of encompassing dominant effects, there is of course no assertion of the literal validity of the Volterra mechanics, any more than that it is claimed that complex molecules are billiard balls and that classical theory can describe quantum phenomena. Rather the position set forth is that in the long term and over the large view of a big multicomponent biological association, the association operates in some measure *as if* actuated first by the rough biological rule incorporated in the Volterra dynamics, observed population oscillations being describable as effectively of Volterra type. 'Predator' and 'prey' of course need not have only literal meaning but, additionally, the significance that one species' losses add up somehow to another's gain, without precise specification of just how.

It is certainly true that the behaviour of many laboratory microcosms indeed shows Volterra's scheme to be nugatory as a general microscopic proposition; characteristically in some of these cases it is probabilistic rather than deterministic results that are manifested: for example, the survival of one of two competing species, but not always the same one. Behind any deterministic scheme such as Volterra's there are surely more basic stochastic elements. This in itself is no bar to useful deterministic theorizing. One may even surmise that the useful deterministic theory, granting its usefulness, must stand as some deducible limit within the complete stochastic theory, a touchstone and a clue.

I have mentioned the simplicity of Volterra's mechanics. Actually the coupled differential equations admit of complete analysis only for two interacting species. Wherein is their general simplicity? It is in two conservation laws: (1) they admit one integral to view and none other is apparent, the quantity

$$G \equiv \Sigma q_i \beta_i (e^{v_i} - v_i)$$

is conserved throughout the motion, $v$ being log $(N/q)$, $q$ the average population level $\bar{N} = (1/T) \int_0^T N(t)dt$; (2) they admit a Liouville's theorem in the many-dimensional space $v_1, v_2 \ldots$ ('phase space'), to the effect that a bag containing a cloud of phase points (each corresponding to different sets of initial values of the $N_i$ or $v_i$ and thereafter being propelled by the dynamical laws) keeps its volume but not its shape unchanged throughout the course of its motion. The parallel with physics is striking: the two-body problem is amenable to full discussion, the many-body problem is not; yet from energy conservation and Liouville's theorem (in the physical phase space of coordinates and momenta of the many particles) progress can be made. The name of this progress is statistical mechanics, its chief author is J. Willard Gibbs. The above-mentioned cloud is a Gibbs ensemble and is the mental instrument for statistically assaying the vastly complicated and practically unknowable motion of any single phase point; it is exactly the amassing of ignorance in the cloud by which knowledge is gained. As one may see, statistical mechanics has in essence nothing to do with mechanics, but is really a statistical theory of differential equations; it simply happens that the Volterra equations, as well as Newton-Hamilton's and possibly others, fit into the scheme.

The statistical-mechanical idea is that despite (or even because of) our inability to unravel the details of motion of very large numbers of variables, we can still draw statistical conclusions about them and can describe through system-type quantities, such as temperature and entropy, important macroscopic features of the whole (for example, entropy tends to increase, and energy tends to flow downward along temperature gradients). Moreover, if we had full knowledge of the details it would be an embarrassment against the necessary poverty of data obtained from observation: necessary because there would not even be enough data books for recording information about all the variables. Thus the full knowledge would come to be reduced to manageable proportions via averaging procedures dictated by the meager data available; statistical mechanics of some type would be invented. The statistical overlay of dynamics, in short, is very nearly compulsory when the dynamics covers large numbers of degrees of freedom. If very accurate laws of population dynamics were known, whether deterministic or stochastic, some means of surveying the mass of them statistically would still be essential to understanding macroscopic observations. It could be argued from this standpoint that the accuracy of the laws destined in part for a statistical hopper might not have to be so great, especially if the observations themselves were not very fine.

The conservative character of Volterra's dynamics, if it have some general meaning, then stands as a curious feature of a system that physically is energetically and materially open and nonconservative. It is an interesting question how this may be comprehensible in the roughest conceptual terms from the underlying biochemical ultimates; one wonders how understandable it is in principle (Elsasser, 1958).

The role of any macroscopic parameter—such as temperature—is somewhat different biologically from what it is physically. Whereas temperature as a system property is directly observed physically, it is only indirectly observable in the biological association, for the association is observed fundamentally through population-time curves of one or a few selected species. The physically comparable observation is that of the displacement along an axis, as a function of time, of a Brownian particle. Both curves bear the stamp of the operation of complex environs; neither is predictable in detail but only in respect to statistical properties such as mean amplitude or mean frequency of oscillation. In short, both are kinds (different kinds) of random noise and, as such, they function as 'thermometers' or other thermodynamic indicators as well as anything else could. The macrofeatures of the large system cannot but reveal themselves through the fluctuations of the microobservables. To say, for instance, that the whole system is in thermodynamic equilibrium is only to say that different long strips of the microcurves exhibit always the same statistical qualities: the strips of noise, taken from different time intervals, look 'the same'. The macroecological world is in some ways bound to be a strange one for us, for we are submicroscopic elements in it—somewhat like electrons in a molecule in a great sea of molecules—and therefore conscious firstly of the fine particularities that impinge on us. Nevertheless, how does the biological world look to an observer far removed from local detail and unable to distinguish even the discrete individuals comprising a population, but seeing things only grossly, as the chemist sees the colour change of an indicator in the course of a titration without so much as a glance at its atomistic structure? Our hyperfine powers of observation and our too-intimate engagement with what we are observing may perhaps be hindrances to a certain order of understanding.

The basic rules of statistical mechanics tell what distribution of phase points in phase space is proper to the description of thermodynamic equilibrium; it is the canonical distribution

$$P(v_1, v_2, \ldots)\, dv_1\, dv_2 \ldots = e^{(\Psi - G/\theta)}\, dv_1\, dv_2 \ldots$$

giving the probability that the $v_i$ are found in the intervals $v_i$, $v_i+dv_i$. Here $\theta$ means 'temperature'', $\Psi$ 'free energy'', the factor $\exp(\Psi/\theta)$ being a normalization constant. If one replaces $G$ by the Hamiltonian of a physical system (that is, the energy written as a function of coordinates and momenta of the component particles) one obtains the chance for finding the coordinates and momenta of all particles in any infinitesmimal range of their possible values. For a gas of weakly interacting atoms this leads to the Maxwell-Boltzmann law

$$P(V_x, V_y, V_z)\, dV_x dV_y dV_z \sim \exp - \left( \frac{\frac{1}{2}M(V_x^2+V_y^2+V_z^2)}{kT} \right) dV_x dV_y dV_z$$

for the probability that any one of them has a velocity in $V_x$, $V_x+dV_x$; $V_y$, $V_y+dV_y$; $V_z$, $V_z+dV_z$ (here $\theta$ stands for $kT$= Boltzmann's constant $\times$ absolute temperature).

In completely parallel fashion for the biologic case, if one at random opens the door on an association and peers in at some species, for example the $k$th, one has a chance

$$P(v_k)\, dv_k \sim \exp - \frac{q_k \beta_k}{\theta}(e^{v_k} - v_k)\, dv_k$$

of finding

$$v_k \equiv \log (N_k/q_k) \equiv \log n_k$$

in the range

$$v_k, v_k + dv_k$$

Or, in $n$-language, one's chance for finding

$$n_k \text{ in } n_k, n_k + dn_k$$

is

$$P(n_k)\, dn_k \sim n_k^{x_k - 1} e^{-x_k n_k} dn_k$$

$x_k$ being short for $q_k \beta_k / \theta$. Substantially this same law was invoked by Corbet *et al.* (1943) for interpreting butterfly- and moth-catch data, with $x_k$ taken as some small parameter (meaning here $\theta \gg q_k \beta_k$).

The way is not yet clear for using the apparatus of Gibbs ensembles to interpret the basic observational datum, the single population-time curve that was likened before to a Brownian particle's displacement-time curve. The answer here is that (ergodic hypothesis)

time averages = ensemble averages

more exactly that the time average of any function of population numbers, $f(N_1(t), N_2(t), \ldots)$, translated into a function of phase variables, $F(v_1(t), v_2(t), \ldots)$ is

$$\frac{1}{T}\int_0^T F(v_1(t), v_2(t), \ldots)\, dt = \int_{\text{(all phase space)}} F(v_1, v_2 \ldots) e^{(\psi - G/\theta)} dv_1\, dv_2 \ldots$$

where we assume that the association has been let run a long time and has attained equilibrium, and where $T$ is long enough to cover very many cycles of population oscillation. Strict proof of ergodicity is difficult to ascertain, but indirect arguments and the weight of statistical-mechanical experience point to its validity.

A principal result stemming from this understanding of averages is that (bars meaning time- or ensemble-averages)

$$\frac{\theta}{q_k \beta_k} = \overline{\frac{(N_k - \bar{N}_k)^2}{\bar{N}_k^2}} = \overline{\left(\frac{N_k}{\bar{N}_k} - 1\right) \log \frac{N_k}{\bar{N}_k}}$$

Thus the meaning of 'temperature' is in effect that it measures the amplitude of oscillation (in two ways) away from the average population level $\bar{N}_k$. Zero temperature is the completely quiescent state $N_k = \bar{N}_k$. We can think of 'heat' exchange (that is, G-exchange) between two associations at different temperatures in a meaningful way; the high-$\theta$ association cools off – all its species experiencing a fall in amplitude of fluctuation – as the low-$\theta$ association warms up, all its members increasing their amplitudes of fluctuation. A heat capacity can indeed be defined analogously to the physical one and, curiously, it has a temperature dependence much like common physical ones. Similarly entropy can be defined; it shows a monotonic rise with rising temperature. However, calorimetry is not thermodynamics. The concept of *work* is missing. To introduce it, the variations of external parameters (such as volume in the physical case, or radiation intensity or physical temperature in the biological case), as they induce changes in the association, need to be considered; the general nature of such variations remains to be studied.

Plainly the calculation of time averages relevant to the observationally basic population variations of one species will be vitally controlled by the statistical parameter $x_k$. More elaborate analysis brings out qualitatively that: (1) at small $x$

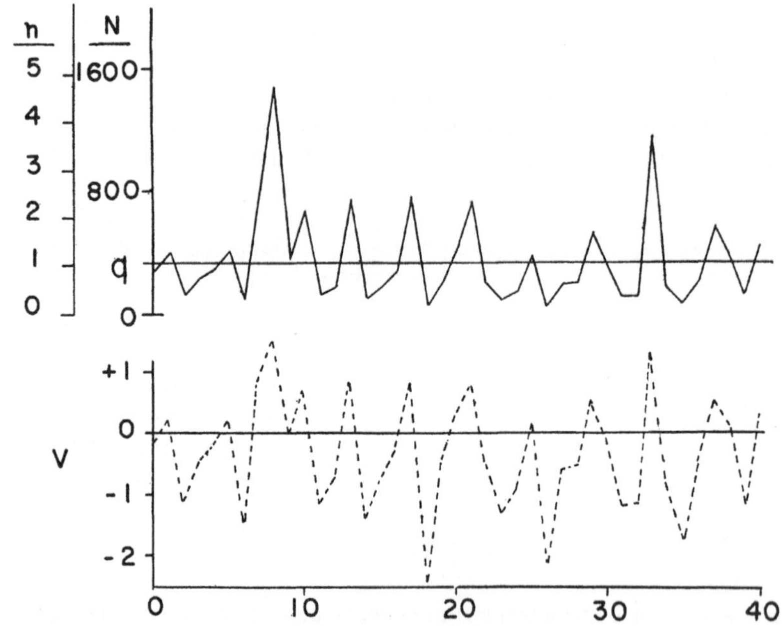


FIGURE 2
Labrador fox-catch for the first 40 years of a 90-year period. Ordinate $N$, upper curve, shows the catch, $q$ being the average catch. Ordinate $n$ is the catch measured in units of $q$. According to Elton, it is reasonable to assume that the catch is proportional to the population. The lower curve shows the reduced variable $v = \log n$. In the statistical theory, such a population-time curve as this is taken to represent a kind of 'noise'. Adapted from Kerner (1959); fox-catch data from Elton (1942)

(high θ) the population spends most of its time at below-average levels, oscillating below in long shallow troughs and above in short high peaks; (2) at large $x$ (low θ) the frequency of oscillation is highest (but is altogether a slowly varying function of $x$) and the upward swings away from the mean level are closely comparable with the downward swings; (3) the amplitudes of oscillation above and below the mean level, taken separately, are not rapidly varying with $x$ except at small $x$ (for example, $x<1$) where the upward amplitude alone increases rapidly with decreasing $x$; and (4) the mean amounts of time spent above and below the average level are similarly slowly dependent on $x$ except at smaller $x$.

The important question is: Can the single parameter $x$ describe the host of statistics that can be ferreted out of the population-time curve? A whisper of an answer is given in figures 2 and 3 and table I, where Elton's (1942) Labrador fox-catch data is analysed and a comparison with theory made (original sources

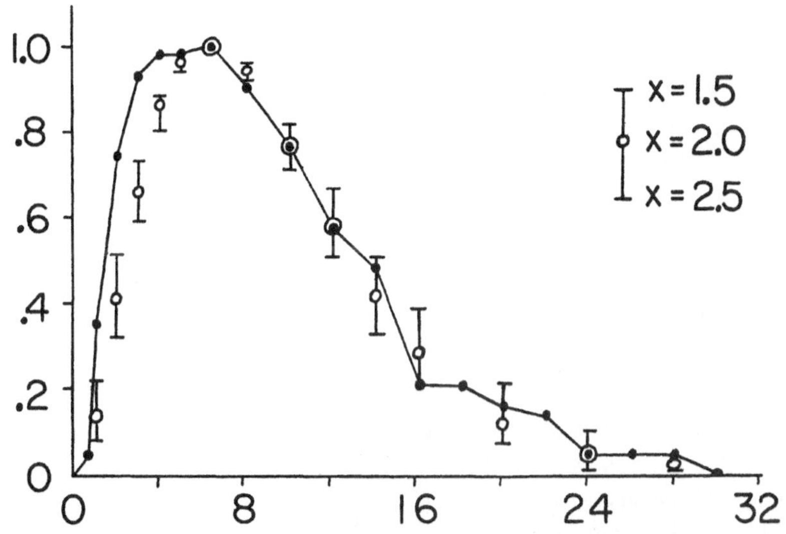


FIGURE 3
Comparison of observed and theoretically expected frequencies with which the fox-catch curve crosses a given axis. In figure 2 draw horizontal lines at intervals of 50 fox catches; the ratio of crossings of the curve on these lines to the crossings on the line $q$ is here plotted as the ordinate against the ordinal number of the line as abscissa. Solid dots are observed values; circles are expected values corresponding to the choice $x=2{\cdot}0$ of the parameter $x$; for comparison, expected values for $x=1{\cdot}5$ and $2{\cdot}5$ are also shown. The statistical errors of the observed values are appreciable, possibly 40 per cent or more. Adapted from Kerner (1959)

TABLE I
Comparison of observed and theoretically expected statistical data for the fox-catch curve of figure 2

| *Quantity averaged* | *Time average* | *x* | *x range* |
|---|---|---|---|
| Fraction of time spent at below-average levels | 0·595 | 1·90 | 0·70–15·0 |
| Amplitude of oscillation of $n$ for $n < 1$ | 0·445 | 2·10 | 1·00–2·75 |
| Amplitude of oscillation of $n$ for $n > 1$ | 0·729 | 1·71 | 1·42–2·05 |
| $(n-1)^2$ | 0·536 | 1·87 | 1·69–2·07 |
| $(n-1)\log n$ | 0·489 | 2·04 | 1·86–2·27 |
| $\log n$ | −0·258 | 2·08 | 2·31–1·91 |

The question at issue is whether the *single* theoretical parameter $x$ can account for a variety of statistics. The first column gives the observed time average, using directly the polygonal line of figure 2; the second column gives the theoretical $x$ resulting in the same value; and the third column gives the range of $x$ corresponding to a ±10 per cent alteration of the observed values. Adapted from Kerner (1959)

may be consulted for details). The whispered answer seems to be, possibly yes.

A final point remains to be mentioned about Volterra's model. Its conservative character stems in a basic way from the evenness of the number of interacting species. This has quite a dubious and artificial air about it. Its general import, however, may not be without real biological meaning, since, when the number of species is odd, what occurs is a tendency of decay into an even association with exhaustion of at least one species. In dynamical terms, the system exhibits some dissipative characteristics. It has been part of ecological and evolutionary thought for a century or more that closely similar species cannot both occupy the same ecological niche indefinitely, but that the slightly more 'successful' species eventually completely supplants the other (Volterra-Lotka principle, see Hardin, 1960). It turns out to be feasible to describe such extinction-domination struggles within Volterra's theory as an odd → even decay. The advantage over the usual description is (1) the avoidance of the unrealistic element, in the latter, of an ultimately static population of the successful species, (2) an accounting of the strugglers in immersion in a larger biological world. Under the hypothesis of close similarity of the competitors, the dissipation still can be viewed through the lens of the conservative theory; a species of secularly drifting statistical mechanics can be constructed; and the short-term oscillations of the strugglers may be seen against the backdrop of the controlling long-term rise of one and fall of the other. A description of the workings of the creation and annihilation of species is not beyond the range of the theory.

## References


Corbet, A. S., R. A. Fisher & C. B. Williams. 1943. The relation between the number of species and the number of individuals in a random sample of an animal population. *J. Animal Ecol.*, **12**: 42–58.

D'Ancona, U. 1954. The Struggle for Existence (E. J. Brill, Leiden, Germany).

Elsasser, W. 1958. *The Physical Foundation of Biology* (Pergamon Press, New York, N.Y).

Elton, C. 1942. *Voles, Mice, and Lemmings* (Clarendon Press, Oxford, England).

Hardin, G. 1960. The competitive exclusion principle. *Science*, **131** : 1292–7.

Hinshelwood, C. 1951. Decline and death of bacterial populations. *Nature*, **167** : 666–9.

Kerner, E. H. 1957. A statistical mechanics of interacting biological species. *Bull. Math. Biophys.*, **19** : 121–46.

Kerner, E. H. 1959. Further considerations on the statistical mechanics of biological associations. *Ibid.*, **21** : 217–55.

Kerner, E. H. 1961. On the Volterra-Lotka principle. *Ibid.* To be published.

Lotka, A. J. 1920. Undamped oscillations derived from the law of mass action. *J. Am. Chem. Soc.*, **42** : 1595–8.

Volterra, V. 1931. *Leçons sur la Théorie Mathématique de la Lutte pour la Vie* (Gauthier-Villars, Paris, France).

Volterra, V. 1937. Principes de Biologie Mathématique. *Acta Biotheoretica*, **3** : 1–36.

# A Statistical Mechanics of Temporal Organization in Cells

Reprinted from Society for Experimental Biology Symposium No. 18, 1965, pp. 301–26

Brian C. Goodwin
Massachusetts Institute of Technology

Running through science there is a duality in the description of systems, which has resulted in some of the deepest problems of scientific method and analysis. This duality arises from the analytical procedure which reduced all complex, macroscopic systems to a set of simpler, microscopic components. Natural processes are thus regarded as having two sides : a macroscopic or phenomenological one, which is observed directly ; and a microscopic one, which may not be directly observable but which is postulated to underlie and in some sense explain the macroscopic process.

Cell biology is currently going through a molecular revolution in which almost daily discoveries are laying bare the molecular organization of cellular processes. However, the macroscopic aspects of cell behaviour, which form the content of cell physiology, are left largely unexplained by these developments. Thus, for example, recent studies in molecular biology have shown that negative feedback control mechanisms operate at the molecular level in cells, but detailed analysis of these control circuits fails to explain the integrated, higher order properties of cells, such as homeostasis, competence, and circadian organization. There remains a gap between molecular biology and cell physiology, which is a direct consequence of the analytical method of science.

The problem of bridging such a gap, of resolving microstructure and macrostructure in systems, is not new in science. It arose first in connexion with classical thermodynamics, which is a phenomenological or macroscopic science, and the kinetic theory of gases, which is a molecular or microscopic theory. The man who set himself the task of deducing thermodynamic laws from the dynamic properties of molecules was Boltzmann. He and his successors, notably Willard Gibbs, succeeded in developing a theory which allows one to do just this. It is called statistical mechanics, and in considerable measure it succeeds in overcoming the duality inherent in scientific analysis, although in so doing it has posed some theoretical problems which to this day remain unresolved.

The purpose of this paper is to study the possibility of forging a direct link between molecular biology and cell physiology by using essentially the same formal, logical procedures which were used in the development of classical statistical mechanics. The starting point will be a study of the dynamic behaviour of molecular feedback control circuits in cells, which will play, in our theory, a role similar to the kinetic theory of gases in statistical thermodynamics; and the goal is the derivation of certain macroscopic variables which are formally analogous to thermodynamic quantities: temperature, entropy, free energy, etc. These macroscopic variables will then be used to describe and analyse the coordinated or organizational properties of cells, such as homeostasis and adaptation. One of our tasks will be to discover the observational or experimental significance of these new thermodynamic-like variables, and to see how they can be used to give analytical precision and quantitative content to our intuitive ideas about the nature of cellular organization.

## THE CONTROL CIRCUITS

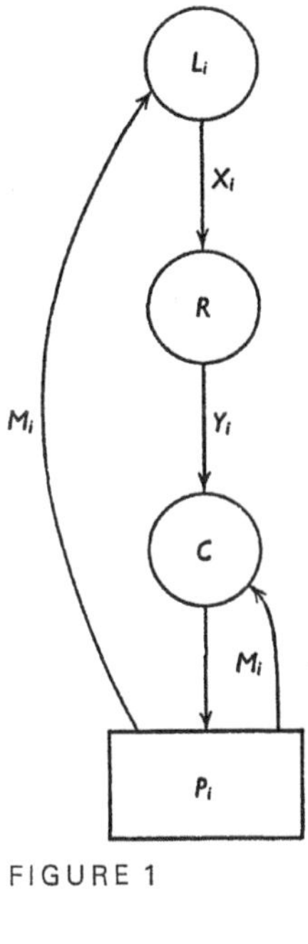


FIGURE 1

The kinetic theory of gases had a ready-made dynamic foundation in Newtonian mechanics. Unfortunately there is no parallel to this for the dynamics of molecular control mechanisms. However, on the basis of the purely qualitative features of cellular control circuits, and the behaviour of negative feedback devices studied in engineering, it is possible to derive a set of differential equations which describe, albeit in very crude form, the dynamics of a class of control mechanisms which on circumstantial grounds seem to have a high probability of occurring in cells.

There are at least two types of feedback mechanism operating in cells to control macromolecular activities. They are known as feedback inhibition and feedback repression, and according to current theory they have the characteristics shown diagrammatically in figure 1. Here $L_i$ represents a genetic locus which produces messenger ribonucleic acid (mRNA) of the *i*th informational species in quantities represented by $X_i$. This specific 'signal' encounters a cellular structure, $R$ (a ribosome), where its activity results in the synthesis of informationally homologous protein in quantities denoted by $Y_i$. The protein then travels to some cellular locus, $C$, where it exerts an influence upon the metabolic state either by enzyme action or by some other means. (We will concentrate in the following on the case where $Y^i$ is an enzyme.) The result of this activity by the enzyme is the generation of a metabolic species in quantity $M_i$, which enters a metabolic pool, $P_i$.

Two different closed control loops are now obtained by a feedback of the metabolite either to the site of enzyme activity, $C$, or to the genetic locus, $L_i$. In the first case the metabolite reduces the activity of the enzyme; this is feedback inhibition (Umbarger, 1956; Magasanik, 1958). In the second case the metabolite reduces the rate of deoxyribonucleic acid (DNA)-mediated synthesis of mRNA at the genetic locus, and this is called feedback repression (Vogel, 1957). There are a great many details and problems which this description leaves out. For example, there is the question of how $M_i$ exerts its effect upon the enzyme (cf. Gerhardt & Pardee, 1962); and the nature of the actual genetic repressor remains unknown, although there is evidence that it is a complex between $M_i$ and some macromolecular species (Monod & Jacob, 1961). Furthermore, there is the possibility that a third level of control exists, wherein the metabolite feeds back to the ribosome-mRNA complex and reduces the rate of protein synthesis.

In the following analysis we will consider in detail only the dynamics of feedback repression, involving the control loop from the gene, through mRNA and protein to metabolite, and back to the gene. The control sequence in this circuit is believed to be the following: $X_i$ controls the rate of synthesis of informationally homologous protein; $Y_i$ controls the rate of production of the *i*th metabolite; and $M_i$ controls the rate of synthesis of mRNA as well as taking part in metabolic reactions. These quantities will therefore be regarded as the essential control variables in the respective biochemical reactions. This involves the assumption

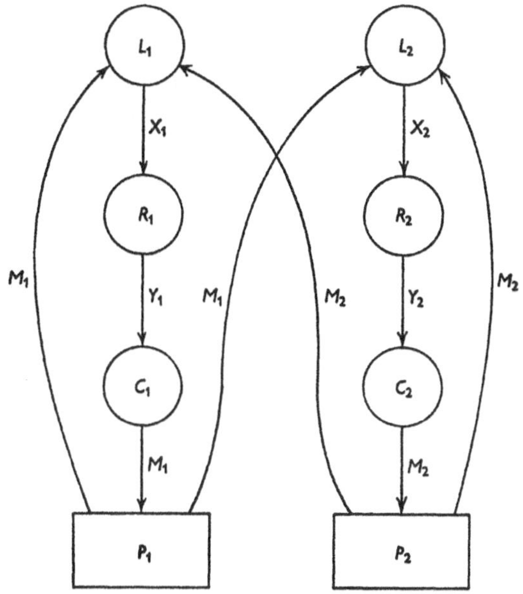


FIGURE 2

that other possible rate-limiting factors, such as size of nutrient and energy pools, concentrations of aporepressors (i.e. the macromolecule postulated to act in conjunction with the metabolite to form the effective repressor complex), etc., can be treated as parameters of the 'motion' of the variables, $X_i$, $Y_i$ and $M_i$. That is to say, these quantities are considered to remain constant or to change very slowly compared with the dynamic motion of the control variables. It is possible to extend this control scheme to cover more complicated circuits, such as when several enzymes form a biosynthetic sequence and are controlled simultaneously by one single metabolic species, the end-product of the sequence (Gorini, 1963). We will not, however, consider these modifications here.

More complicated situations will also be analysed, in which repression occurs between different control components, so that we get the situation shown in figure 2. Here a metabolite controlled by $Y_i$ interacts by repression with another genetic locus, $L_2$, while a reciprocal interaction occurs from $L_2$ to $L_1$, through $M_2$.

THE CONTROL EQUATIONS

We will now establish a set of very coarse functional relationships between the variables $X_i$, $Y_i$, and $M_i$, which describe the essential dynamic features of the type of control system which we have in mind. If the differential equations thus obtained can serve as the dynamic basis for the construction of a statistical mechanics and 'thermodynamics' of cellular processes, then the crudeness of the initial equations will not be a severe limitation in the macroscopic analysis. This is because it is the nature of a statistical mechanics that many of the microscopic details are smoothed out, as it were, and only the fundamental dynamic properties are retained. Therefore in spite of the approximations necessary in the microscopic description of the control circuits, we may nevertheless get some idea of the macroscopic quantities relevant to a general description of cell behaviour, and some suggestion of experimental procedures for controlling and observing these macroscopic quantities.

Consider first an equation for the rate of synthesis of the $i$th species of protein, when the controlling variable in this process is $X_i$, the concentration of the corresponding species of mRNA. The general form of the equations to be considered is

$$\frac{dY_i}{dt}=f_i(X_i,Y_i,M_i)-g_i(Y_i,M_i) \qquad \text{....(1)}$$

where $f_i(X_i, Y_i, M_i)$ is a function describing the rate of synthesis of protein, while $g_i(Y_i, M_i)$ relates to the rate of its degradation. The simplest conceivable functions which satisfy the control requirements of our model are given by:

$$f_i(X_i,Y_i,M_i)=\alpha_iX_i, \quad g_i(Y_i,M_i)=\beta_i$$

Then $$\frac{dY_i}{dt}=\alpha_iX_i-\beta_i \qquad \text{....(2)}$$

Here $\alpha_iX_i$ represents the rate of mRNA-controlled protein synthesis, while $\beta_i$ is the rate of protein degradation, assumed to be a constant. Since protein synthesis is an almost completely irreversible process, no terms for the reverse reaction on the template are included.

In equation (2) the constant $\alpha_i$ is a composite parameter containing a rate constant for the template synthesis of the $i$th species of protein and concentration terms for activated amino acids, the precursors of protein synthesis. The simplification involved in using such a representation for what is clearly a very complex biochemical process may seem to invalidate our analysis from the start. The one feature of the real process which is incorporated in this equation is control of protein synthesis by messenger RNA. Since it is the dynamic consequences of this control which we seek to investigate, we will go no further than equation (2) at present. The more general form of equation (1) allows certain modifications to be included if they prove to be essential.

CONTROL EQUATION FOR mRNA SYNTHESIS

The equations which we consider for messenger RNA synthesis will be of the general form

$$\frac{dX_i}{dt} = \phi_i(X_i, Y_i, M_i) - \psi_i(X_i, Y_i, M_i) \qquad \ldots.(3)$$

Here $\phi_i(X_i, Y_i, M_i)$ is a function describing mRNA synthesis, and $\psi_i(X_i, Y_i, M_i)$ represents the rate of its degradation. It is assumed that the kinetics of repression of mRNA synthesis by the metabolite $M_i$ are essentially the same as those of enzyme inhibition. This means that we are dealing with a surface-binding phenomenon wherein the repressing molecule or complex combines reversibly with the DNA template and so interferes with its synthetic activity. The template also combines reversibly with the precursors for RNA synthesis.

There are many details which must be considered before even a crude expression for control of mRNA synthesis by metabolite can be obtained. These include the kinetics of the reaction between DNA templates and activated nucleotides on the one hand and repressors on the other, and the relationship of the repressing complex to the metabolite concentration. It also requires certain assumptions about the 'storage capacity' of the metabolic pool. When the pool size is very small, one would expect very little or no metabolite to 'spill out' of the pool and repress the activity of the genetic locus. But when the pool is large, then a considerable fraction of the metabolite would be expected to serve a repressive function. These and other considerations are treated in detail elsewhere (Goodwin, 1963), and unfortunately there is too little space to present them here. I will therefore pass directly to the simplest possible equation which can be derived on the basis of our assumptions for the control of mRNA synthesis by metabolite, which is

$$\frac{dX_i}{dt} = \frac{a_i}{A_i + m_i M_i} - b_i \qquad \ldots.(4)$$

Here the rate of degradation of mRNA, $b_i$, is assumed to be a constant. The other parameters, $a_i$, $A_i$, and $m_i$ are quite complicated functions of more elementary constants. The storage capacity of the metabolic pool is an important quantity in determining the size of $A_i$ and $m_i$.

In equation (4), the main characteristic is the appearance of $M_i$ in the denominator of the expression for mRNA synthesis. This is a consequence of the assumption that repression is a surface adsorption phenomenon. Similar expressions were obtained by Szilard (1960) in his very interesting studies of control processes in cells.

#### CONTROL EQUATION FOR METABOLITE SYNTHESIS

There is one more step to take before the equations reach their final form. Since the concentration of metabolite was assumed to be controlled by the concentration of enzyme, $Y_i$, we can write down one more control equation showing how these variables are related. The simplest kinetic scheme for the synthesis of $M_i$ is an equation of the form

$$\frac{dM_i}{dt} = r_i Y_i - s_i M_i \qquad \ldots.(5)$$

The parameter $r_i$ represents a composite constant which includes the rate constant for the enzyme, the concentration of its substrate, and the Michaelis constant for the reaction whose product is $M_i$. The term $s_i M_i$, $s_i$ a constant, implies that $M_i$ is drawn off from the pool at a rate dependent upon its own concentration, i.e. that this process is primarily metabolite-controlled.

At this point in the argument we use a device which is frequently employed in kinetic studies. Because enzyme-catalysed reactions are much faster than template-directed macromolecular synthesis, their ratio being about 1000 : 1 (cf. Goodwin, 1963), it can be assumed that the variable $M_i$ will be always at, or very close to, a steady-state value in relation to the comparatively slow rates of change in the variables $X_i$ and $Y_i$, which are macromolecular quantities. We can therefore write

$$\frac{dM_i}{dt} = r_i Y_i - s_i M_i = 0 \qquad \ldots.(6)$$

and then solve for $M_i$ in terms of $Y_i$. This puts the equation (4) into the final form

$$\frac{dX_i}{dt} = \frac{a_i}{A_i + k_i Y_i} - b_i \qquad \ldots.(7)$$

where $k_i = \dfrac{m_i r_i}{s_i}$.

We thus end up with a pair of differential equations in $X_i$ and $Y_i$, (2) and (7), which represent the dynamics of a particular type of closed feedback control circuit. In the derivation of these equations many assumptions have been made which may prove to be invalid, and an extremely complex system has been reduced to one of almost absurd simplicity. However, the hope is that the major qualitative features of the control system have been included in the model; and it can be made much more complicated and sophisticated without altering its essential dynamic properties.

In all this we have not mentioned the units of the variables, which are in some sense concentrations. The units which have actually proved to be most useful are rather unorthodox; they are simply molecules per cell. The reason for this choice is that the usual chemical units for homogeneous solutions, moles per litre, have little meaning for macromolecular populations in the heterogeneous conditions of the cell interior. Furthermore, it has been estimated that for mRNA, the size of the populations can be very small in certain cells, amounting to perhaps 100 molecules of a particular informational species. There are many other questions which cannot be considered in detail here, and I must refer to a more complete treatment in my book *Temporal Organization in Cells.*

THE DYNAMICS OF THE CONTROL SYSTEM

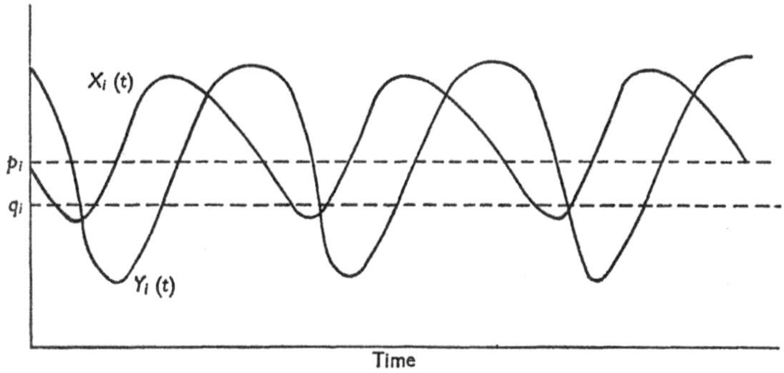


FIGURE 3

Figure 3 shows the dynamic behaviour of the variables $X_i$ and $Y_i$ in the control circuit of figure 1, whose motion is governed by the equations

$$\left.\begin{aligned} \frac{dX_i}{dt} &= \frac{a_i}{A_i + k_i Y_i} - b_i \\ \frac{dY_i}{dt} &= \alpha_i X_i - \beta_i \end{aligned}\right\} \qquad \ldots.(8)$$

The variables undergo continuing oscillations as functions of time, varying about fixed, steady-state values $p_i$ and $q_i$, which values are obtained by setting the equations equal to zero and solving for $X_i$ and $Y_i$. Thus the major dynamic characteristic of the control circuit is the occurrence of oscillations in system variables. It is a common experience among engineers that negative feedback control systems have a strong tendency to oscillate. Engineers usually try to avoid these 'parasitic' oscillations, as they are called, and design their control systems to prevent, if possible, the oscillations occurring. It is a central assumption however, in this study that, in the course of evolution, cells have not selected against dynamic oscillations in their control circuits, but have made use of them to organize the staggering complexity of cellular dynamics into a well-ordered, rhythmic sequence of biochemical processes. The oscillations are thus regarded as the dynamic basis of temporal organization in cells.

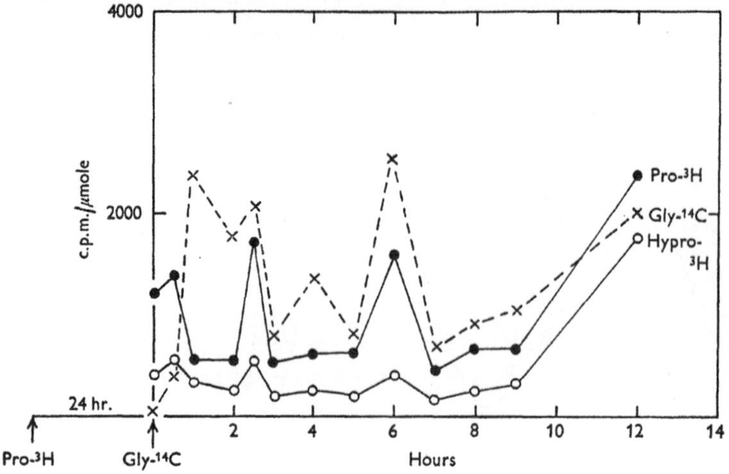


FIGURE 4
Dynamics of the appearance of proline, hydroxyproline, and glycine in extractable collagen of embryonic chick bone tissue. (After Tanzer & Gross, 1963)

There is much indirect evidence to support the idea that oscillations are an intrinsic feature of molecular dynamics in cells, most of which has arisen in connexion with the study of rhythmic phenomena, particularly circadian rhythms (cf. Harker, 1958; Pittendrigh, 1960, 1961; Halberg, 1960). Recently, however, some very remarkable experimental observations have been made on the dynamics of proline incorporation into collagen in embryonic chick cells by Tanzer & Gross (1963) and by Jackson (1963), which provide direct evidence for this belief. These investigators have observed very marked fluctuations in the specific activity of the proline pool of chick embryos after injection of radioactive proline, and synchronous fluctuations in the specific activity of labelled collagen. In figures

4 and 5 some typical results are reproduced, with the kind permission of the authors. Figure 4 shows the dynamics of the appearance of three different amino acids, proline, hydroxyproline, and glycine, in extractable collagen of embryonic chick bone tissue after administration of labelled amino acid to the embryo. Jackson's observations on free proline in the chick embryo are shown in figure 5. Here we have observations on a system undergoing large and fairly regular fluctuations, which can hardly be ascribed to random noise in the embryo or the experimental procedure.

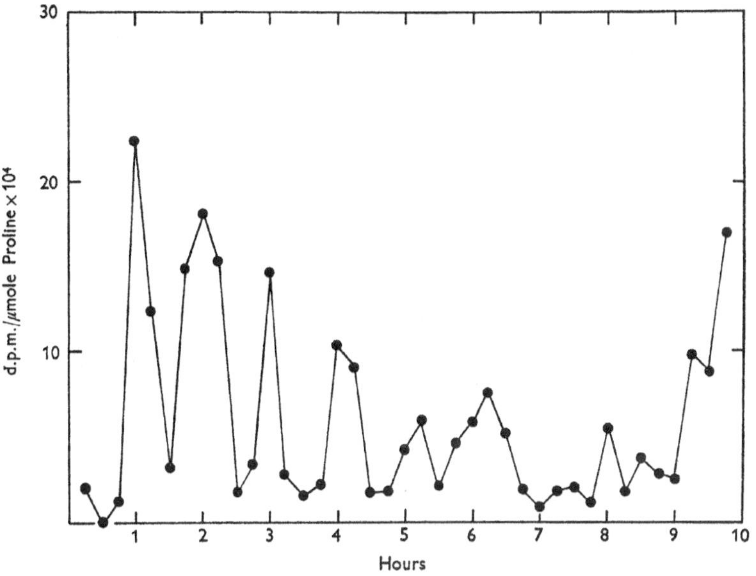


FIGURE 5
Free proline. (After Jackson, 1963)

Further evidence that the observed variations represent a non-random cellular variable is given in figure 6. It is possible to eliminate the fluctuations by a well-defined modification in the experimental procedure. The addition of a chaser of cold proline to the chick embryo after the hot injection causes the variations to disappear, and the dynamics of the system are smoothed out. This result finds a ready interpretation in the present theory of oscillating control circuits. Proline, a non-essential amino acid, is synthesized in chick cells from precursors via a biosynthetic sequence which according to our assumptions is controlled by the end-product, proline. Only as long as the amino acid is being produced by this endogenous, self-regulating system will oscillations occur. (That the variable $M_i$, in this case taken to be proline, undergoes oscillations is an immediate consequence of equation (6), since $Y_i$ oscillates.) If proline is added exogenously in quantities which saturate the control system, then the oscillations will cease, and in fact the endogenous production of proline will be shut off. The theory predicts that any oscillating circuit should be damped out in this manner by adding saturating quantities of the feedback molecule, the end-product, provided of course that the cells are permeable to it.

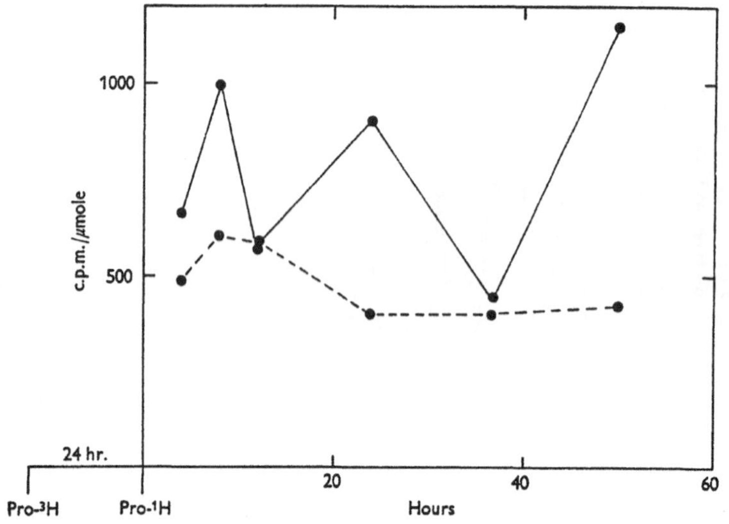


FIGURE 6
Hydroxyproline: specific activity. (After Tanzer & Gross, 1963)

Furthermore, the level at which the cut-off occurs should be at maximum pool size, i.e. at or near the bottom of the oscillating specific activity curve. This is also seen in figure 6. More work is required before it can be definitely concluded that what is being observed in these studies is a periodic variation arising from oscillations of the kind described in this paper. The absence of strict periodicity can be explained by the fact that not all the chicks used for the study are in synchrony at the level of intracellular metabolic oscillators. Indeed, it is remarkable that periodicity is even suggested by the investigations, since this implies a high degree of synchrony between cells synthesizing collagen, and between chicks of the same developmental age. These studies do suggest, however, that oscillatory behaviour is a characteristic of molecular dynamics in cells, and they provide a direct experimental approach to the study of temporal organization in cellular control systems.

### THE INVARIANT INTEGRAL

The property of equations (8) which allows us to construct a statistical mechanics is the fact that they have an integral. This we see by combining the two equations into one and writing

$$(\alpha_i X_i - \beta_i)\frac{dX_i}{dt} + \left(b_i - \frac{a_i}{A_i + k_i Y_i}\right)\frac{dY_i}{dt} = 0$$

This expression can now be integrated, with the result that we get a constant of the motion (an invariant integral):

$$G_i(X_i, Y_i) \equiv \frac{\alpha_i X^2}{2} - \beta_i X_i + b_i Y_i - \frac{a_i}{k_i}\log(A_i + k_i Y_i) = \text{constant} \qquad \ldots.(9)$$

This integral is analogous to the energy integral in physics, and it allows us to proceed to a thermodynamic-like description of cell behaviour, arising from the oscillating dynamics of molecular control mechanisms.

It is convenient to transform to new variables defined by

$$x_i = X_i - p_i$$

$$1 + y_i = 1/Q_i(A_i + k_i Y_i)$$

where $Q_i = A_i + k_i q_i$, and $p_i$, $q_i$ are the steady-state values of $X_i$ and $Y_i$ which make $dX_i/dt$ and $dY_i/dt$ simultaneously zero. The equations thus become

$$\left.\begin{aligned} \frac{dx_i}{dt} &= b_i\left(\frac{1}{1+y_i} - 1\right) \\ \frac{dy_i}{dt} &= c_i x_i \end{aligned}\right\} \qquad \ldots.(10)$$

where $c_i = \dfrac{\alpha_i k_i}{Q_i}$.

The integral now takes the form

$$G_i(x_i, y_i) \equiv \frac{c_i x_i^2}{2} + b_i[y_i - \log(1 + y_i)] = \text{constant} \qquad \ldots.(11)$$

Now as a first approximation we can regard the cell as being made up of a large number of single control circuits of the type shown in figure 1. All these units are coupled together biochemically within the cell because they draw upon common metabolic pools of activated precursors for macromolecular synthesis. We call this type of coupling 'weak interaction', a concept which plays a very important role in the statistical mechanics. The very complex array of interactions occurring in the biochemical space in which the deterministic control components function is not explicitly defined in our theory, for we are largely ignorant of its details. However, we cannot ignore these interactions, for they constitute an integral part of cell structure. In the statistical mechanics these undefined interactions are represented dynamically as a 'noisy' biochemical space in which specific control processes take place, and through which dynamic motion is transmitted from one oscillator to another. It is thus that random processes and distribution theorems enter the dynamics of the control systems, one consequence of which is a maximum 'entropy' theorem.

The crude model which we thus obtain for the cell is a large number, say $n$, of individual, self-regulating control circuits immersed in a biochemical space which supports their existence. Each control circuit has an integral of the same type, and so for the whole system there is a general integral

$$G(x_1, x_2, \ldots, x_n; y_1, \ldots, y_n) = \sum_{i=1}^{n} G_i(x_i, y_i) = \text{constant}. \qquad \ldots.(12)$$

This is the integral which will play a role in the statistical mechanics similar to the Hamiltonian or total energy integral in physics. The dynamic system made up of the set of $n$ control circuits with a total of $2n$ variables $(x_i, y_i; i=1, 2, \ldots, n)$ will be referred to as the epigenetic system of a cell.

When we consider coupled control circuits of the type shown in figure 2, a somewhat altered set of equations is obtained by using essentially the same assumptions and procedure which led to equation (8). The equations are:

$$\left.\begin{aligned} \frac{dX_1}{dt} &= \frac{a_1}{A_1 + k_{11}Y_1 + k_{12}Y_2} - b_1 \\ \frac{dX_2}{dt} &= \frac{a_2}{A_2 + k_{21}Y_1 + k_{22}Y_2} - b_2 \\ \frac{dY_1}{dt} &= \alpha_1 X_1 - \beta_1 \\ \frac{dY_2}{dt} &= \alpha_2 X_2 - \beta_2 \end{aligned}\right\} \qquad \ldots.(13)$$

These lead to a new integral wherein quadratic terms appear, i.e. terms like $X_1X_2$. These terms reflect the presence of what are called 'strong interactions' in the system, and they lead to complicated dynamic effects characteristic of coupled non-linear oscillators. It is these strong interactions which lead to complex but stable relationships between oscillators, so that a definite temporal ordering of biochemical processes can be achieved. These effects include synchronous locking or entrainment between oscillators, wherein two coupled components lock together at the same frequency; and subharmonic resonance or frequency demultiplication, wherein a stable but asymmetric relation arises between two coupled oscillators, so that in a certain time-interval they complete a number of oscillations which is a rational fraction of those completed by the uncoupled (free-running) components. Both these phenomena appear to occur in cells or organisms with diurnal rhythms (cf. Pittendrigh & Bruce, 1957). The properties of the molecular control circuits studied in this paper are regarded as providing the dynamic foundation for such behaviour in cells.

### THE STATISTICAL MECHANICS

The central mathematical content of a statistical mechanics is a probability distribution which allows one to use powerful statistical techniques for evaluating mean values of various functions which can be used to study the general, macroscopic properties of the complex, integrated system. The most useful distribution for our present purposes is that defining a construct known as the canonical ensemble. Once again lack of space prevents a full discussion of the significance of this theoretical concept, and reference must be made to a fuller treatment (Goodwin, 1963). The probability distribution is defined as

$$\rho = e^{(\psi - G)/\theta} \qquad \ldots.(14)$$

where $G$ is the general integral of the system, defined by equation (11) in the case where there is no strong interaction in the system, and by a more complicated integral for a system with coupling between control components. In this expression $\theta$ is a quantity analogous to temperature in physical systems, while $\Psi$ is analogous to free energy. These thermodynamic expressions will be prefixed in the following by the adjective talandic, deriving from the Greek ταλαντωσις, meaning oscillation. Thus we will speak of talandic temperature, talandic free energy, etc., in referring to the macroscopic quantities which define the 'thermodynamic' properties of oscillating control systems of the type considered.

By using the procedures of statistical mechanics, the following result is readily obtained:

$$c_i\overline{X_i(X_i-p_i)}=\theta=\frac{b_ik_i^2}{Q_i}\frac{\overline{Y_i(Y_i-q_i)}}{A_i+k_iY_i} \qquad \ldots.(15)$$

In these expressions the bar signifies that the mean value of the quantity is referred to.

This shows us that $\theta=0$ when $X_i=p_i$ and $Y_i=q_i$, i.e. when the variables are at their steady-state values and no oscillations occur in the system. Thus the point of zero talandic temperature is where there is no dynamic motion, analogous to the zero point of temperature in physics, i.e. where there is no kinetic motion of molecules. It may prove to be the case that this point of absolute zero with no oscillating motion in the control circuits is just as unattainable in living cells as its analogue is in physical systems, and that it is reached only with the death of the cell.

The condition of large $\theta$ corresponds to a very excited state in the oscillators, with the variables undergoing oscillations of large amplitude. It is of fundamental significance in this analysis that when $\theta$ is small, the oscillations are small and have the character of sinusoidal oscillations, i.e. they are effectively linear. However, as $\theta$ increases, the oscillations get larger in amplitude and progressively more non-linear. In the system with strong coupling, it can be shown that the complex interactions characteristic of non-linear oscillators, such as frequency demultiplication and synchronous locking, cannot occur when $\theta$ is very small, and can arise only at elevated $\theta$-values. Thus the talandic temperature of the system is found to be directly related to its degree of non-linearity, and may be regarded as providing in some sense a measure of the organizational capacity of the system in the time domain. We will shortly discuss this question in greater detail.

It is interesting to consider the actual units of $\theta$ and the order of magnitude required for significant non-linearity in the dynamics of the control circuits. In equation (14) the parameter $c_i$ is $a_ik_i/Q_i$. Here $a_i$ has the units of a rate constant, $1/\text{time}=1/T$. The units of $k_i$, an equilibrium constant, are $1/\text{concentration}=1/C$. $Q_i$ has no units, and $b_i$ is a rate, $C/T$. Thus we get from (15):

$$\frac{1}{T}\cdot\frac{1}{C}\cdot C^2=\frac{C}{T}=\frac{C}{T}\cdot\frac{1}{C^2}\cdot C^2$$

The talandic temperature scale thus has the units $C/T$.

The magnitude of $\theta$ in the epigenetic system is determined by the microscopic parameters of the control circuits as well as by their oscillatory motion. Some of these parameters can be calculated on the basis of current estimates for macromolecular synthetic rates, but other parameters can only be guessed at. Thus the studies of Loftfield & Eigner (1958) and Dintzis (1961) show that the protein synthetic time in the cells of higher organisms is of the order of a very few minutes; and investigations by Penman, Scherrer, Becher & Darnell (1963) suggest that the mean lifetime of mRNA in He La cells is several hours. These and other results allow one to estimate the values of the parameters $\alpha_i$, $\beta_i$, and $b_i$. However, a quantity such as the storage capacity of a cell for a particular metabolic species can only be guessed at. It is nevertheless possible to make some reasonable suggestions for the values of the microscopic parameters in the equations (8), and the results obtained are (cf. Goodwin, 1963):

$$a_i = 200,\ b_i = 5/12,\ \alpha_i = \tfrac{1}{5},\ \beta_i = 20,\ k_i = 24,\ Q_i = 480$$

These are to be regarded as order-of-magnitude estimates for a control circuit in the cell of a higher organism.

Using these values, it has been calculated that an oscillation of considerable magnitude, with the population of mRNA of a particular species varying about a steady-state value of 100 molecules/cell with an amplitude of roughly 50 molecules, requires that $\theta$ has a value of approximately 100. The period of such an oscillation is about 5 hours. In general the estimates suggest that $\theta$ will be in the range from 25 to $10^3$, giving oscillations with periods in the range of 2–14 hr. With $\theta$ less than about 1/10, the system approaches linearity and higher-order dynamic behaviour characteristic of non-linear oscillators is unlikely to occur.

### THE CELL AS A RESONATING SYSTEM

In general this analysis of intracellular dynamics leads one to the point of view that the cell is a kind of resonating system which cycles constantly through a set of states. It is not a quiescent, passive system which moves or changes state only in response to stimuli from the environment, but rather is a vibrant entity with intrinsic dynamic activity which is oscillatory in nature. This oscillatory activity is a type of biological energy, and the suggestion made here is that cells should be considered as resonant systems with properties somewhat analogous to those which characterize conjugated biochemical species such as the purines and pyrimidines and confer upon them a position of unique biochemical importance. The many resonant modes which, according to this analysis, arise from the oscillatory behaviour of intracellular negative feedback control circuits and their interaction could explain the remarkable intrinsic coherence and stability of biochemical processes in cells. They could also account for the great lability which cells have for adaptive response to the physical environment or to other cells. The introduction of frequency and resonance considerations into the analysis of cell-cell interactions in developing embryos provides a new dimension for the investigation of such phenomena as embryonic induction, competence, and individuation (i.e. the self-organizing features of interacting embryonic cell populations).

The particular class of control circuits which has been considered in this analysis and called 'epigenetic' could account for relatively slow cellular rhythms, with periods of several hours. However, shorter control loops such as those arising from feedback inhibition would be expected to result in oscillations with shorter periods, and one should be prepared to find a wide spectrum of frequencies occurring in cells. Furthermore, the phenomenon of subharmonic resonance, well known as a consequence of interactions between non-linear oscillators, greatly extends the frequency spectrum over that of the free-running or uncoupled control circuits. Thus, for example, if one of the control circuits of figure 2 shows, separately, oscillations with a period of about 3 hours and the other a period of about 6 hours, then the whole system when appropriately coupled can show oscillations with periods ranging from 3 to 24 hr, and greater. As the order of the subharmonic increases, the amplitude of the oscillation also increases, so that relatively small free-running amplitudes can be greatly magnified when oscillators are coupled. Thus, for example, cells and organisms can generate slow rhythms, such as tidal and circadian rhythm, by the proper interaction of control processes whose fundamental frequencies are considerably greater than 1 or 2 oscillations per day. There is evidence (Pittendrigh and Bruce, 1957) that this is in fact how organisms generate frequencies which keep them in time to terrestrial periodicities.

Another phenomenon commonly observed in non-linear systems is for two oscillators with different frequencies to get locked together at the same frequency when they interact. This is known as synchronous locking of oscillators, and the common frequency can be at or anywhere between the free-running frequencies of the interacting components, depending upon the coupling. The phase relations between locked oscillators are also widely variable with the parameter values. It seems likely that stable temporal relations of this kind between different control processes in cells could play a very important role in keeping different biochemical activities always in a correct phase or time relationship with one another (Halberg, 1960).

The stability of the temporal relationships which can arise from the interaction of non-linear oscillators is strongly dependent upon the 'amount' of non-linearity in the system. In the statistical mechanics which has been described briefly in this paper, the general parameter $\theta$ is, in a rather direct sense, a measure of the non-linearity present in the system. The talandic temperature level is therefore a major macroscopic quantity in determining the stability of the time structure which can emerge in strongly interacting control processes. Thus the frequency spectrum of a cell and its resonance characteristics, in the sense suggested by this study, are intimately related to θ, which emerges as a general system parameter of central importance for the analysis of temporal organization in cells.

### TALANDIC TEMPERATURE AND HOMEOSTASIS

The type of stability with which we are concerned in this study is a stability of temporal relations between biochemical processes. The above considerations on synchronous locking and subharmonic resonance allow one to glimpse the richness of time structure which can arise in the dynamics of oscillating control systems when strong interactions between control units are introduced, compared with the relatively simple dynamics of uncoupled oscillators. They also emphasize the importance of θ in relation to temporal ordering between the interacting components. Even if there is a rich pattern of strong interactions in the system, they may result in very little temporal organization if θ is small. Only at elevated talandic energy levels (large θ) can the dynamic consequences of non-linear interactions emerge.

We can thus begin to see how talandic temperature relates to such physiological notions as homeostasis. When θ is large, small disturbances will not cause synchronously locked oscillators to drift out of phase, for example, because they will spontaneously lock again after the disturbance has ceased. But when θ is small, the forces of dynamic interaction between oscillators are greatly reduced, and the oscillators will not be ordered by stable relations to one another. Random disturbances will cause the components to drift without any mutual coherence, and any time structure in the system will rapidly decay. The importance of θ as a major system parameter in relation to the homeostatic capacity of cells in time thus becomes evident.

### ADAPTIVE ASPECTS OF TALANDIC PHENOMENA

The emphasis implicit in the present theory is on rhythmic properties of cells which are stationary in time, and are neither growing nor differentiating, but simply maintaining themselves. The homeostatic properties of coupled oscillators discussed briefly above relate to an internal stability of relations in strictly periodic systems, which cycle through the same state at regular intervals. These ideas are readily extended to include the relations which arise between cells and periodic behaviour in environmental variables. Periodicity is a dominant feature of a terrestrial environment, so that adaptation to such an environment involves rhythmic variation in the biochemical activities of cells. The most obvious rhythms in organisms are those which are linked to the light-dark cycle of the planet, and the homeostatic aspects of these diurnal rhythms have been discussed by Harker (1964). We may observe that the phenomenon of entrainment is extremely important in ordering cellular activities relative to environmental variables. Thus once again the talandic temperature of oscillating control systems becomes an important variable in considerations of the capacity which an epigenetic system has for adaptive and stable response to environmental signals. A cell in a very low talandic energy stage can respond only passively to a periodic environmental signal; it has very little 'oscillatory energy' of its own to use in adapting its activities to environmental cycles, so that its response depends largely upon the intensity of the exogenous stimulus. However, a cell in a high talandic energy state has a high capacity for adaptive response to periodic environmental signals by means of nonlinear interactions. It can also organize its biochemical activities with greater autonomy since the exogenous stimulus need be used only as a clue to set the endogenous oscillators relative to the environment, the oscillators then maintaining a coherent rhythmic state by means of the high talandic energy of the system.

### TALANDIC FORCE AND TALANDIC WORK

So far we have considered only stationary or conservative behaviour in control systems, and that is the natural province of the present theory. The equilibrium condition of the epigenetic system occurs when all the steady-state values $(p_i, q_i; i=1, \ldots, n)$ are fixed and the talandic energy is equally distributed over all components so that the talandic temperature is uniform throughout the system and equations (13) are satisfied. However, it is possible to move away from this equilibrium point and to consider the forces which are at work to change the state of the epigenetic system. The correct quantitative approach to this question is through the analogues of force, work and free energy in the present theory. The concept of force enters a statistical mechanics in association with quantities which are called external parameters. These are the environmental variables which act upon the epigenetic system, such as light, temperature, inducers, mutagens, etc. These stimuli cause the system to move to new states by altering the values of the microscopic parameters. Considering for simplicity the case of control systems without strong interactions, the generalized force conjugate to an external parameter, $s_r$, is defined as

$$F_r \equiv -\frac{\partial G}{\partial s_r} = -\sum_{i=1}^{n}\left\{\frac{\partial c_i}{\partial s_r}\frac{x_i^2}{2} + \frac{\partial b_i}{\partial s_r}[y_i - \log(1+y_i)]\right\}$$

The canonical mean of this force can be shown to be equal to

$$\bar{F}_r = -\sum_{i=1}^{n}\left\{\frac{\partial \Psi_{pi}}{\partial c_i}\frac{\partial c_i}{\partial s_r} + \frac{\partial \Psi_{qi}}{\partial b_i}\frac{\partial b_i}{\partial s_r}\right\} \qquad \ldots.(18)$$

where $\Psi_{pi}$ and $\Psi_{qi}$ are the talandic free energy functions for the variables $x_i$ and $y_i$, respectively. The work done by the stimulus, acting over a period $t_0$ to $t_1$, is defined as

$$W = \int_{t_0}^{t_1} \bar{F}_r \frac{ds_r}{dt} dt$$

$$= -\sum_{i=1}^{n}\int_{t_0}^{t_1}\left\{\frac{\partial \Psi_{pi}}{\partial c_i}\frac{\partial c_i}{\partial s_r} + \frac{\partial \Psi_{qi}}{\partial b_i}\frac{\partial b_i}{\partial s_r}\right\}\frac{ds_r}{dt}dt$$

If now $\theta$ is held constant and the process is a reversible one, we can write

$$W = -\sum_{i=1}^{n}\left\{\int_{c_i(t_0)}^{c_i(t_1)}\frac{\partial \Psi_{pi}}{\partial c_i}dc_i + \int_{b_i(t_0)}^{b_i(t_1)}\frac{\partial \Psi_{qi}}{\partial b_i}db_i\right\}$$

$$= \Psi_0 - \Psi_1$$

Thus we see that the amount of talandic work done in a reversible process is equal to the change in the talandic free energy of the epigenetic system. This is strictly analogous to the situation in classical thermodynamics.

### TALANDIC TEMPERATURE AND GROWTH

Now the direction of 'spontaneous' change in a thermodynamic system is that defined by a decrease in the free energy function. Figure 7 shows the family of curves obtained when $\Psi_{pi}$ is plotted as a function of $c_i$, for various values of $\theta$. We see immediately that $\Psi_{pi}$ decreases as $c_i$ decreases. (Similar curves are obtained for $\Psi_{qi}$ as a function of $b_i$.) Since $c_i = (\alpha_i k_i)/Q_i$, $Q_i = A_i + k_i q_i$, and $b_i = a_i/Q_i$, a decrease in $c_i$ and $b_i$ means generally that the steady-state values $p_i$ and $q_i$ increase. The direction of spontaneous change in the system is therefore an expansion: the macromolecular populations tend to increase if the constraints in the system are such that some movement away from equilibrium is possible. Thus we may say that the dynamic system we have been studying has a spontaneous tendency to grow.

Another property evident from figure 7 is that when $\theta$ is large there is a much greater tendency for this expansion to occur than when $\theta$ is very small, for the talandic force causing the system to expand is directly proportional to the slope of the talandic free energy curve, as we see from (18). When $\theta$ is small the slope

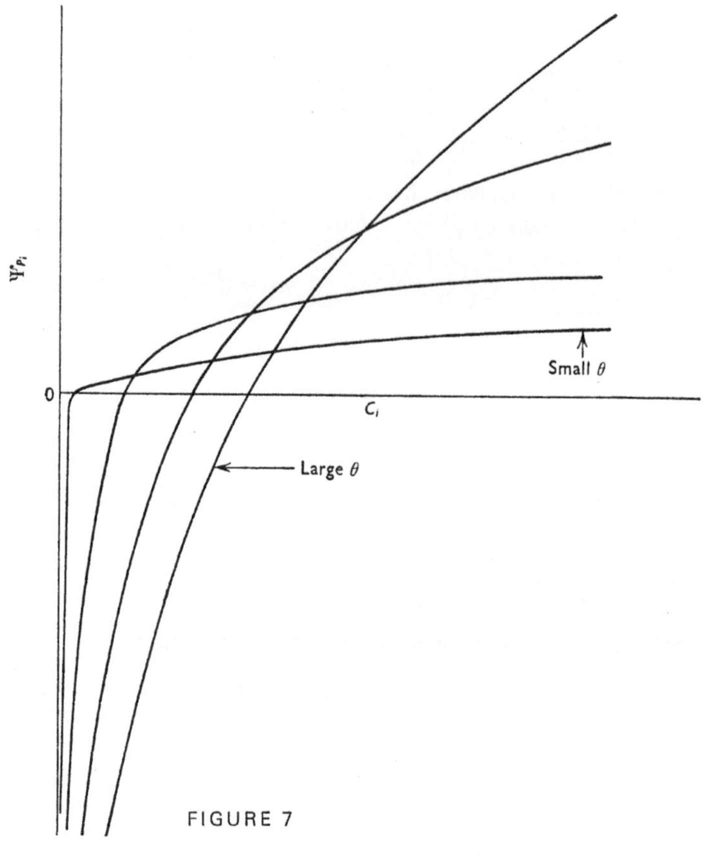


FIGURE 7

of the free energy curve in figure 7 is small over most of the range $c_i$; but with θ large, the slope is uniformly large, and the system will change its state rapidly when parametric constraints are loosened. These considerations suggest how one may understand cell growth rates, and lay a quantitative foundation for their analysis in terms of the dynamic properties of cellular control systems. We also have here a possible basis for analysing the adaptive responses of cells to external stimuli such as inducers, in terms of the concepts of talandic force and work.

### TALANDIC PROPERTIES OF EMBRYONIC SYSTEMS

Turning briefly now to an embryological problem, the ideas of the present theory may have some relevance in relation to the property of competence in embryonic cells. There is good evidence that the length of time during which a cell or tissue is competent to respond to a particular inductive stimulus is determined by processes within the cell, and is to a considerable extent independent of its environment (Waddington, 1934, 1936; Holtfreter, 1938). Some kind of timing mechanism appears to be operating in embryonic cells which fixes their period of competence, and the notion immediately presents itself that the fundamental dynamic property which underlies such 'clocks' is once again oscillatory behaviour in cellular control systems. One might visualize the development of a particular competence and its duration in analogy with the setting of an alarm clock. At particular developmental stages in the differentiation of a cell, biochemical clocks might be started which go off at some later time. The ringing of the clock would be the attainment of a certain level of activity at certain genetic loci, with the occurrence of particular protein species at critical concentrations in the cell. The alarm would ring for a while, and then stop; the high activity would last for some time, and then drop off again, as occurs in cells with circadian time-structure. However, these postulated clocks in embryonic cells would not be truly cyclic, in the same way that an alarm clock is not cyclic; it goes off once, and then has to be set again. Embryonic processes are manifestly non-cyclic and irreversible in this sense, and this creates a difficulty for a theory in which the equilibrium condition is defined by steady states. There appear to be ways of extending the theory to cover irreversible processes in the embryological sense. One could then approach the problems of developmental canalization and homeorhesis (Waddington, 1957), questions which are still outside the province of the present theory. The suggestion we would like to make here, however, is that the phenomenon of embryological competence may well find an explanation in terms of cellular timing mechanisms which are generated by non-linear interactions between biochemical oscillators of the type considered in this study.

There is a further aspect in which considerations of temporal organization in cells may be of significance in relation to embryonic processes. Periodicities in time can generate periodicities in space by leaving a physical trace, as it were, of the rhythmic activity. In connective tissue there occurs a rather remarkable orthogonal layering of collagen fibres, one layer of fibres lying at 90° to the next. It is very difficult to imagine how this could occur during embryonic development if there is a continuous, steady release of collagen from chondrocytes. However, if collagen is released from these cells periodically, with a well-defined rhythm, then one has a new variable with which to explain the physical periodicity. We have already seen that there is experimental evidence for oscillations in the synthesis of collagen in the results of Tanzer & Gross (1963) and Jackson (1963), shown in figures 4–6. Gross (1961) has discussed the structural layering of collagen fibres in some detail from this point of view, and he suggests that oscillatory behaviour in embryonic cells may frequently underlie spatial periodicities in the morphology of the adult. The possibility of using time structure to generate space structure gives to the developing organism a further dimension for morphogenesis. The homeostatic properties of interacting biochemical oscillators in cells then apply directly to connective tissue formation as a homeorhetic process.

### EXPERIMENTAL ASPECTS OF THE THEORY

A theory of biological organization has little value if it cannot be given an experimental foundation. In this last section we will consider how talandic temperature, the central macroscopic variable to have arisen in this study, may be controlled and measured experimentally in cells. Successful experimental control of $\theta$ would amount to an explicit test of the theory; whereas if such control cannot be realized, then the theory would have to be abandoned, for it would then have practically no predictive value.

The main idea which has arisen in this study is that a system organized microscopically in the manner described by equations (8) can exist in many different talandic energy states without any change occurring in the steady-state values of the system variables, $p_i$ and $q_i$. In terms of the epigenetic system of cells, this means that with all microscopic parameters fixed so that the steady-state levels of the different species of molecular and macromolecular populations do not change, it should be possible for the cell to be in any one of a large number of different states in the sense of its level of oscillatory activity or excitation. The question of how to control $\theta$ thus becomes the problem of exciting or damping the oscillators experimentally and observing these changes of talandic energy in some way.

One procedure might be as follows: consider a culture of tissue with well-defined rhythmic behaviour, such as the circadian respiratory periodicity in cultured adrenal glands reported by Andrews & Folk (1963). Let the culture be maintained on a defined medium, in which there is a limiting quantity of an essential amino acid, so that protein synthesis is controlled by the concentration of this component. Suppose that the culture is given a small pulse of the limiting amino acid, so that protein synthesis is briefly stimulated. The pulse should be small enough for the added quantity to be used up in, say $\frac{1}{2}$–1 hr after it is added ; or else there should be a flow of medium through the culture vessel such that the concentration of amino acid can be constantly regulated, and pulse times closely controlled.

The effect of the pulse should be a temporary stimulus of protein synthesis. This one would anticipate, both because the protein synthetic system will be temporarily relieved of amino acid limitation, and because of the inductive effect which amino acids appear to have on mRNA synthesis ( Stent & Brenner, 1961 ). The pulse should therefore cause a transient shift in the oscillating trajectories of the different protein species. The direction in which such a disturbance will shift an oscillatory trajectory, whether towards or away from the steady-state value, will depend upon what part of the trajectory the system is on when the disturbance begins. If it is on that part of the trajectory which lies below the steady state, then the stimulus will shift the system 'up' towards it, thus decreasing the amplitude of the oscillation. But if the protein concentration is above the steady-state axis when the stimulus begins, then the shift is away from the axis and the amplitude of oscillation will be increased.

We come now to an observation which is critical to the argument. The oscillations occurring in the feedback control mechanisms studied in this work are distinctly asymmetrical. In figure 3 it can be seen that the variables undergo greater excursions above their steady-state values than below, and it can be shown that this asymmetry increases as $\theta$ increases ( Goodwin, 1963 ). The asymmetry is particularly marked in the case of oscillations in protein concentrations, and the fundamental dynamic reason for this behaviour is the occurrence of $Y_i$ in the denominator of the expression for the rate of mRNA synthesis. Our assumptions about the dynamics of feedback control of macromolecular synthesis in cells thus have an important consequence : protein and mRNA concentrations spend more time above the steady-state axis than below. It follows therefore that a transient increase in protein and mRNA synthesis will be more likely to occur on that part of the cycle which is above the steady-state axis, and hence should cause the trajectories to move away from this axis. That is to say, the small pulse of limiting amino acid will have a greater probability of increasing the amplitude of the oscillation, hence increasing the talandic temperature, than of decreasing it.

However, a single pulse is not likely to cause a permanent change in $\theta$. What is required to bring about such a change is a repetition of the pulse at intervals of perhaps 2–3 hr over a fairly long period of time, say 2 days. The idea is that the pulsing should not cause any permanent change in the microscopic state of the system ; the pool sizes, synthetic rates, etc., returning to their initial values after the pulse ceases.

## Temporal organization in cells

We now suggest that the way to observe changes in θ is through the rhythmic behaviour of the cells. It can be shown in the theory that the period of the oscillations is directly proportional to the square root of the talandic temperature when θ is fairly large (giving well-defined temporal behaviour in cells, such as circadian rhythms). A basic assumption in this study is that circadian periodicity is directly related to oscillations in biochemical control circuits. We are thus led to the conclusion that the above-described pulsing treatment of cells with well-defined rhythmic behaviour should result in a slowing down of the biological clock, since the periods of the primary oscillations in the cells should increase. As the pulsing treatment continues, the prediction is that the clock period will get progressively longer.

There are many complications which may be expected to arise in such an experimental design. However, the basic principle suggested for the experimental control and observation of θ is periodic, transient disturbances of a rhythmic system, designed to excite the dynamics of the control circuits without permanently changing the microscopic parameters. It is clearly essential that some experimental procedure be devised whereby θ can be controlled and observed, otherwise it can hardly serve a useful purpose in the study and analysis of cell behaviour. A more direct observation of oscillatory motion, and hence a more direct measurement of talandic temperature in cellular control systems than via biological clocks, would certainly be desirable. The experimental methods of Tanzer & Gross (1963) and Jackson (1963) seem to approach this goal, but in this case continuous observation is not possible since the cells are destroyed during the analytical procedure. It is possible that further refinements in such optical techniques as microspectrophotometry and fluorometry will allow one to make measurements on the 'motion' (in the sense of changing concentrations) of molecular populations in single, living cells, and thus measure θ directly. Some experimental foundation is clearly essential for a theory which seeks to replace the intuitive notions which we have concerning cellular organization by precise, quantitative concepts. The analysis presented here is still in a very exploratory state, but the hope is that it may lead to a theory which has real content and can contribute to the development of a theoretical biology which is firmly grounded in experiment.

## References

Andrews, R.V. and Folk, G.E. (1963). Respiratory circadian periodicity in cultured hamster adrenal glands. *Fed. Proc.*, **22**, 382.

Dintzis, H.M. (1961). Assembly of the peptide

chains of hemoglobin. *Proc. nat. Acad. Sci., Wash.*, **47**, 247–61.

Gerhart, J. C. and Pardee, A. B. (1962). The enzymology of control by feedback inhibition. *J. biol. Chem.*, **237**, 891–6.

Goodwin, B. C. (1963). *Temporal Organization in Cells* (London: Academic Press).

Gorini, L. (1963). Control by repression of a biochemical pathway. *Bact. Rev.*, **27**, 182–90.

Gross, J. (1961). Collagen. *Sci. Amer.*, **204**, (5), 120–30.

Halberg, F. (1960). Temporal coordination of physiologic function. *Cold Spr. Harb. Symp. quant. Biol.*, **25**, 289–310.

Harker, J. E. (1958). Diurnal rhythms in the animal kingdom. *Biol. Rev.*, **33**, 1–52.

Harker, J. E. (1964). Diurnal rhythms and homeostatic mechanisms. *Symp. Soc. exp. Biol.*, **18**, 283–300.

Holtfreter, J. (1938). Veränderungen der Reaktionsweise in alternden isolierten Gastrulaektoderm. *Roux. Arch. EntwMech. Organ.*, **138**, 163–96.

Jackson, D. S. (1963). Personal communication.

Loftfield, R. B. and Eigner, E. A. (1958). The time required for the synthesis of a ferritin molecule in rat liver. *J. biol. Chem.*, **231**, 925–43.

Magasanik, B. (1958). The metabolic regulation of purine interconversions and of histidine biosynthesis. In *A Symposium on the Chemical Basis of Development*, pp. 485–94. (Ed. McElroy, W. D. and Glass, B.) (Baltimore: Johns Hopkins Press.)

Monod, J. and Jacob, F. (1961). General conclusions: Teleonomic mechanisms in cellular metabolism, growth, and differentiation. *Cold Spr. Harb. Symp. quant. Biol.*, **26**, 389–401.

Penman, S., Scherrer, K., Becher, Y. and Darnell, J. E. (1963). Polyribosomes in normal and poliovirus-infected He La cells and their relationship to messenger-RNA. *Proc. nat. Acad. Sci., Wash.*, **49**, 654–62.

Pittendrigh, C. S. (1960). Circadian rhythms and the circadian organization of living systems. *Cold Spr. Harb. Symp. quant. Biol.*, **25**, 159–84.

Pittendrigh, C. S. (1961). On temporal organization in living cells. *The Harvey Lectures*, Series 56, p. 93 (New York: Academic Press).

Pittendrigh, C. S. and Bruce, V. G. (1957). In *Rhythmic and Synthetic Processes in Growth*. (Ed. Rudnick, D.) (Princeton: University Press.)

Stent, G. S. and Brenner, S. (1961). A genetic locus for the regulation of ribonucleic acid synthesis. *Proc. nat. Acad. Sci., Wash.*, **47**, 2005–14.

Szilard, L. (1960). The control of the formation of specific proteins in bacteria and in animal cells. *Proc. nat. Acad. Sci., Wash.*, **46**, 277–92.

Tanzer, M. L. and Gross, J. (1963). Collagen metabolism in the normal and lathyritic chick. *J. exp. Med.* (In the press.)

Umbarger, H. E. (1956). Evidence for a negative-feedback mechanism in the biosynthesis of isoleucine. *Science*, **123**, 848.

Vogel, H. J. (1957). Repressed and induced enzyme formation: A unified hypothesis. *Proc. nat. Acad. Sci., Wash.*, **43**, 491–6.

Waddington, C. H. (1934). Experiments on embryonic induction. I. The competence of the extra-embryonic ectoderm in the chick. *J. exp. Biol.*, **11**, 211–17.

Waddington, C. H. (1936). The origin of competence for lens formation in the amphibia. *J. exp. Biol.*, **13**, 86–91.

Waddington, C. H. (1957). *The Strategy of the Genes* (London: Allen and Unwin).

# New Thoughts on Bio Control[1]

A. S. Iberall
General Technical Services Inc., Pennsylvania

My colleagues, notably Dr Cardon, and I have been pursuing a closely coordinated novel view of the complex biological system. This is contained in nine basic reports written for NASA and a number of derived papers that we have given or reported on at various meetings [1]. I will attempt to give you some sketchy outline of our central ideas up to date. (It is only fair to the reader and ourselves to state that Dr Cardon and I are physical scientists only recently concerned with biology.)

Starting in our first report–available as a N.Y. Academy of Science review of September 1964–we sought to define the regulating and control functions of the system as a whole.

First we sought the central theme of the biological system and found it well enunciated in the biological literature in the form of Bernard's and, later, Sechenov's and Cannon's [2–4] insistence on the primacy of the interior milieu and of its regulation–homeostasis–as the significant theme of operation of the complex system. We then found and accepted H. Smith's view of the kidney [5] as the master regulator of the system, for in its function, more than anywhere else, does the preservation of the character of the watery internal milieu lie. As physical scientists–basically in the tradition of the modern chemical engineer with a view of flow charts, unit processes, and regulating and control functions–we thus had a core model of chemicals in a watery environment with a master flush valve at the exit to preserve the watery environment and regulate concentration contents.

It was clear, next, that ever since Wiener [6] the internal electric system had to be related to possible regulating and control functions. However, within the spirit of discovery that we were seeking, it appeared not so much that Wiener had developed key knowledge about the electrical system–Adrian and Sherrington [7, 8] were much more influential and important from a biological point of view–but that he called attention to a control (cybernetic) point of view for the system.

At this point it was necessary to put some mechanisms, systems, and processes into the bath. Here we had to start with what would be specialized and peculiar to physical theory and its dynamic analysis. If one were to look for regulating and control chains, it would be first necessary to dispose of any self-sustained

[1] This report was substantially presented, by invitation, for the local Biomedical engineering section–IEEE–Phila., Sept. 1966; Dept. Physiology–U. of Pittsburgh, Dec. 1966; Serbelloni conference on Theoretical Biology, Lake Como, Aug. 1967.

oscillatory or so-called limit cycle phenomena. These are indicative of unstable linear chains that have unknown causality and 'purpose' to begin with in the analysis of a new unknown system. We proposed to tease these out and lay them aside and then to continue to look for regulators and controllers.

To those who might consider this idea vague, it really consists of the following: what is unique in the description of the dynamics of a system—whether by equations, physical chains, or verbal chains—is the set of singular states of motion of a system and the kinds of operating stability that one finds around these singularities. It is thus that the actual 'stable' dynamic motion of a system is cast. This is the topological overview of non-linear mechanics in considering the dynamics of systems. As Minorsky [9] puts it so well, the non-linear limit cycles of Poincaré are the stable non-linear steady states of operation of a system, just as the damped oscillation is the stable linear steady state of operation of a system. Whether a system be linearly stable (at rest) or non-linearly stable (in limit cycle oscillation) depends upon the specific character of its singularities. In an *a priori* unknown system, they can only be discovered experimentally.

We therefore went searching, and began to tease out a broad spectrum of oscillators, which may also be called engine cycles, rhythms, DC to AC converters, or non-linear limit cycles. In fact, after a while, it dawned on us that we had become keepers of biological macrospectroscopy, the dynamic spectrum of biological effects, in which the chains are biochemical, bioelectric, and biomechanical. (When we were introduced to Brian Goodwin's work [10], we regarded him as a keeper of microspectroscopy.) We found so many such chains that we were led then to believe a key idea: that what the biological system consisted of was a great collection of limit cycle oscillators; basic regulating and control functions were to be found in the mediation of the stability of these oscillators, mainly by inhibition; that the slowly varying DC parameters of the milieu mediate the operating point—in amplitude, or frequency, or in actual stability margin—of these oscillators so as to achieve desired regulation.

I can name some of these 'major' physiological oscillators for clarity and suggest their nominal time scale.

| | | |
|---|---|---|
| bioelectric nervous waves (spikes, etc.) | 0·1 | sec. per cycle |
| heart beat complex | 1 | sec. |
| ventilation | 4 | sec. |
| blood circuit flow | 10 | sec. |
| blood flow oscillations | 30 | sec. |
| metabolic oscillations | 100 | sec. |

| | |
|---|---|
| vasomotor oscillations | 400 sec. |
| many fast endocrine oscillations | 300–1000 sec. |
| gas exchange oscillations | 2000 sec. |
| metabolic fuel oscillations | 5000 sec. |
| heat balance oscillations | 3 hours |
| circadian rhythms | 24 hours |
| water cycles | $3\frac{1}{2}$ days |
| many longer range endocrine rhythms | 20–40 days |

Since Cannon's concept of homeostasis, on careful reading, was a concept of a regulating process or chain, but with the implication (in his examples) of quasi-static regulation [4], it seemed appropriate to propose a change in name to indicate a more dynamic nature of the regulation. For this, the term homeokinesis was offered, for it is by manipulation of the kinematic variables–of space and time alone–that this dynamic regulation takes place.

By our fourth report [1] we were then prepared to propose a new operational definition of life.

'Thus life is tentatively defined as any compact system containing a complex of sustaining non-linear limit cycle oscillators, and a similar system of algorithmic guiding mechanisms, that is capable of regulating its interior conditions for a considerable range of ambient environmental conditions so as to permit its own satisfactory preservative operation; that is capable of seeking out in the environment and transferring and receiving those fluxes of mass and energy that can be internally adapted to its own satisfactory preservative operation; that is capable of performing these preservative functions for a reasonably long period of time commensurate with the "life" of its mechanical-physical-chemical elements (i.e. clocks made of parts that should wear for hundreds of years that run for two seconds are not crickets); and–likely as a luxury part of the definition–that are capable of recreating their own system out of materials and equipment at hand (one will have to note in the future how much of the biological system can be rebuilt, or one can recall the story of the amorous young bridegroom who has watched his bride remove most of her apparent charms before his eyes–teeth, eyes, hair, wooden leg, etc.).

'The purpose of this definition is to guide the physical search for "explanations" of the operation of the biological system; and to leave the physical scientist closer to some physical base by which he can model, "build", or assess systems that resemble naturally "living" systems by suitable operational definitions; and a clue that "life" does not have to be explained by only one mechanistic scheme per system, but may involve many possible types of successful operation.'

We view the function of such a definition, in an operational Bridgman sense, that it should give us a hunting licence in the form of a basic concept of what we are hunting for. In this definition it is declared to be dynamic regulating chains. These may be fully unstable and oscillatory, or may be marginally unstable and aperiodic. We would hope that by this means we can have forced the intrusion of thinking about dynamic processes into biology. The significance of dynamic AC analysis of systems does not have to be stressed to any group who has been influenced by modern electrical engineers.

Further, to give the system its primary functional keynote, its 'purpose,' we have proposed the following catch-phrase:

*The biological system is an intermittently self-actuated motor system operating in both short and long term, 'that seeks to sustain the metabolic reaction: fuel + oxygen → carbon dioxide + water, plus sufficient power so that it can continue its self-actuated motor activity to seek to sustain its metabolic activity.*

Shorter still: *the system eats and moves about, so that it can continue to eat and move about.*

Thus, for example, in our second summary paper at the 1965 IFAC Symposium in Tokyo, we could represent our overview of the biological system two years after our N.Y. Academy paper. As an excerpt:

'The experimental survey of a complex biological system like the human discloses a large number of autonomous oscillators continuously operating in the system. A partial frequency spectrum in man would consist of primary neural frequencies in the range of 5 to 50 cps, a muscle motor unit frequency at about 10 cps, a heart beat about 1 cps, breathing rate about 1 cycle per 4 seconds, several eating cycles per day, circadian rhythms of 1 cycle per day, sex urge approximately 1 cycle per few days and menstruation about 1 cycle per 30 days. To these may be added cycles, demonstrated within the past few years, in ventilation rate, local skin temperature, and metabolism of approximately 100 seconds, 400 seconds, 30 minutes, and 3 hours. The first appears to be an engine cycle, primarily in skeletal muscles, in which the major heat production takes place. The 400 second cycle appears to represent a vasomotor action which partitions blood flow among the major organ circulations. The 30 minute cycle may be a gas exchange cycle, likely representing a total body carbon dioxide equilibrium, and the 3 hour cycle is probably an overall thermal balance. A 3-day cycle has been found in body weight. It has been tentatively identified as a water balance cycle.

'It has become increasingly apparent that the many oscillators in the biological system are not incidental characteristics of the system, but represent the working components of the system. In summation, they are the biological system. In accordance with this view, it is proposed that homeostasis, Cannon's organizing biological concept of a complex regulation characteristic of the system, is obtained as a result of shifting the stability of these intrinsic non-linear oscillators. The oscillators are likely modulated or shifted in operating point by electrical and chemical signals. In our view such action is an illustration of dynamic regulation. It is possible that the biological system is not able to operate in any other way. In fact, it is likely that the same type of instability mediation is the foundation for all automatic control theory. However, the more general thesis is beyond the scope of discussion in this paper.

'To illustrate the mechanisms of homeostasis, a number of biological systems will be discussed. A first case in point is that of heat production in mammals, which is of interest as part of the system of human thermoregulation. Local skin temperatures, ventilation rate, and oxygen consumption rate all show sustained limit cycle oscillations with a prominent component near 100 seconds. We have postulated that this is a heat engine cycle; that the engine consists essentially of skeletal muscles; that the level of operation of local muscle engines is regulated by an oxygen choke which limits the oxygen supply and thus the rate of local oxidation; that regulation of oxygen rate is achieved in a 100 second cycle of capillary red blood cell flow, mediated by the vasodilatory effect of adrenaline. That adrenaline is the hormone mediator of the cycle is suggested by its calorigenic action, its vasodilatory effect primarily in skeletal muscles, its effect of increasing muscle activity and a time of action for transient effects in a range between one and two minutes.

'The mechanism of the adrenaline action is not known.'

The essence of the matter, as this last thought intends to imply, is that having now proposed all of these grandiose systems views of the complex mammal, the human, the burden of proof is now upon us to make the description stick. This, to us, has meant that we must identify and run down these dynamic regulating chains. This is what we are now beginning to try to do. However, there are so many of them that we have to choose carefully and for the most hard-hitting value. Our resources are really quite limited. However, to add some more flavour to the presentation, I think I can add what we had in addition grasped more recently and expressed in our sixth report, August 1966 [1].

In the metabolic reaction:

| Time elements | Fuel + | A matrix Oxygen → | $CO_2$ + | $H_2O$ |
|---|---|---|---|---|
| 1 | | | | |
| 2 | | | | |
| 3 | | | | |
| 4 | | | | |
| 5 | | | | |

we basically consider that there seems to be a matrix of regulation levels and effects, a hierarchy of regulation functions that seem to pile up under each item; the regulation of fuel, oxygen, $CO_2$ and $H_2O$. Many of these are quite independent and tend to appear redundant. What we expressed in the past, coming more nearly into focus in this sixth report, is that when the ecology provides cueing sensitivities of a temporal type which may have survival value, and there is a suitable cyclic time fracturing (or phase locking) commensurate with this period provided by possible physical-chemical chains, these chains will become a dominant dynamic regulating element in the system. Thus, the spectrum of effects we have identified seem to fit this scheme. However, beyond this, we have finally vaguely seen that the spectrum tends to be even more limited. We have further said in our sixth report:

'The apparent competition between autonomous physiological oscillators and the environmental cues (day, month, year) is to be resolved by recognizing that cycles must fit and thus be entrainable as small numbers with all such cues. We eat, defecate, urinate, sleep a few times per day. These are useful adaptations. An animal (call him a dinosaur) that must chase food for too long to make one meal can't make it. He will not satisfactorily entrain. Time is against him. When time is ample, then the system instabilities will lead to an orbital entrainment in such cycles that can fit the time comfortably. This is one added thought of how the patterning richness is regularly reinforced. The second thought emerges even more strongly from endocrine considerations.

'Each endocrine gland seems to put forth a spectrum of hormones, and it has been a little discouraging to attempt to exhaust, by long lists, their apparent multiplicity of functions. Yet major functions emerge, and the spectrum covers functions that are usefully adaptive. However, it also emerges that more than one gland may put out hormone components that collaborate at common functions. In reviewing time constants, it appeared fascinating that, while the spectrum is rich, it tends to appear finite in number. Thus the concept of time fracturing of physical-chemical chains seems to roost on a more specific perch. Any particular ductless gland may contribute hormonal elements. However, most of them will be involved in chains with a small number of time constant ranges. (For example, we might propose some typical numbers, 0·05–0·1 second, 0·5–1 second, 5–10 seconds, 20–40 seconds, 60–120 seconds, 300–500 seconds, 20–40 minutes, 3–4 hours, 20–8 hours, 3–4 days, 20–40 days.) It is not the case that they must be involved in such time ranges. However, those that can be entrained in chains that have considerable adaptive value have greater survival value for their species. Thus the hormone patterns are interlocked–from gland to gland–to form a matrix of chain function (sugar metabolism, for example) and time. It is this highly locked-in matrix that exists in the endocrine system, the nervous system, etc., that provides the underlying functional structure of the system. The patterns then form and wrap themselves around these more permanent poles. The final ingredient, of course, is that these salient polar times fit, by small numbers, and interact with the AC cueing of the environment. The night–day cycle is a most powerful polarizer. The breakdown of time ultimately to the 1–2 minute or the 0·05–0·1 second cycle time is of course more subtle. The 28-day menstrual cycle drive is quite real, and the breakdown into 4 weeks is reasonable. It is disconcerting to consider cues for a 3½ day water cycle. Which is causal for which ?' (That is, the week for the water, or the water for the week ?)

'The existence of such time cycles does not mean that all animals must share them. Development in each species can have occurred independently. It is likely that similar chemical cycles may be arrived at, but this arises because the biochemical chains, in general, are not so specific, although particular ones may be sharply deterministic.'

Thus far, we have sketched out our scheme for the system via its more 'involuntary' physiological oscillator chains. Now I will discuss a little our proposal that this is extendible to the less 'involuntary' psychological or behavioural system.

We have postulated that what is most characteristic of behaviour is an orbital synchronization with various elements in the milieu. This model has been subconsciously influenced by a quantum mechanical model of free electron conduction in metals. In Frenkel's *Wave Mechanics* [11] there is a delightful modelling of the electron motions as follows: The electrical field causes an accelerated motion of free electrons with a relaxational phase on collisions. The free electron path is thus defined as a sequence of mean free path relaxations and mean relaxation time between collisions with what are basically not crystal lattice points but lattice dislocations. However, at any point the electron is not locked up for one oscillation but it makes a number of pinwheel rotations. The human–it is proposed–acts similarly. A signal DC or AC in the milieu–for example, the flick of the skirt of a passing comely female, or the pulsations in the stomach, or a painful nerve signal, etc., each mediate the oscillator complex, and lock the system 'posture' for that 'moment' into an orbital synchronous path. The command algorithm (an extremely large code book) was not all determined at birth, or did not emerge at various maturational points. It was partially 'learned' and developed out of the matrix of experience. Such development, I was taught a long time ago, represents an epigenetic process.

To give behaviour some picturable substance, consider the bank of physiological oscillators as not being rigidly closed but with partially open inputs, and quite a few in number. Let the oscillators resemble the various resonators of a piano or organ, each of which can have its pitch, amplitude, and timbre mediated. Now let the input winds of the milieu–both internal and external–play over the bank of open inputs. They will make melodic patterns in the organ. Thus far, we only have an instrument like a passive Aeolian harp–even if the individual resonators are active elements.

However, let these oscillations also drive the motor system into space-time orbits. Then the system will be urged to drift down the path of life, however not continuously. It will lock up, circle quite a few times, pass on, etc. There are open portions in this path motion.

Continuity of behaviour is provided by the learned part of the algorithm. Mother, in the mother-child relation, teaches the child various patterns that are fairly adequate to provide the range of needs that will saturate the physiological oscillators over time.

To make this more comprehensible, there is likely a two-state oscillator complex–this is Freudian derived–representing a satisfying or 'euphoric' state, and an 'anxious' or 'dysphoric' state. In this view we find Henry Stack Sullivan's formulations [12] most nearly fitting what we were looking for. Any first infant experience may be satisfactory or not. However, all that can emerge in time is an alternation of such states. In the infant brain, the major signalling complex is the oral one. We consider that this signalling complex, pounding into the plastic brain, develops correlates for satisfying and unsatisfying states. Later, as the nervous system develops and the more detailed outlining of the system, descending toward the anal and genital regions, gradually helps set up a conflict of signalling into the brain, there is a tendency to separate out and segregate these signalling interfaces into the brain, becoming represented as analogue imprints of these system interfaces in the brain. The summation of these signalling complexes, as represented by their imprints in the more primitive structures in the brain, we think of as representing the structure of Freud's id (or *es*, the *it* of the system, as he named it in his own language). It is this measure which somehow is to lead us to anxiety-provoking or euphoric states. We then visualize that the coordinated summation of this oscillator complex, projected into higher centres, represents the ego sum of the system. A more nearly unitary measure of the buzz of information to ego centres is provided by a part sexual flux of libido. Its state presents a measure of the existing, satisfying, operating state of the system. Guided by this measure, system operators fall into an anxious or euphoric state, represented as a two-state operative complex.

As another integrative measure taken by the brain, unification of discreet signal information into an integrated analogue of the information represents the gestalt of the analogue. The brain is thus compacted with a large number of unified analogues to various signalling complexes and motor responses. It is such analogue packages that the mother helps develop in the maturing child together with a program routine of a tolerably satisfying nature that threads through the system hungers.

The missing ingredient is Freud's super-ego, the ego ideal, which represents the image of a satisfying ego state. This, added to the developmental algorithmic content on how to achieve a satisfying ego state, provides the motion for the behavioural patterning that emerges. The conflicts from all the signalling interfaces are not to be satisfyingly resolved by moving in some direct path. The super-ego is developed as a rule book on what motor actions will preponderately move in the direction of a more satisfying state. The speed of motion will also depend on the oscillatory two-state operator system—whether overall anxious or euphoric.

Thus modulated by the two-state oscillator, under the driving guidance of the super-ego algorithm, the system will hurl itself into motor activity that will synchronize in orbital paths with specific oscillator patterns.

In the larger developmental picture there are a number of large-scale flaring instabilities of a maturational nature. Two of these we can be certain of physiologically, and the others, if not yet clear physiologically, are certainly clear observationally, even under comparative anthropological study.

The first is the oral interface of infancy. The primacy of food seeking and of establishing routines for food acquisition may be fairly detailed.

The second is the great flaring instability of adolescence. The physical characteristic changes are, of course, noticeable. However, equally noticeable is the interest in the opposite sex and the preparedness of the system behaviourally and physiologically for mating–the maintenance of the metabolic reaction through reproduction of the species. The adaptive value of this aperiodic flaring of an entire sexual cycle is obvious.

The third newly-emergent flare is Sullivan's chum or peer stage in which the child is integrated sufficiently for it to suddenly break its tie to the mother's orbit and discover its own kind, its peers and chums.

Each of these periods is marked by the continuing development of the guiding algorithm, within the framework of interpersonal relations and the surrounding milieu. A useful thought is that man is an instrument to be used by and for man. It carries with it the idea that the system gain at zero frequency is indeterminate. (This means that the system is unstable at rest.)

The basic command algorithm for behaviour can be considered to be a continued adaptation to make the ego image agree with the ego ideal by means of the sequence of practised analogues that are stored within the brain.

The marginal stability of the system is such that environmental signalling is always blowing over the inputs and putting the analogue systems into continued melodic lines of response. The system is always practising its repertoire. It is this continued response that makes a follower-type characteristic melodic line appear for each and every questing and questioning environment.

In a successful biological species there is a satisfactory patterning of behaviour which threads all of the system hungers. In an unsuccessful individual it may lock up or become too wild. I have presented these ideas less sketchingly in our eighth report as a joint effort with Warren McCulloch. The report became available just at the time of this conference. An additional ingredient that emerged in that report is the foundation for interpersonal forces, the keynote of behaviour stressed by Sullivan. If, in the interior, there is this well-coordinated body image of both inside and outside the system, with its many vector dimensions, then upon meeting a member of a like (or similar) species in a reacting situation, there is an exchange of body image. The individual imagines the nature of his body image projected into the other individual as the other person's apparent actions emerge. If the image and actions are concordant, then an 'empathy', a binding 'force', emerges. The basic two elements of concordance, likely, are complementarity and congruence.

By these means, roughly, we have attempted to bring a dynamic concordance through the biological system by its spectrum of dynamic oscillator chains. Our problem now is to demonstrate the reality of these chains.

We can provide a more rudimentary illustration. We have taken a large motor system—the hind limb of the complex mammalian animal—as a suitable laboratory to test some of these ideas. For example, we are attempting to follow the detailed metabolic reaction in that system.

So far, in the gross animal we have demonstrated appreciable dynamic cycles in:

respiration rate
heat production rate
temperature.

From their temperature concordance, we have postulated that their operative dynamics are to be associated with pulsing dynamics in the cardiovascular system at the level of the microcirculation. We have demonstrated corollary dynamics in the femoral system (for the first 3) and in capillaries (for the 4th):

blood sugar oscillation
blood oxygen oscillations
blood $CO_2$ oscillation
red blood cell oscillations.

We have traced a line of evidence from Krogh, Sir Thomas Lewis, among modern microcirculationists, and others.

To illustrate briefly, we postulated that there should be a chemical engine cycle faster than a thermal relaxation time of three hours. We found it in thermal, metabolic, and ventilatory cycles at 100 seconds, 400 seconds, and greater. We have chosen to highlight the 100 second cycle as the engine cycle because of its large magnitude of variation. For example, we have found a running variation of near two to one in metabolism and ventilation at the 100 seconds level, ranging as much as peaks that are 5–10 times larger than minima in a 5 hour observation period [13].

We have decided that the large cycles must stem from major mesodermic organ systems such as muscle, liver, heart, brain, etc. We have sought and found allusion to such time scales in the microcirculation and expected to find it in capillary opening and closing. Instead we found the cycle in the red blood cells flow. We postulated that the engine cycle could either be run by metering of fuel, oxygen, or a combustion by-product. The high level of regulation posed for fuel and the rapid follower regulation and storage capacity for $CO_2$ ruled these two out, whereas the more limited storage of oxygen suggested that the engine cycle must be run by an oxygen choke. In this view, the muscles were considered to be an unstable system capable of metabolic conversion at any level to which they were supplied with oxygen. The fluctuation of red blood cells was an additional step in the evidence for this concept. Early in 1966 another step was found in the experiments of another investigator (Whelan), who showed that oxygen tension in the tissue was low and oscillatory–in the range of 0–5 mm Hg, with a period of the order of 100 seconds.

We are continuing in our effort to show an adrenaline involvement in setting this local system into its oscillatory chain. We carry with us the provocative thought that the brain sets a motional pattern, but not the resisting load. Then it must implement its choice by providing the needed oxygen carburettor to run the muscle engine.

In any case, the story and proof of such chains is exciting business. Yet it is beset by the following situation.

We have measured and found ventilation to be oscillatory time and time again. It has been verified by a few other investigators, for example, Goodman [14], now at NIH; more recently, Lenfent [15]. Yet we can find no other ventilation physiologists who report these results.

We have found normal heart rate to be oscillatory. This has essentially not been reported by other observers.

We have found that Anderson [16] and others (for example Hansen [17]) have found blood sugar to be considerably oscillatory and we have verified these findings; yet many physiologists deny these findings.

Thus, many of the observations of fundamental dynamic findings are themselves in doubt. We have sufficient confidence in our experimental findings to make even the consideration of the validity of our theses to hang on the validity of our dynamic observation. If the others are right and we are wrong in measurements, then it is dubious that our dynamic concepts should receive much attention. However, if measurements in normal unanaesthetised animals held normally quiescent or normally active demonstrate such dynamic oscillations, then we feel that our ideas may warrant the attention that we believe they deserve.

## References


1. A. Iberall and S. Cardon. (1) *Ann. N.Y. Acad. Sci., 117* (1964) 445.
A. Iberall and S. Cardon. (2) NASA CR-141, Jan. 1965.
A. Iberall and S. Cardon. (3) NASA CR-219, May 1965.
A. Iberall and S. Cardon. (4) NASA CR 129, Oct. 1964.
A. Iberall, S. Cardon and T. Jayne. (5) Dec. 1965, Interim Report.
A. Iberall, M. Ehrenberg and S. Cardon. (6) Sixth Report to NASA, Aug. 1966.
A. Iberall. (7) *Math. Biosciences, 1* (1967), 375.
A. Iberall and W. McCulloch. (8) *Currents Mod. Bio., 1* (1968), 337.
E. Young. (9) NASA CR-990, Dec. 1967.
A. Iberall and S. Cardon. In *Proc. IFAC Symp. on Syst. Eng. for Control Syst. Design* (Sci. Council Japan, Toyko, 1965).
E. Young, A. Iberall, M. Ehrenberg and S. Cardon. *Proc. Ann. Conf. Eng. in Med., Bio.,* Vol. 8, 1966.
M. Ehrenberg, C. Oestermeyer, E. Bloch and S. Cardon. *Microvasc. Res.* (forthcoming).
A. Iberall, M. Ehrenberg, S. Cardon and M. Simenhoff. *Metab.* (forthcoming).

2. C. Bernard. *An Introduction to the Study of Experimental Medicine,* Schuman (N.Y., 1949).

3. I. Sechenov. *Selected Physiological and Psychological Works* (For. Lang. Pub. House, Moscow, 1952).

4. W. Cannon. *The Wisdom of the Body* (Norton, N.Y., 1939).

5. H. Smith. *From Fish to Philosopher* (Ciba, N.J., 1959).

6. N. Wiener. *Cybernetics* (Wiley, N.Y., 1961).

7. E. Adrian. *The Mechanism of Nervous Action* (U. Penn. Press, Pa., 1932).

8. C. Sherrington. *The Integrative Action of the Nervous System* (Yale U., Conn., 1961).

9. N. Minorsky. *Nonlinear Oscillations* (Van Nostrand, N.Y., 1962).

10. B. Goodwin. *Temporal Organization in Cells* (Academic, London, 1963).

11. J. Frenkel. *Wave Mechanics* (Dover Repr., N.Y., 1950).

12. H. Sullivan. *Collected Books* (Norton, N.Y., 1953).

13. A. Iberall. Trans. ASME Series D, J. Basic Eng., *82* (1960) 92, 103, 513.

14. L. Goodman. Trans. IEEE, Biomed. Eng., *13* (1966) 67.

15. C. Lenfent. *J. App. Physiol., 22* (1967). 675.

16. G. Anderson, Y. Kologlu, C. Papadopoulos. Metabolism *16* (1967), 586.

17. K. Hansen. *Acta. Med. Scand. Suppl., 4* (1923), 27.

## Cellular Oscillations and Development
### Comment on the papers of Iberall and Goodwin

by C.H. Waddington
University of Edinburgh

The argument that a fundamental characteristic of cellular biosynthesis is a tendency to oscillation and the establishment of limit-cycles, opens a new theoretical approach to many classical problems, including those of differentiation. This is an aspect of biology in which our understanding is, in my opinion, at a very much less satisfactory level than many optimistic devotees of molecular biology pretend. The Jacob-Monod account of how single structural genes are turned on in prokaryotic organisms goes only a minimal distance towards explaining the phenomena of eukaryotic cellular differentiation ; and an appeal to histones as a *deus ex machina* is as imprecise theoretically as it is shaky experimentally. In these circumstances it is, I think, excusable, and may even prove profitable, to publish speculative ideas which still lack the empirical confirmatory evidence which would be demanded in an area in which we had a firmer grasp of the outlines of the logical structure within which effective theories must be framed. In this spirit—of throwing bread upon the waters—I should like to make some comments on the 'Goodwin-oscillator, limit-cycle' model of biological systems. I have already briefly alluded to some possible applications of these new ideas [1].

▸ *Differentiation of oscillating systems* There I suggested that we can now contemplate a previously unsuspected type of inductive interaction between cells or tissues, in which one side of the reaction acts as a 'temporal template' and entrains the oscillations of other elements to its own frequencies and phases. I should like now to add the point that one of the major difficulties in the theory of differentiation would be greatly simplified if we could interpret embryonic determination in terms of a fixation of certain frequencies in an oscillatory system. This is the difficulty which arises from the fact that during development general characteristics become determined earlier than specific and detailed ones. These general characteristics are often of a kind which it is very difficult to interpret in terms of the activities of single genes. This, which is the part of the situation that offers the real challenge to understanding, does not, of course, always occur. For instance, it is well known that in the early development of the amphibia, part of the dorsal ectoderm of the gastrula may, by induction, become determined as neural tissue, when it will proceed to differentiate into some part or other of the nervous system, though it is left for later processes, such as interaction with the neighbouring mesoderm, to determine whether this will be part of the fore-, mid-, or hind-brain, the spinal cord, or peripheral nerves. At the time when the cells are determined simply as 'neural tissue of some kind' the only changes known to occur are the visible alterations of cell shape involved in the formation of the columnar epithelium of the neural plate. It is not wholly implausible to suggest that these changes are due to the coming into activity of one or a few genetic loci; and one might argue that the first step in the differentiation of the neural tissue was therefore the 'activation' of a few genes, which would be followed later by activation of others, which convert the cells into the various parts of the brain, spinal cord, etc.

But consider another case. It is a well-established fact (see for instance [2]) that small groups of cells in the limb buds of the chick embryo are determined as fore-limb as opposed to hind-limb (or vice versa) well before they are determined as proximal versus distal (thigh versus toe, or upper arm versus fingers), let alone determined as muscle, cartilage, bone, etc. For instance, Saunders transplanted a small piece of tissue from the region which would develop into the thigh into a position near the future wing tip. The transplant developed into toes and claws! That is, it retained its character of 'something to do with the hind-limb', but was still flexible as regards its distal-proximal character, and, finding itself in the distal region of a limb, developed in the parts appropriate to the distal region of a hind-limb.

I have always found it extremely difficult to see how to comprehend results like this in terms of gene activation. It seems to me totally implausible to suggest that there is a gene, or even a small group of genes, 'for' hind-limb, and another one 'for' fore-limb. We could only make such a supposition if we could envisage there being some substance (gene-produced protein) which characterized all tissues in the hind-limb and distinguished them from those of the fore-limb.

A picture which is intellectually much easier to accept (though experimentally it might be difficult to approach) would emerge if we could consider that the characteristic feature of these early determined states of developing cells is a certain pattern of oscillations. We would not have 'a neural plate substance, a fore-limb substance, a hind-limb substance', etc., but neural plate, fore-limb or hind-limb oscillatory patterns, which could be regarded as analogous to musical themes or chord sequences. The later phases of differentiation, in which the cells of the fore-limb bud become differentiated into the various cartilages, bones, muscles, etc., must certainly involve the 'activation' of different structural genes controlling the proteins in these different sorts of cells; but we could interpret these changes as similar to the development of the initial theme according to the conventions of some school of classical musical composition – I suppose the analogue of what jazz musicians do to a chord sequence in a jam session would be some sort of cancer!

I confess that I have not yet seen a way in which one could *prove* that frequency of oscillation is involved in embryonic differentiation. But most embryological phenomena seem easily interpretable in such terms. For instance, in the determination of the neural system in amphibia it is generally agreed that two processes are involved: one which induces competent ectoderm cells to become neural tissue, in the first instance of a type characteristic of the fore-brain; and another which interacts with the former in such a way as to produce the more posterior parts of the nervous system, such as spinal cord. There is still argument whether the former process must act earlier and the latter later, or whether the temporal order of them is unimportant (see [3]), but both schools of thought agree that when you combine 'fore-brain' with 'spinal cord' you find that you get all the intermediates between these geographical extremes, such as hind-brain, ear region, and so on. This is compatible with an interaction of frequencies; but also with several other theories, e.g. of gradients of chemical concentrations. What we need is an experimentally decidable question which could discriminate between the *a priori* plausible possibilities.

▸ *Morphogenesis in oscillatory systems.* Several new theories of morphogenetic processes could be developed in relation to systems consisting of numbers of mutually interacting oscillators. For instance, at the Third Symposium Wolpert and Cohen sketched out a theory in which the relation between a transient peak and some other oscillation, which served as a time base, was invoked as a possible way of encoding the 'positional information', which, Wolpert argued, is necessary to specify the behaviour of a given cell within a mass of tissue. The type of theory I suggested in 1965 was quite different. I was thinking of the formation of fields of vibration within delimited areas. The well-known 'Chladni figures', in which the nodes of such fields are exhibited by strewing the vibrating surface with light powder, often have a character which reminds one strongly of biological forms – for instance, in the degree of precision with which the curves are drawn, the degree of geometrical complexity of the curves, and the way in which a given pattern may suffer a slight, graded overall distortion when one of the parameters is changed. Although the elementary theory of simple examples of such vibratory fields has been well explored mathematically, systems in which there are variations in the nature of the material, or in which a stochastic element plays a part, are difficult

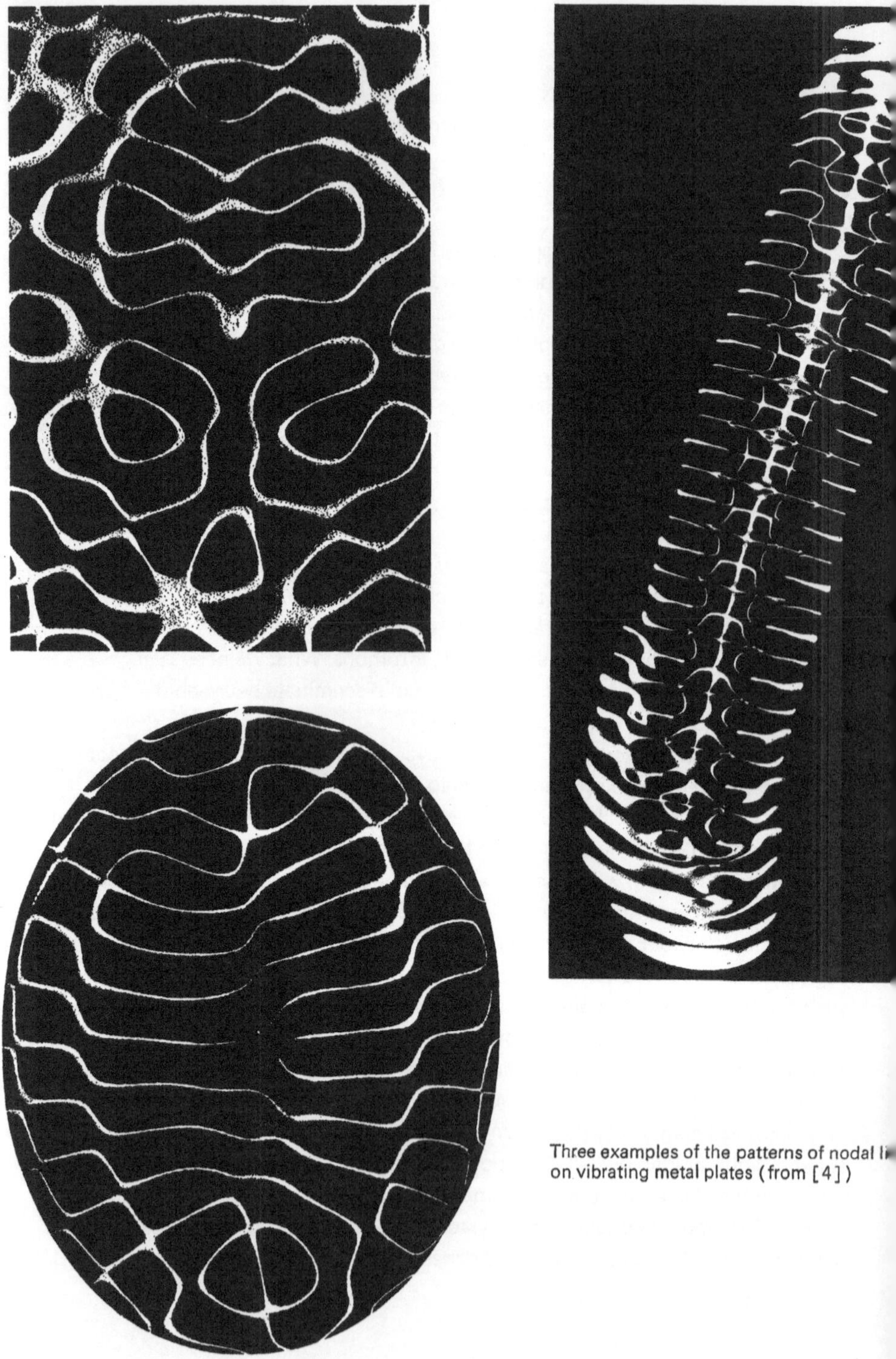

Three examples of the patterns of nodal li
on vibrating metal plates (from [4])

to handle analytically. A doctor living near Basel, Hans Jenny, has made a hobby of exploring this field, which he calls Cymatics, by a variety of experimental procedures. To illustrate the 'biological look' of the forms produced, I have copied three figures selected from the many in his recent, very beautiful book [4].

## References

1. Waddington, C.H. *J. Theoret. Biol. 8* (1965) 367.

2. Saunders, J.W., Jnr., Cairns, J.M. and Gasseling, M.T. *J. exp. Zool. 135* (1957) 503.

3. Yamada, T. *Adv. Morphogen. 1* (1961) 1.

4. Jenny, Hans. *Cymatics* (Basilius Presse, Basel, 1967).

# A Physicochemical Basis for Pattern and Rhythm

*A shorter version reprinted from Intracellular Transport © 1966 Academic Press Inc., New York*

John I. Gmitro and L. E. Scriven

Department of Chemical Engineering, Institute of Technology, University of Minnesota, Minneapolis, Minnesota

## INTRODUCTION

Any general principles governing the origins of regular patterns in space and rhythmic oscillations in time seem very likely to find application, directly or indirectly, at many levels of biological science—beginning with the intracellular, supramolecular level. A basic problem in the physical sciences which is increasingly attracting mathematicians and engineering scientists is the explanation of how specific dynamic patterns and rhythms can arise in spatially uniform, steady-state situations. Of course, macroscopic systems always suffer some sort of low-level noise, but how can chaotic, weak disturbances have no effect in some circumstances, yet trigger development of strong, regular pattern and rhythm in others?

The genesis of dynamic patterns depends on the coupled effects of transport processes and transformation processes. The study of both lies at the heart of engineering science today. Their application to multicomponent, chemically reactive systems is the special concern of chemical engineers. This is one area in which chemical engineering and cell physiology run parallel; probably both could profit from closer communication and perhaps even active collaboration. The report that follows is offered as an example of current research in engineering science which may be of interest in connection not only with intracellular transport but also with other biological phenomena that may be better known to the reader than the authors.

We begin by mentioning a few strictly physical examples of dynamic pattern and rhythm. From these we attempt to abstract the key factors and to identify a set of specific problems which can be precisely formulated from the viewpoint of physical science—that is, in mathematical terms. The first several of these are then formulated and solved for a prototype class of situations involving simultaneous diffusion and chemical reaction in a variety of geometric configurations. The bearing of the results on signal propagation, pattern and rhythm generation, and mechanical movement is discussed in more qualitative terms which we earnestly

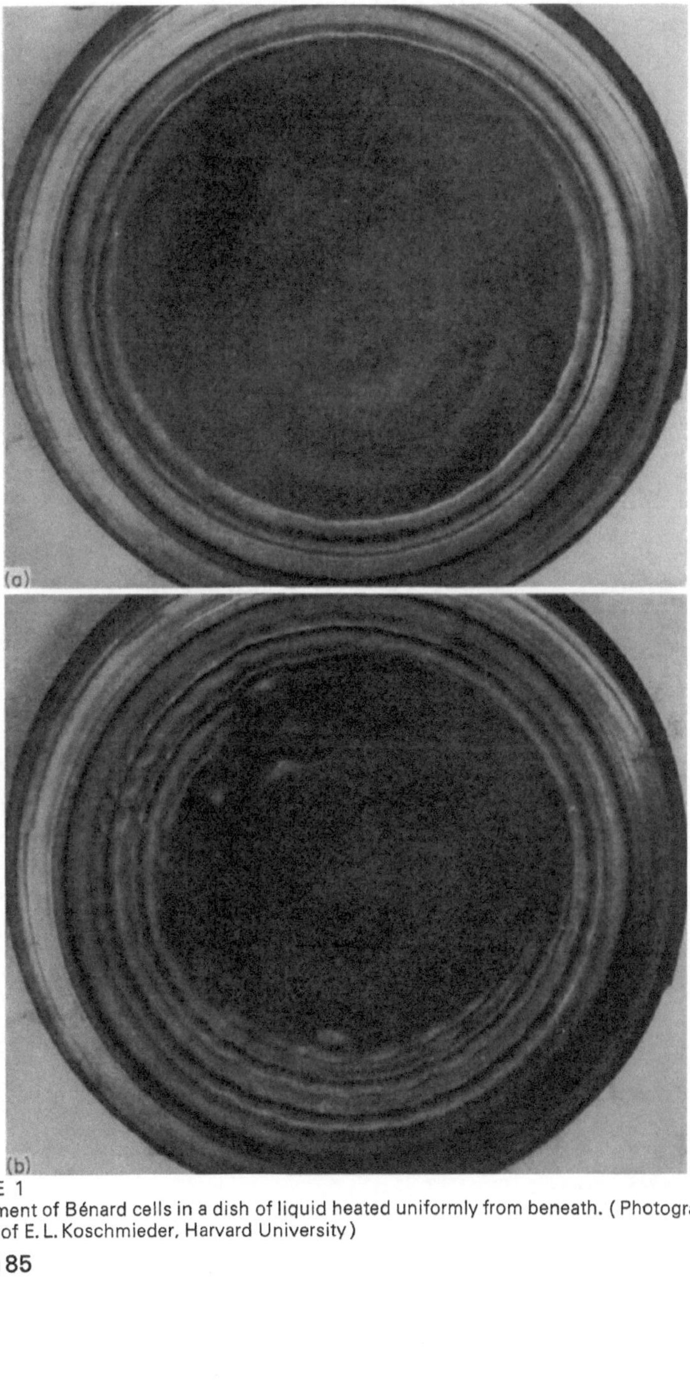

FIGURE 1
Development of Bénard cells in a dish of liquid heated uniformly from beneath. ( Photographs courtesy of E. L. Koschmieder, Harvard University )

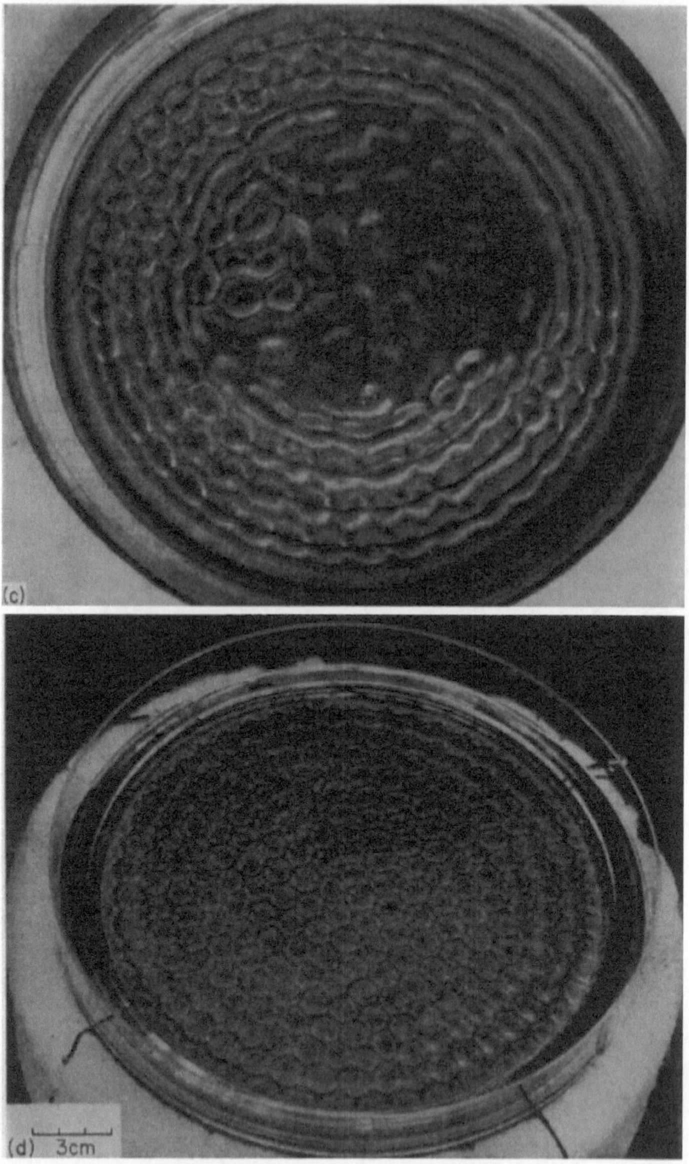


FIGURE 1 (continued)
Development of Bénard cells

hope will be informative to the reader who chooses to skip over the mathematical language of the sections on Formulation, and Instability and Wave Propagation.

Figures 1 (a–d) are photographs showing the development of a dynamic flow structure known as Bénard cells in a shallow dish of ordinary liquid that is being uniformly heated over its bottom side. Almost the same flow patterns can be brought about by two different physical mechanisms, one stemming from the dependence of surface tension on temperature, the other from the dependence of density on temperature (Scriven and Sternling [14]). In the latter case, the hotter, buoyant fluid at the bottom of the dish tends to rise and the colder, denser fluid at the top to sink. Such a turnover would lower the potential energy of the system and render it more stable until the heating from below re-established the unstable density profile. In the turnover itself, hot rising columns would necessarily exist somewhere alongside cold sinking columns, and in this situation of lateral velocity and temperature gradients there would be viscous forces opposing the flow and heat conduction reducing the buoyancy differences responsible for the flow. Thus there are two competing tendencies: one toward establishment of dynamic pattern, the other toward its destruction once it is formed (Sani and Scriven [13]). In fact, the rate of heating from beneath must exceed a certain critical value before the anabolic process can surpass the catabolic process sufficiently to establish flow, which tends to settle down in the steady Bénard-cell pattern if the critical value is not too greatly exceeded. The photographs indicate that the presence of the side of the dish favors concentric ring cells at first; these ultimately break up to give the strikingly hexagonal planform of Bénard

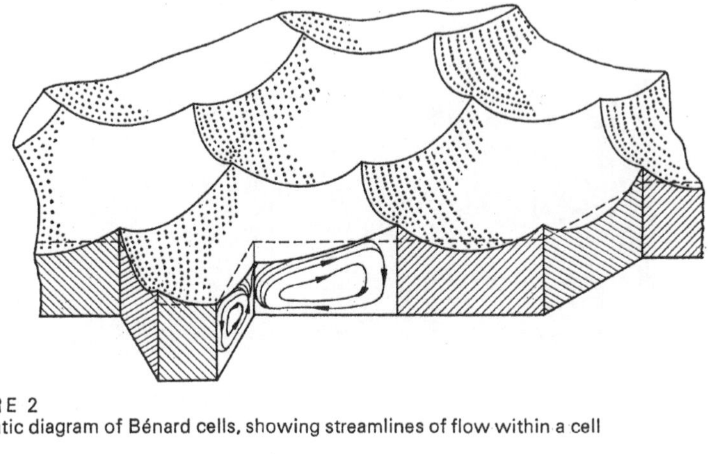

FIGURE 2
Schematic diagram of Bénard cells, showing streamlines of flow within a cell

cells, provided the dish diameter is much larger than the natural cell size. Flow within a cell is diagramed in figure 2, where streamlines are shown. The boundaries between cells are simply symmetry planes across which there is no flow of fluid. They are purely dynamic.

Bénard cells remain fixed in location and the flow within them is steady. They are an example of *stationary* convection. If a dish of liquid mercury is spun fast enough about its axis, a second type of convection, called *oscillatory*, occurs. The cellular planform becomes a little more complicated although still basically hexagonal, while the flow within cells may diminish and reverse periodically or the cellular pattern itself may translate through the liquid. In either case the net result is an oscillating, or rhythmical, flow at each point in the liquid.

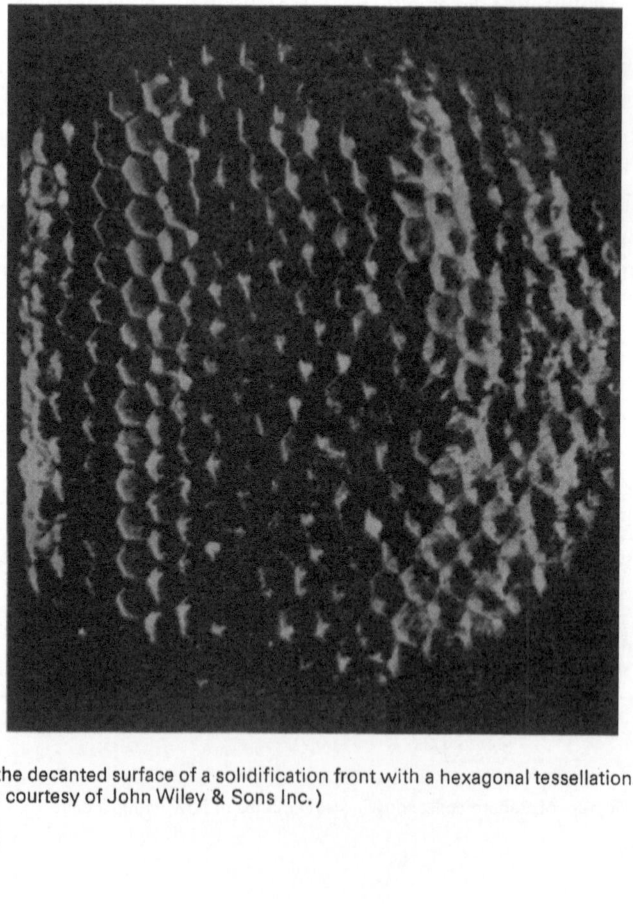

FIGURE 3
Top view of the decanted surface of a solidification front with a hexagonal tessellation. (Photograph courtesy of John Wiley & Sons Inc.)

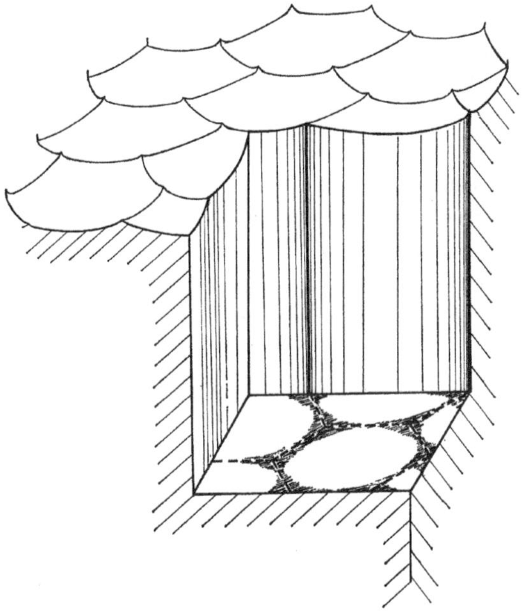

FIGURE 4
Concentration patterns frozen into the solid as hexagonal columns behind a solidification front. Broken lines indicate boundaries of the unit pattern. Regions of high solute concentration are shaded

The values of the critical heating rate for the onset of flow have been very successfully predicted by the theory of convective instability (cf. Chandrasekhar [7]). The analysis below is patterned after that theory.

Another remarkable instance of dynamic structure occurs under certain conditions of freezing of solid out of molten solution. The transformation process is solidification with its accompanying heat release. Both heat and solute are transported. There is a tendency for solute to be redistributed in a regular hexagonal pattern at the freezing face of the solid. This tendency, which stems from local supercooling in the melt, is opposed by thermal transport and reduced by diffusional transport. When these catabolic processes do not prevail, a regular concentration pattern may form over the freezing face; this results in the tessellated profile pictured in figure 3 and a permanent, hexagonal-column concentration pattern frozen into the solid behind as diagramed in figure 4 (Chalmers [6]). The static structure is merely a partial record of the dynamic processes by which it has been produced—this is the important point.

## A physicochemical basis for pattern and rhythm

Many other well-studied examples of dynamic structure in physical systems could be cited from fluid mechanics, meteorology, geophysics, and astrophysics. As for biological examples, we prefer to leave these and the biological implications of what follows to the biologically expert reader.

In all cases the key factors appear to be three. First, transformation processes: changes in physical state, as by phase transition, or in chemical state, as by chemical reaction. Second, transport processes: changes in location, as by convection or diffusion. Third, coupling of the two types of processes together: both must proceed simultaneously and affect each other. For our present purposes chemical reactions play the part of transformation while simple diffusion plays that of transport.

Analysis of these factors can be logically organized around the following set of problems:

1. Origin of pattern and rhythm from a uniform and steady state of transformation in systems in which departures from uniformity give rise to transport processes (equilibrium systems are a particular case). A natural adjunct of this item turns out to be
2. Signal transmission by propagation of small local disturbances in an initially uniform and steady-state system. With solutions to these problems, one can study
3. Control of pattern size, rhythmic period, propagation velocity, and wavelength, especially the dependence of possibilities on the complexity of the system—here, the number of participating chemical species. Beyond this lie more difficult questions of
4. Evolution and stability of particular patterns, rhythmic variations, and waveforms, in which so-called nonlinear effects are likely to be dominant. Before answering these questions one can investigate
5. Effect of pre-existing pattern on spontaneously developing pattern and rhythm and disturbance propagation, as most simply exemplified by these processes in homogeneously compartmentalized systems. If at the outset chemical effects alone are considered, as is done here, a parallel problem is
6. Coupling of chemical patterns and waves to electrical and mechanical stress fields in the material and thereby to forces, accelerations, and movements. Ultimately this coupling and the accompanying convective transport should be included in the first problem.

This is a large undertaking and we restrict ourselves here to the first three and last items, focusing on the physicochemical side of the overall problem. For a variety of reasons, some of which may become evident, it has seemed desirable to study pattern and rhythm in *surfaces* or *membranes*, and *lines*, or *fibres*.

#### FORMULATION

The basic system under consideration is diagramed in figure 5. It consists of a membrane or thread, uniform across its thickness, within which various chemical species are reacting and diffusing along its length. At the same time some or all of the participating species are exchanging with the surrounding media. The number of participating species is left open, for an important question to be answered is what influence the number of species has on pattern and rhythm. Here reaction is the only transformation process and diffusion the only transport process inside the system ; modifications necessary to account for convective, electrical, and other effects can be made subsequently.

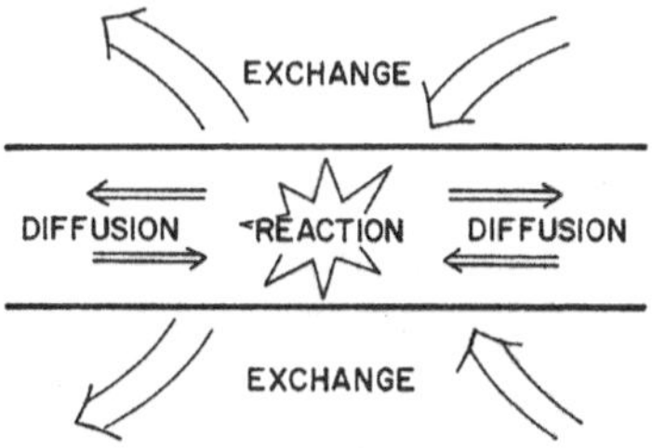


FIGURE 5
Diagram of basic system considered in the text. Reaction and diffusion take place within or on a membrane or thread. There is also exchange with the surrounding media

*Equations of change*

At every location within the system each chemical species obeys the conservation equation,

| | | | | | | | |
|---|---|---|---|---|---|---|---|
| Rate of accumulation within the system | = | net rate of production by chemical reaction within the system | + | net rate of influx by diffusion along the system | + | net rate of input by exchange with the surroundings | ....(1)* |

In mathematical symbols the equation becomes

$$\frac{\partial c_i}{\partial t} = R_i + J_i + Q_i \qquad \text{....(2)}$$

* For the purposes of this publication the mathematical treatment has been reduced to a brief series of notes, which it is hoped will provide some indication of the approach used. Fuller details are printed in the original article. Numbers of equations, tables and figures have been taken from the original.

where $c_i$ is the local concentration of the $i$th species and $t$ is time. The system is not really described until constitutive relations specifying the rate processes present have been substituted for the rate symbols $R_i$, $J_i$, and $Q_i$.

. . . .

*Excursions from uniform steady state*

At steady state the concentrations of participating species are constant in time, by definition. If the state is uniform as well, there are no gradients in the system, and the equation of change reduces to a statement that rates of production and output of each component must just cancel:

$$0 = R_i^s - \sum_{j=1}^{N} H_{ij}^s (c_j^s - c_j^\circ) \qquad \ldots.(7)$$

Here $c_j^s$ stands for the concentration of the $j$th component in the steady state. The corresponding values of the rates of chemical production and of the mass-transfer coefficients are $R_i^s = R_i(c_1^s, c_2^s, \ldots, c_N^s)$ and $H_{ij}^s = H_{ij}(c_1^s, c_2^s, \ldots, c_N^s)$ respectively.

Unsteady states can be represented by concentrations that are sums of steady-state values and excursions from the steady state:

$$c_i = c_i^s + x_i \qquad \ldots.(8)$$

. . . .

With matrix notation and the rules for matrix multiplication the equation of change of excursions, obtained by substituting Eq. (8) and Eq. (7) in Eq. (2), takes the compact form of a standard linear partial differential matrix equation:

$$\frac{\partial [x]}{\partial t} = [K]\,[x] + [D]\nabla^2[x] \qquad \ldots.(12)$$

Equation (12) states that if a steady state of the system is slightly perturbed, the rate at which the excursion grows or decays is controlled by the competition between (*a*) first-order processes depending on chemical reaction within the system and mass exchange with the surroundings, and (*b*) diffusive processes governed by 'bumpiness' of the concentration distributions within the system.

Whether an excursion grows, whether it periodically oscillates, and what size of pattern is likely to emerge as it develops, are questions that can be answered by solving Eq. (12).

*Solution in terms of elementary patterns*

Equation (12) can be solved by the methods of harmonic analysis, which rest on a remarkable theorem going back to Fourier: Any spatial pattern may be expressed as a weighted sum of members of a suitably chosen set of *elementary patterns* (Bell [2]; Tolstov [18]). Mathematically, the elementary patterns are a complete set of characteristic functions, or eigenfunctions, appropriate to the geometric configuration of the system of interest.

. . . .

These are eigenfunctions of the Laplacian operator and satisfy the equation

$$\nabla^2 F_k(\mathrm{r}) = -k^2 F_k(\mathrm{r}) \qquad \ldots.(14)$$

The constant $k$ is known as the characteristic parameter, or eigenvalue, corresponding to the eigenfunction $F_k$. Geometrically, $k$ describes the mean size, or wavelength, of the corresponding elementary pattern. In fact, the mean pattern size, $l$, is $2\pi/k$. (If $k$ is complex, as for wave propagation, then $l=2\pi/k_r$; the imaginary part $k_i$ gives the rate of attenuation of pattern or wave with distance.)

. . . .

Standard methods are available for determining eigenvalues and eigenvectors of a given matrix (Frazer *et al.* [8]). The eigenvalues are themselves the roots of the determinantal equation

$$\det([\mathrm{K}] - k^2[\mathrm{D}] - \lambda[\mathrm{I}]) = 0 \qquad \ldots.(18)$$

where [I] is the identity matrix. For every value of $k$ there are as many eigenvalues, $\lambda_{k,1}$, $\lambda_{k,2}$, etc., as participating chemical species, in general.

A solution of the equation of change in an excursion, Eq. (12), is

$$[\mathrm{x}] = \sum_{k=0}^{\infty}\Big(\sum_{n=1}^{N} a_{k,n}[\mathrm{A}]_{k,n} e^{\lambda_{k,n}t}\Big) F_k(r) \qquad \ldots.(19)$$

Before taking up the interpretation of this result it should be pointed out that there may be other solutions, solutions that amount to propagating waves in particular. For example, the function representing simple harmonic waves propagating in the x-direction, viz.*

$$[\mathrm{x}] = [a] e^{i(k_r x - \omega t)} e^{-k_i x} \qquad \ldots.(20)$$

is found by substitution to satisfy Eq. (12) provided the wave-number $k$ $(=k_r + ik_i)$ and frequency factor $\omega$ $(=2\pi f)$ together satisfy the determinantal Eq. (18) with $\lambda$ replaced by $-i\omega$.

Equation (18) for the eigenvalues of the transformation-and-transport matrix is the pivot on which the whole theoretical analysis turns. . . .

* The excursion [x] is a set of real quantities which may be taken equally well as the real part or as the imaginary part of the right-hand side.

To sum up with regard to instability: Any uniform, steady-state system of chemical reaction and diffusion suffers low-level noise, either transmitted from surroundings or arising internally in molecular fluctuations. Appearing more or less continually and randomly throughout the system as small concentration perturbations, the disturbances are without effect or trigger development of pattern and rhythm, according to the nature of sets of coefficients that characterize reaction, exchange, and diffusion. What matters are the eigenvalues of matrices comprising these coefficients; in particular, the eigenvalue containing the largest growth factor. Because the eigenvalues through the coefficients may depend, as remarked above, on factors other than the participating species, it is entirely possible that a previously stable steady state may turn unstable owing to changes in such factors. Factors other than the participating species can also control whether the ensuing instability is stationary or oscillatory, how rapidly it develops, and the pattern size and rhythmic frequency. Once the number of participating species is known and values of the coefficients are available, one can determine all of these features from the solutions of Eq. (18) for every admissible value of the size factor, $k$.

. . . .

EXAMPLES

In the preceding sections we have shown how the origin of regular patterns and rhythms in the type of system under consideration, and chemical signal transmission as well, can all be thought of as immediate consequences of Eq. (18), which involves sets of parameters that describe chemical reaction, exchange with the surroundings, and internal diffusion in small excursions from the uniform steady state originally present. We also outlined the derivation of the pivotal equation from basic physicochemical principles. Our object now is to show by means of examples what some of the implications of the analysis are, particularly in regard to control of pattern size, rhythmic period or frequency, propagation speed, and wavelength.

To do this we examine cases of one, two, and three participating species. In the first two we can write the characteristic equation for the eigenvalues, $\lambda_{k,n}$, explicitly in terms of the system parameters and the pattern factor $k$; in the third case it is scarcely worth while attempting to do so (a cubic equation must be solved). From the characteristic equation, whether or not in explicit form, the relationships can be found which the system parameters must satisfy, as functions of $k$, in order for a given type of instability to occur—stationary, oscillatory, marginal. The

Routh-Hurwitz criteria for eigenvalues of matrices and Routh's algorithm for polynomials with real coefficients are very useful (Gantmacher [9]).

. . . .

DISCUSSION

Without examining thoroughly all of the new possibilities of control of pattern size, rhythmic period, growth factor, wave propagation speed, attenuation factor, and wavelength which accompany a third participating species, we can see that richer ranges of possibilities come with each additional compound and the reactions into which it enters. Instability behaviour just possible with fewer compounds can be realized in a variety of ways; entirely new behaviour can be produced. The same is true with regard to propagation of chemical waves when the number of participating species is increased.

Chemical concentration waves could provide large numbers of parallel signal-transmission channels. In small-scale systems these might for some purposes be competitive with electrical transmission means. A steady-state reaction system can be arranged to propagate signals at far faster speeds than diffusional movements, and without the attenuation that also limits purely diffusional processes, as shown above. Concentrations of participating species can be exceedingly low; indeed, concentration could be reinterpreted as the probability density of finding a molecule of the species in a given locale.

In an engineering sense these observations mean that more species broaden the range of alternatives and may permit a better optimum solution to any particular design problem. Indeed, a new alternative may provide a solution where no acceptable solution was available before. The point is that increased numbers of species in a multicomponent system of transport and transformation may have decisive influence on the appearance and nature of pattern and rhythm.

Linearized stability theory illuminates the origin of pattern and rhythm as well as the propagation of small disturbances. But although its predictions of dominant pattern size or wavelength are likely to be close to the values for resultant steady states or limit cycles, the particular pattern shape and variation of amplitude with time or waveform which evolve are invariably determined by nonlinear effects that have not been considered here. Until an analysis of these effects has been completed, many patterns can be put forward as candidates.

*Representative geometric configurations and patterns*

The analysis and examples that have been presented actually pertain to a whole gamut of line-like and surface-like configurations. All that is necessary is to place appropriate interpretations on the elementary pattern functions, $F_k(r)$, and the pattern-size factor, $k$. The most important cases are listed in Table II; representative examples are shown in figure 11. Some mechanical deformations that might be produced by the chemical patterns are also shown, and will be discussed.

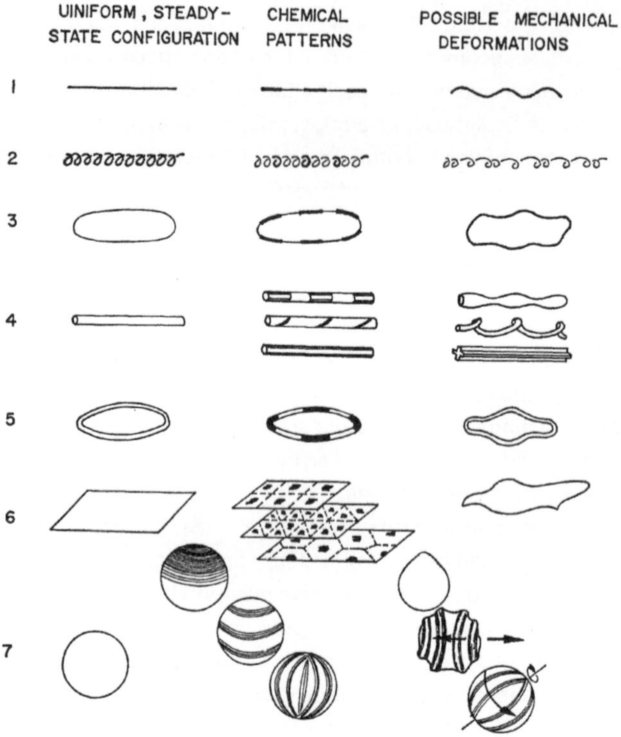


FIGURE 11
Representative geometric configurations with some regular chemical patterns that can arise spontaneously. Concentration dependent stress could produce the corresponding mechanical deformations

If the system is unbounded, or at least very extensive compared to significant pattern sizes, then continuous ranges of pattern size $l$ and the factor $k$ must be considered. If, on the other hand, the system is closed, as are loops, rings, and spheres, then these factors can take on only certain discrete values. The reason is that the basic unit of pattern, whatever it is, must repeat itself in the system an integral number of times. Rarely can the fastest-growing among all pattern sizes be

TABLE II
Elementary pattern functions for representative geometric configurations*

| *Configuration* | *Coordinates* | *Elementary pattern functions* | $k^2$ |
|---|---|---|---|
| Line | $x$ | $e^{i2\pi x/L}$ | $4\pi^2/L^2$ |
| Circle | $\phi$ | $e^{im\phi}$ | $m^2/R^2$ |
| Cylinder | $x, \phi$ | $e^{i(2\pi x/L+m\phi)}$ | $\frac{m^2}{R^2}+\frac{4\pi^2}{L^2}$ |
| Ring | $\theta, \phi$ | $e^{i(n\theta+m\phi)}$ | $\frac{m}{R_1{}^2}+\frac{n^2}{R_2{}^2}$ |
| Plane | Rectangular: $x, y$ | $e^{i(2\pi x/L_x)+2\pi y/L_y)}$ | $4\pi^2\left(\frac{1}{L_x{}^2}+\frac{1}{L_y{}^2}\right)$ |
| | Polar: $r, \phi$ | $J_m(ar)e^{im\phi}$ | $a^2$ |
| Sphere | $\theta, \phi$ | $P_n{}^m(\cos\theta)e^{im\phi}$ | $\frac{n(n+1)}{R^2}, \quad -n \leqslant m \leqslant n$ |

* Position is measured by standard length and angle coordinates.

so accommodated, in which case the nearest admissible pattern size is likely to be dominant.

A cylindrical surface is two-dimensional and can support circumferential as well as longitudinal wave patterns. A rich assortment of combinations can be made, e.g. the helically wound concentration pattern diagramed in the fourth row of figure 11. Moving patterns might provide means for facilitating diffusion of selected species along a thread, or along a membrane for that matter.

Of all the concentration patterns that can exist on flat surfaces (and certain others) only three are strictly regular—those in which the basic unit is a square, triangle, or hexagon, as sketched in the sixth row of figure 11. Pattern functions that are combinations of elementary pattern functions are known for all three (Chandrasekhar [7]). There is no way of predicting from the linearized analysis whether one of these three or some less regular candidates having the same pattern size will be established when a large expanse becomes unstable. The presence of a boundary or edge nearby may favour one pattern over another in a fairly predictable way, however. The configuration of the source in cases of wave propagation may also favour one pattern over another. Figure 12 serves as a reminder that a point source produces circular waves on a homogeneous and isotropic plane; a line source, lineal waves.* Many, many other wave patterns can of course be generated by interference of multiple sources.

*These are described by functions listed for the plane in Table II, with $m=0$ and $L_y=0$ respectively.

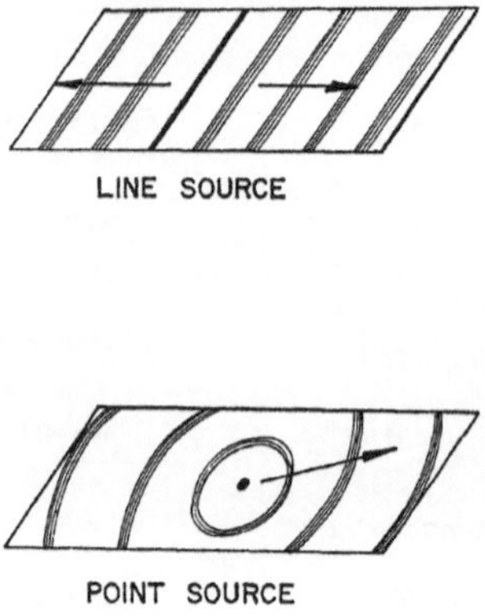


FIGURE 12
Lineal and circular waves propagating on a plane

In the case of the closed spherical surface, linear theory provides a little more information. Before describing it we should point out some of the patterns possible on a sphere: two axially symmetric zonal patterns ($n=1$ and $n=4$; $m=0$) and one sectorial pattern ($n=10$, $m=10$) are shown in the seventh row of figure 11. There are only five strictly regular patterns: these correspond to the five regular polyhedra, the Platonic solids, as shown in figure 13. Of these, the cube and

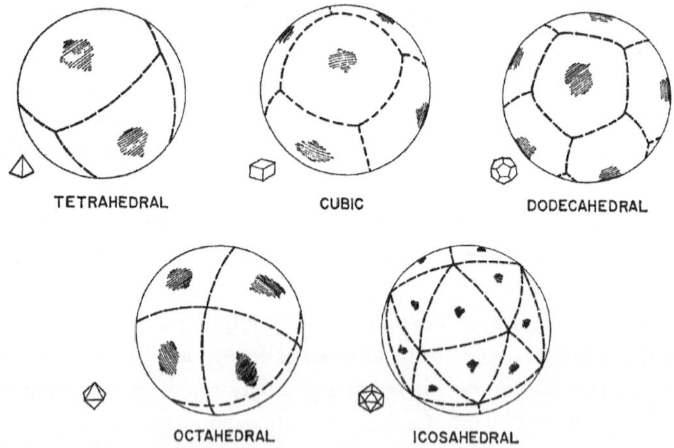


FIGURE 13
Strictly regular patterns on the surface of a sphere, corresponding to the five regular polyhedra

octahedron are a conjugate pair, and so are the dodecahedron and icosahedron; for if the concentration maxima form one pattern, the minima must form the other (the same is true for the triangle-hexagon pair on simple surfaces). Further distinctions can be drawn on the basis of multicomponent concentration patterns, but we need not go into these here. Pattern functions that are combinations of elementary spherical harmonics are known for all five.

Because the surface of a sphere is closed, the pattern-size factor, $k$, can take on only certain discrete values, which happen to be given by $k^2=n(n+1)/R^2$, where $n$ is any non-negative integer. Types of pattern corresponding to values of $n$ from 0 to 10 are listed in Table III. It is significant that for $n=1$ there is a unique pattern

TABLE III
Spherical patterns for $n=0$ to $n=10$*

| $n$ | *Type of pattern* |
|---|---|
| 0 | Completely uniform |
| 1 | One axisymmetric pattern, nonregular |
| 2 | One biaxial pattern, nonregular |
| 3 | Only strictly regular pattern corresponds to the tetrahedron |
| 4 | Only strictly regular pattern corresponds to the cube and octahedron |
| 5 | No regular patterns |
| 6 | Only strictly regular pattern corresponds to the dodecahedron and icosahedron; one complex polyhedral pattern with cubic/octahedral symmetry |
| 7 | One complex polyhedral pattern with tetrahedral symmetry |
| 8 | One complex polyhedral pattern with cubic/octahedral symmetry |
| 9 | One complex polyhedral pattern with tetrahedral symmetry |
| 10 | One complex polyhedral pattern with cubic/octahedral symmetry; one complex polyhedral pattern with dodecahedral/icosahedral symmetry |

* $k^2=n(n+1)/R^2$.

and likewise for $n=2$. Among the basic patterns for which $n=3$ the only one that is strictly regular corresponds to the tetrahedron. Those corresponding to the cube-octahedron pair and dodecahedron-icosahedron pair enjoy the same distinction at $n=4$ and $n=6$ respectively. Now, since the dominant value of $k$, and hence $n$, is under the control of reaction, exchange, and diffusion, the linear theory indicates clearly how by adjusting the rate coefficients of these processes ($a$) a simple axisymmetric pattern can be selected and ($b$) one of the strictly regular patterns could be selected if some tendency for regularity existed. Whether a strictly regular pattern could actually be obtained depends on nonlinear effects under study.

Chemical patterns on a closed loop and, cursorily, on a sphere were considered by the mathematician Turing [19] in a pioneering paper entitled 'The Chemical Basis of Morphogenesis'. Turing's paper stands alone in the biological literature and is independent of the mainstream of stability analysis in the physical sciences. It has been one of the factors motivating our work, which connects his ideas to the main stream, enlarging on them and extending them in ways he unfortunately did not live to pursue.

Besides a continuous closed loop, Turing analysed the origin of pattern in a ring of cells, that is, a compartmentalized loop. Now a logical question to raise is what effect has pre-existing nonuniformity on the appearance of new patterns in general. In our view the simplest way to attack this question is to analyse a regularly compartmentalized system—identical uniform regions separated by permeable barriers of negligible thickness, like Turing's cells. Incidentally, such barriers could be conceived of as arising at nodal lines of a regular pattern that might have previously been established in an originally uniform system. Results of our analysis of regularly compartmentalized systems will be presented elsewhere.

*Coupling to mechanical stress and movement*

The chemical state of a material influences its mechanical state—first the state of stress, and then the state of strain. Temperature-induced deformations commonly occur in everyday materials and are the subject of the well-developed theory of thermoelasticity (Nowacki [12]). Quite analogously, composition changes can induce elongation, swelling, shearing, bending, and twisting; the responses of wood products to humidity changes are familiar examples. Moving chemical patterns can in general produce moving patterns of mechanical stress which bring about movement. We focus here on such mechano-chemical coupling, although the chemical state of a material also influences its electromagnetic state and may do so sufficiently to activate electro-kinetic effects.

Moving stress patterns produce net forces and accelerations and thereby moving patterns of strain-displacement, which is simply mechanical motion. Thus, systems of coupled chemical reaction and diffusion of the sort we have been considering can give rise to pulsations, rippling, streaming, and so on, by what we call 'active-stress' mechanisms. An *active* stress is one depending on variables other than strain in a way that permits chemical energy and other forms of energy to be transformed directly to mechanical energy.

There are no reasons known to us to doubt that active stresses can arise in every category of mechanical stress. In line-like structures the only meaningful category is tensile stress, which of course includes contractile stresses; in rods and tubes: tensile, bending, and torsional (twisting) stresses; in surfaces and membranes: isotropic tangential stress (often called membrane or surface tension), pure shear tangential stress, and internal torsional stress. In films, sheets and plates: the pertinent categories are these plus various types of bending and torsional stresses. In continuous and internally structured bulk phases there are the three basic types: isotropic stress, or pressure; pure shear stress, occasionally referred to as 'shifting' or 'shearing' forces; and antisymmetric stress, or internal torsional stress. Of these, only so-called contractile forces and two or three others seem to have been considered in the literature on mechano-chemistry.

By invoking various of these possibilities we can envisage chemical patterns producing the mechanical deformations shown on the right of figure 11. The efficacy of isotropic stress, or pressure, is firmly established by findings in thermo-elasticity (Nowacki [12]; Bupara [4]). In all cases it is easiest to deal first with situations of 'weak' stress-coupling, wherein the motions produced by active stress mechanisms do not affect appreciably the underlying chemical patterns and rhythms. The motions can have rather interesting consequences, however, from the points of view of hydrodynamics and, perhaps, cellular and organelle motility.

Fibres can displace themselves along their length through viscous surroundings by worming and squirming, including bending in planar and helical waves, as shown by G. I. Taylor [17] and Hancock [10]. Translation of a cylinder through liquid and transport of liquid contents of a tube by peristaltic waves are implied by Taylor's analysis of self-propulsion of a sheet by means of progressive bending waves [16]. That self-propelled rotation of a cylinder and a sphere can be produced by progressive waves of surface deformation has been established (Bupara [4]). So also has self-propelled translation of a sphere by means of axisymmetric zonal waves of the sort suggested in the seventh row of figure 11 (Lighthill [11]; Bupara [4]).

The theoretical analyses, prior to Bupara's recent one, simply postulated surface-deformation waves without going into their origin and control. He demonstrated that such waves can be caused by concentration-dependent pressure that is weakly coupled to moving chemical patterns. Analysis of systems with strong stress coupling is feasible though considerably more complicated. But the mechanical motions permit convective transport which may influence appreciably the other transport and transformation processes, and so should be included from the outset. Mathematically, the convective transport terms should be returned to the equations of change, from which they were omitted here [Eqs. (1)–(6)], and an equation of momentum and stress must be brought in. Convection along and to a fibre or membrane gives rise to additional means of controlling the emergence and properties of pattern and rhythm.

SUMMARY

Elucidation of principles governing the spontaneous appearance of pattern and rhythm is an important problem in contemporary physical and engineering science, and one that surely relates to many levels of biological science. The key factors evidently are transformation processes and transport processes—one of each, at least—coupled together. There are many purely physical samples, but we have chosen to study a class of physicochemical systems of coupled chemical reactions and diffusion. They are open systems in that every point on a surface or a line is adjacent to the surroundings; in geometric configuration they may be either open or closed.

We find that a uniform steady state of almost any such system, beginning with the simplest, is potentially able to propagate chemical signals or to give birth to progressive chemical waves or standing chemical patterns. Possibilities in the simpler systems are drastically limited, but as the number of participating chemical species goes up there is an ever-increasing richness of possibilities for accomplishing any of these things, and for doing so within narrower design specifications, so to speak.

Analysing the possibilities involves sets of dependent concentration variables and a multiplicity of reaction rate, exchange, and diffusion coefficients. The mathematics of ordered sets known as matrices proves to be a powerful tool; the concepts and language are useful in thinking about still more complex systems.

Composition affects the local state of stress in a material, and thereby can develop net forces, acceleration, and motion. In this way travelling concentration patterns supply a physicochemical basis for spontaneous movements. Convective transport, which accompanies movement, raises additional possibilities of 'feed-back' and regulation of spontaneously appearing pattern and rhythm.

ACKNOWLEDGMENTS

This work was supported by Air Force Office of Scientific Research Grant No. AF AFOSR-219-63. A travel grant by the American Institute of Biological Sciences made possible the participation of L. E. Scriven in this symposium.

## References

1. Adler, R. B., Chu, L. J. and Fano, R. M. 'Electromagnetic Energy Transmission and Radiation.' (Wiley, New York, 1960.)
2. Bell, E.T. 'Mathematics—Queen and Servant of Science.' (McGraw-Hill, New York, 1951.)
3. Bellman, R. 'Introduction to Matrix Analysis.' (McGraw-Hill, New York, 1960.)
4. Bupara, S. S. 'Spontaneous movements of small round bodies in viscous fluids.' Ph.D. Thesis in Chemical Engineering, Univ. Minnesota, 1964.
5. Carson, J. R. 'Electric Circuit Theory and the Operational Calculus.' (McGraw-Hill, New York, 1926.)
6. Chalmers, B. 'Principles of Solidification.' (Wiley, New York, 1964.)
7. Chandrasekhar, S. 'Hydrodynamic and Hydromagnetic Stability.' (Oxford Univ. Press, London, 1961.)
8. Frazer, R. A., Duncan, W. J. and Collar, A. R. 'Elementary Matrices.' (Macmillan, New York, 1946.)
9. Gantmacher, F. R. 'Matrix Theory.' 2 vols. (Chelsea, New York, 1959.)
10. Hancock, G. J. '*Proc. Roy. Soc. (London) Ser. A217*, 96 (1953).
11. Lighthill, M. J. *Commun. Pure and Applied Math., 5*, 109 (1952).
12. Nowacki, W. 'Thermoelasticity.' (Addison-Wesley, Reading, Massachusetts, 1962.)
13. Sani, R. L. and Scriven, L. E. 'Convective Instability.' Unpublished Report on Grant AF AFOSR-219-63 (1964).
14. Scriven, L. E. and Sternling, C.V. *J. Fluid. Mech., 19*, 321 (1964).
15. Sokolnikoff, I. S. and Redheffer, R. M. 'Mathematics of Physics and Modern Engineering.' (McGraw-Hill, New York, 1958.)
16. Taylor, G. I. *Proc. Roy. Soc. (London) Ser. A209*, 447 (1951).
17. Taylor, G. I. *Proc. Roy. Soc. (London) Ser. A211*, 225 (1952).
18. Tolstov, G. P. 'Fourier Series.' (Prentice-Hall, Englewood Cliffs, New Jersey, 1962.)
19. Turing, A. M. *Phil. Trans. Roy Soc. (London) Ser. B237*, 37 (1952).

# Self-reproducing Automata–some implications for Theoretical Biology

Michael A. Arbib
Stanford University

*Abstract* In Section 1 we review various models of self-reproducing automata which share with life the property of repetitive production of ordered heterogeneity. In Section 2 we note that these systems also share with living systems a dependence on a separate stable description of the reproducing organism, and we discuss a number of reasons why this condition may be necessary in more complex organisms. In Part 3 we face up to the fact that our automata are extremely prone to large malfunction as a result of small damage, and make a first step towards rectifying this deficiency by exhibiting an automaton model of a worm which, upon being damaged, will readjust its composition so that the front third is head, the next third is body, and the final third is tail. Part 4 concludes the paper with a brief discussion of how our model may be extended to deal with problems of evolution.

▸ *Introduction.* Life has been defined as *the repetitive production of ordered heterogeneity* [1]. Elsasser [2] has noted the importance of the term 'heterogeneity', since repetitive production of order *per se* could describe the operation of physical laws in a lifeless universe. Since many biologists would agree with Hotchkiss [1], it is of no small interest that the self-reproducing automata we present in Section 1 satisfy his definition. This suggests that automata theory may help refine theoretical biology by allowing us to completely describe systems which share more and more of the properties we ascribe to living systems, so that we may avoid over-simplifications which are almost inevitable at the level of verbal discourse. Note that our strategy may be a fruitful one whether or not one holds the reductionist view that all vital processes can be reduced to physicochemical terms, i.e. whether one believes that one is converging to an understanding of the nature of life, or sharpening an inevitable distinction between living things and man-made machines. I personally believe reduction is possible [3], but cannot share the confidence expressed by Crick in its imminence [4], since I believe that we are at the very beginning, and require immense breakthroughs in the mathematical theory of automata and other complex systems, before we can hope to really understand the hierarchy of processes that extends from molecule to man. It seems to me that the notion of DNA→RNA→enzyme transduction is of as vital importance to understanding life as the conversion of decimal numbers to binary notation is to understanding digital computers. However, just as computers may be built to operate in non-binary mode, so may there be life without DNA; and we are no more entitled to say we understand embryology when we understand DNA than we are entitled to say we understand computation when we understand radix two. Perhaps our automaton models may break us of too parochial a view of the goals of theoretical biology.

▸ *1. Summary of Basic Results on Self-reproducing Automata.* It should be emphasized that the theory of automata is usually concerned with devices which transform information from an input string to an output string, changes in the automaton being regarded as incidental. It was von Neumann in his Hixon symposium lecture [9] who shifted emphasis to the way in which initial information serves to regulate the growth and change in structure of an automaton. He noted that we associate with machines used for construction a certain degenerating tendency—we expect an automaton to build an automaton of less complexity. However, when organisms reproduce we expect their offspring to be of complexity at least equal to that of the parent. In fact, due to long-term processes of evolution, we even expect to see increases in complexity during reproduction. In view of this apparent conflict, von Neumann felt it worth while to see what could be formulated rigorously in the way of construction theorems for automata.

To appreciate von Neumann's ideas we should briefly recall the basic work of Turing [10]. A. M. Turing was one of the people who, in the thirties, were worried about evolving a precise notion of effective computation. We are all familiar with the idea of an algorithm; one has a recipe such that if one follows it one always gets the right answer—if there is one. One may run on indefinitely if there isn't an answer. Turing produced a formalization as follows: Let us consider a box which has finitely many states, say $q_1$ up to $q_n$, and which operates on a tape divided lengthwise into squares, each of which can bear any one of a finite number of symbols, $x_1$ up to $x_m$. We start with a finite tape on which are printed the initial data, the machine is started in state $q_1$, and operates synchronously. At each time ($t = 0, 1, 2, 3, \ldots$) the symbol it scans and the state it is in determine that the machine stops *or* determine three things: the new symbol (perhaps the same old one, or it may be a blank) it prints on the scanned square, the moving of the tape at most one square, and the change of state. Then it is ready to repeat the cycle, obeying the instruction keyed by the new state-symbol pair. If it ever comes to an end of the tape, then it will add on a new square.

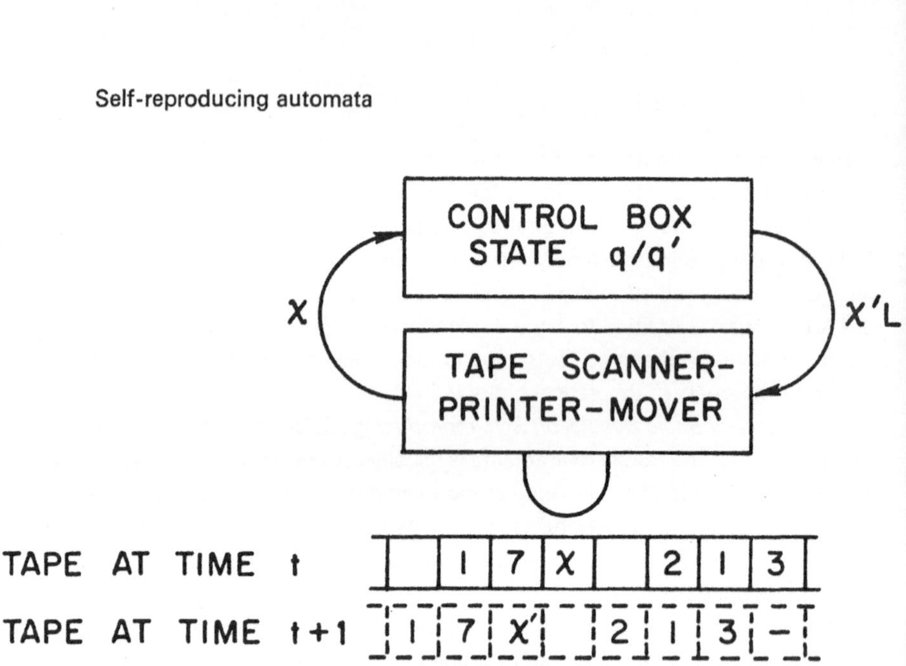


FIGURE 1
A Turing machine, whose program contains the quintuple qxx 'Lq'.

Now suppose we have an algorithm for effectively producing one integer from another. In the past, such a rule has always been transcribable into a set of instructions for a Turing machine, and, of course, it is quite clear that a Turing machine computation is in itself a recipe, an effective process. At the present time, the usual thing is to identify the intuitive notion of an effective computation with the formal notion of computation which can be carried out by a Turing machine.

For our present purposes, the important thing that Turing discovered in his paper of 1936 was that you could build a 'universal' Turing machine. How does it work ? Usually, given a Turing machine, Z, we use it to compute a function, f, by placing the number x coded on the tape and letting the machine run. If and when Z stops the result on the tape is decoded to give us $f_Z(x)$. Every such machine Z is given by a finite list of instructions, one for each state-symbol pair. Thus a Turing machine Z can be represented by a string of symbols e(Z) which encodes its program. Turing gave a program for a Turing machine U which was universal in that if you wrote on U's tape the ordered pair e(Z) and x, then, at the end of U's computation, it would produce $f_Z(x)$—precisely what Z would have computed with input x. If Z wouldn't have stopped, of course U won't stop.

Turing's result that there exists a universal computing machine suggested to von Neumann that there might be a universal construction machine, that is an automaton A which, when furnished with the description $I_N$ of any other automaton N in terms of appropriate functions, will construct a copy of N. In what follows, all automata for whose construction the facility A will be used are going to share with A the property that their description will include the specification of a place where an instruction I can be inserted. We may thus talk of 'inserting a given instruction I into a given automaton'. The reader may at this stage think of the automaton A with the description $I_A$ inserted into it. This entity will proceed to construct a copy of A. But note that this does not make A self-reproducing, for A with appended description produces A without an appended description—it is as if a cell had split in two with only one of the daughter cells containing the genetic message. Such a consideration suggested to von Neumann that the correct strategy might involve 'duplication of the genetic material'. He thus introduced an automaton B which can make a copy of any instruction I that is furnished to it—I being an aggregate of elementary parts and B just being a 'copier'.

Von Neumann then combined the automata A and B with each other and with a control mechanism C which does the following. Let A be furnished with an instruction I. Then C will first cause A to construct the automaton which is described by this instruction I. Next C will cause B to copy the instruction I referred to above and insert the copy into the automaton referred to above which has been constructed by A. Finally, C will separate this construction from the system A + B + C and 'turn it loose' as an independent entity.

Let us then denote the total aggregate A + B + C by D. In order to function, the aggregate D must have an instruction I inserted into A. Let $I_D$ be the description of D, and let E be D with $I_D$ inserted into A.

E *is* self-reproductive. Note that no vicious circle is involved. The decisive step occurs in E when the instruction $I_D$, describing D, is constructed and attached to D. When the copying of $I_D$ is called for, D exists already, and it is in no wise modified by the construction of $I_D$. $I_D$ is simply added to form E. Thus there is a definite chronological and logical order in which D and $I_D$ have to be formed, and the process is legitimate and proper according to the rules of logic.

We thus see that once we can prove the existence of a universal constructor for automata constructed of a given set of components, the logic required to proceed to a self-reproducing automaton is very simple. [Though there is something somewhat whimsical in the idea of a universal constructor, as if a mother could have offspring of any species, depending only on the father. While this may be appropriate to Greek myths, it does not seem appropriate to biological modelling. We shall come back to this matter in later sections.] Our concern now is to examine the difficulties involved in actually providing a universal constructor. Von Neumann did not do this in his original paper, and the task involves hundreds of pages of his book on *The Theory of Self-Reproducing Automata* [11] which was published in 1966 on the basis of a manuscript left at von Neumann's death in 1956. The problem is essentially this. A Turing machine is only required to carry out logical manipulations on its tape, sensing symbols, moving the tape, printing symbols, and carrying out elementary logical operations. A universal computer only has to carry out these operations. But a universal constructor must also be able to recognize components, move them around, manipulate them, join them together. Thus, presumably, constructors of Turing machines require more components than do Turing machines themselves. We are immediately confronted with the possibility of an infinite regress. Given a set of components $C_1$, to construct machines which build all the automata made of components from $C_1$, you may need a bigger set of components $C_2$. To build all machines constructed of components $C_2$, you may need machines put together from a bigger set of components $C_3$. The question is: 'Is there a fixed point ?' Can we find a set of components C such that all automata built from components of C can be constructed by automata built from the same set C ? I have called this [12] *the fixed point problem for components.* This is the fundamental problem in the theory of self-reproducing automata. Once we have found a set of components C in which for each automaton A there can be found an automaton c(A) which constructs A, it turns out to be a fairly routine matter to prove the existence of a universal constructor. We then know from von Neumann that it is a simple matter to prove *the construction fixed point theorem,* namely that there exists a self-reproducing machine U which can construct a copy of U. There have been several procedures following on von Neumann's to exhibit a set of components which satisfy the component fixed point theorem. Von Neumann [11] used 29-state components and gave an elaborate construction taking about 200 pages. James Thatcher [13] used the same components but gave a much more elegant construction taking less than 100 pages. E.F. Codd [14], with remarkable ingenuity, showed that a construction similar to von Neumann's could go through using components with only eight states. I showed [12, 15] that the construction could be done with great simplicity in a matter of eight pages, if one allowed the use of much more complicated components. My rationalization for this use of complex components is that if one wishes to understand complex organisms one should adopt a hierarchical approach, seeing how the organism is built up from cells rather than from macromolecules. We might add that Myhill [16] has given an axiomatic theory of self-reproduction, but that the axioms are formulated in a way which does not allow them to be directly applied to different sets of components but only allows one to generate theorems about self-reproduction when one already has theorems about universal constructors. However, Myhill's paper shows that results in recursive function theory can lead to rather startling conclusions about finite programs containing the possibility of infinite improvements in successive generations of offspring without requiring any randomness in the mutations.

Rather than go into any details of my construction I shall just briefly present three pictures which give some idea of the basic notions involved. Figure 2 shows a CT-machine which under the control of a program in its logic box can read and write on a one-dimensional tape in just the way a Turing machine does, and which can write but not read on a two-dimensional tape. The idea is that the two-dimensional tape is to be thought of as a construction area, and the writing of a symbol is to be thought of as equivalent to the placing of a component. Our task is to find a set of components from which we can build tape, logic box, and construction area. Such a component as shown in Figure 3 is a finite-state module which can contain up to 22 instructions from a rather limited instruction set. We are to think of a two-dimensional plane in which these cells are repeated in a sort of Cartesian array *ad infinitum*. An automaton is then represented as an activated configuration of these cells. I might mention that the little boxes marked W are weld registers which serve to 'weld' squares together so that a number of squares may be 'welded' into a one-dimensional tape in such a way that when any one square of that tape is instructed to move, all cells will move in the indicated direction. The assumption of such a weld operation greatly simplifies our programming. In Figure 4 we see an overall plan of an embedded CT-machine. We see that the logic box has been broken into two pieces, a one-dimensional tape which contains the program and two cells which form a control head. The idea is that on activation by the control head squares of the program tape may either be used to guide the control head in manipulating the computation tape in a Turing machine fashion or else may be used to place selected components in the constructing area and move welded blocks of components around. We

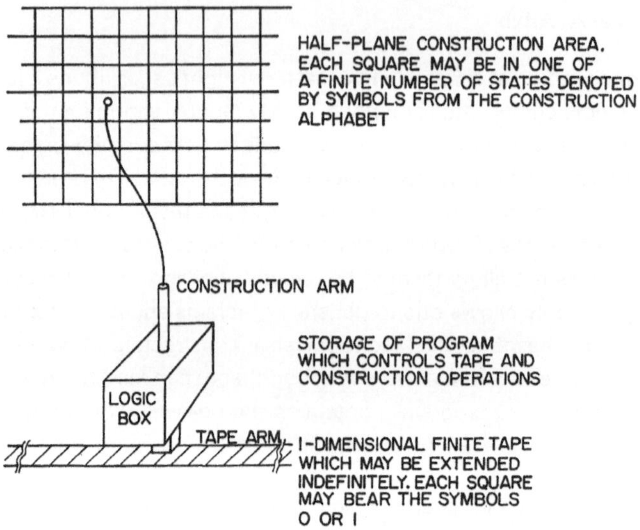


FIGURE 2

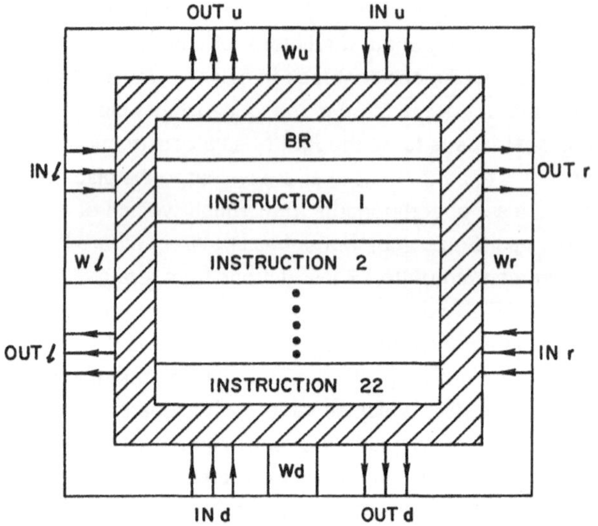


FIGURE 3

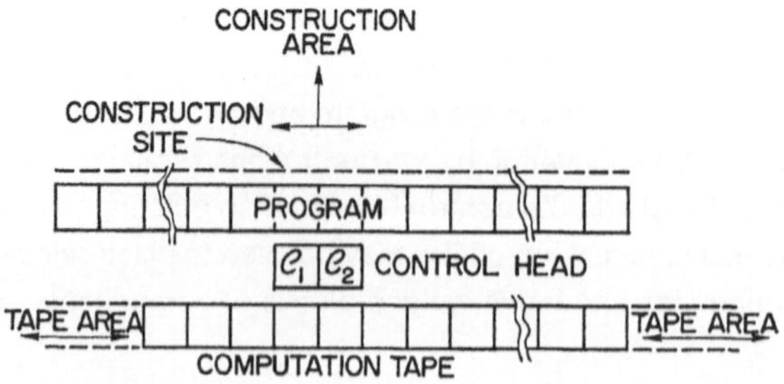


FIGURE 4

show [12, 15] that in fact an instruction code can be specified for our basic modules so that it is not only possible to embed arbitrary Turing machines in the array of those components in the indicated fashion, but in fact to program these machines so that they can construct other such machines in the construction area and to go on from there to show that there exists a universal constructor made of these components. It is then a standard procedure, following von Neumann's argument, to present an actual self-reproducing machine.

Having thus presented the basic logic of the theory of self-reproducing automata, our aim in the rest of this paper will not be to present new results in the theory–save for an amusing example of a regenerating automaton in Section 3–but rather to try and list some of the problems posed for the theoretical biologists by the attempt to compare the mode of reproduction employed in biological systems with that employed in our self-reproducing automata.

▸ *2. What is the role of descriptions?* It is a striking fact that in both real biological systems and in our self-reproducing automata we see a distinction between a description of the system and the active portion of the system. This goes back to the Weissman assumption underlying the current form of the theory of natural selection–namely that one needs to distinguish the genotype (a set of instructions) from the phenotype (their functioning embodiment) and that, whereas different environments produce different phenotypes from a given genotype, this change does not itself produce a change in the genotype. However, changes in genotype do cause marked changes in phenotype, and it is the genotype changes that propagate, whereas phenotype changes do not. Thus selection is of genotypes, but it acts on phenotypes. In Section 4 we shall briefly consider what steps might be taken to fit our automaton-theoretic considerations for a study of evolution, but in the present section we wish to consider the question (whose importance was emphasized to me by Howard Pattee): 'Is a *description* of the object a necessary complication to obtain an interesting self-replicating object?' In other words, must an interesting self-replicating object A made out of our elementary units contain within itself an object e(A), a description of A also made out of these units?

Von Neumann [11] distinguished two methods of self-reproduction. In the passive method the self-reproducing automaton contains within itself a passive description of itself, and reads this description in such a way that the description cannot interfere with the automaton's operations. This is the method used in our self-reproducing automata, and if we identify DNA with an encoding, it appears to be the method used in living organisms. We may contrast this with the active method, in which the self-reproducing automaton examines itself and thereby constructs a description of itself. Von Neumann has suggested that this method would probably lead to a logical paradox. However, I am not at all convinced of this at the time of writing. It would seem to me that DNA does indeed replicate itself by this active method and we must ask not is there a logical paradox inherent in the active method, but rather is there some well-defined cut-off point at which the active method is no longer applicable ?

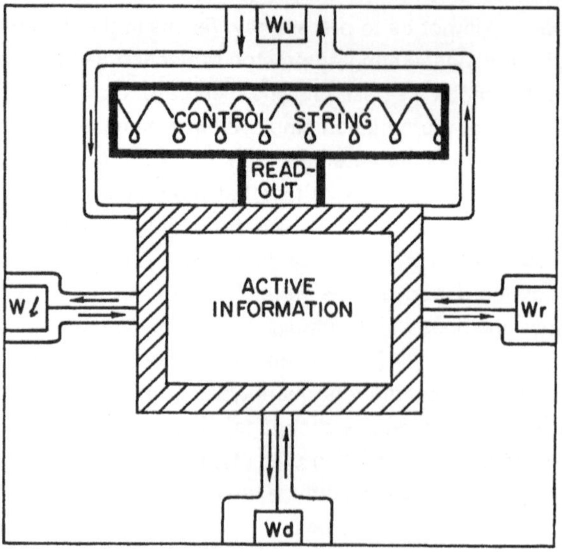


FIGURE 5

Pattee [8] sees the central biological aspect of hereditary evolution as the fact that 'the process of natural selection acts on the actual traits or phenotypes and not on the particular description of this phenotype in the memory storage which is called the gene'. He sees this as essential biologically because 'it allows the internal description or memory to exist as a kind of virtual state which is isolated for a finite lifetime, usually at least the generation time, from the direct interaction which the phenotype must continuously face'.

Goodwin [17] has pointed out that 'in a study of the dynamic properties of a certain class of biological phenomena . . . it is necessary to extract a manageable number of variables from the very large array which occurs. . . . Variables which are major to the phenomenon being investigated become the quantities that define the systems which one intends to study, while [minor] variables become either parameters of the system, thus defining its environment, or . . . noise.'

'The *relaxation time* of a system is, roughly speaking, the time required for the variables to reach steady state after a "small" disturbance. The significance of this concept is the fact that if two systems have very different relaxation times, then relative to the time required for significant changes to occur in the "slower" system, the variables of the "faster" one can be regarded as always being in a steady state. . . . On the other hand, the variables of the "slow" system will enter into the equations of motion of the "fast" one as parameters, not as variables.'

The description, then, is presumably a portion of the system whose relaxation time is so great relative to the dynamics of the system that it may serve as a permanent record insulated from the vagaries of the existence of the organism. In terms of this notion of insulation from the vagaries to which the organism is subjected, we might note that Michie and Longuet-Higgins [18] compare the *germ plasm* (DNA specifications) of a cell to the *program* of a user of a large computer system, and they compare the *soma* (cellular machinery of implementation) to the *monitor* of the computer system (the program which controls input/output devices, assigns priorities to different users, translates programs using various compilers, etc.). They suggest that the genotype-phenotype distinction may be compared to the segregation of user's software and system's software. If a user writes a program which could modify the monitor during execution, difficulties could be created for users of other programs using the same monitor. Perhaps similar difficulties could arise if germ plasm and soma were not segregated.

Perhaps there is even something in the horribly naive thought that if an object is genuinely three-dimensional it cannot be copied in space without completely cutting it apart. Thus we copy it in the four dimensions of space-time instead, by using a genotype to *grow* a replica.

We end this somewhat inconclusive discussion of the nature of descriptions by noting that Thatcher [19] and Lee [20] have shown that there are Turing machines which can print their descriptions on their tapes when appropriately triggered although their description is not explicitly stored within their program. Perhaps this means that an automaton need not contain within itself an explicit separate description at all times, but need only retain the ability to generate such a description when it is necessary. In other words, we may visualize an organism in which the genome is *potential* rather than *actual.* The big question here, then, is one of reliability. If we do not explicitly segregate the genotype from the phenotype, can we guarantee that damage to the phenotype will not yield dangerous alterations to the potential genotype, as suggested by Michie and Longuet-Higgins ?

While our self-reproducing automata share with biological systems the use of a completely separate description, we should note that they are abiological in that the automaton which is being reproduced is completely passive until it is completed. It might be worth analysing why complete passivity is used and what might be done to avoid it.

▶ *3. Morphogenesis.* We examined [12] some of the ways in which our model diverged from biology and indicated the way in which one of these flaws could be corrected, namely we showed how the model could be modified so that the biological program became a string stored in *each* cell, rather than a program embedded in a *string* of cells. This Mark II module is shown in Figure 5. We have kept the tessellation structure, side-stepping the morphogenesis of individual cells. The control string is segmented in words which correspond to possible internal instructions of the original module. The whole control string corresponds to the overall program in our original model. Only a small portion of the control string can be read by an individual cell. *Every cell in an organism has the same control string.* Individual cells differ only in the portion of the control string which can currently be read out. *The change in activation of portions of the control string is our analogue of differentiation. The increase in the number of cells in a co-moving set is our analogue of growth.* Rather than rehearse here that discussion we shall turn to another criticism of the automaton-theoretic approach, namely that it gives us no insight at all into the reaction of embryos to damage. The program of the self-reproducing automaton is completely explicit, specifying the automaton in every detail. Any damage to the program will yield corresponding damage in the automaton. Our purpose here is to indicate a way in which such an array model of an automaton can exhibit some of the embryonic properties of resistance to damage. The problem is similar to the counting problem raised by Maynard Smith [21] and the French Flag problem raised by Wolpert [22], and we may phrase it here in the following form: 'How may we design a cell for our array so that a string of any length whatsoever of these cells, all identical, will so organize itself that the first third of these cells are in one state, we may say they form a head, the second third of the cells are in another state, we may say they form a body, and the final third are in yet another state, we may say they form a tail ?' Furthermore we demand that if this array is cut in any way, each portion of the array will rearrange itself so that the portion will after a certain period of time have a recognizable head, body, and tail. In other words, how can an array count up to three irrespective of the length to which it grows or the length to which it is reduced by damage ?

We shall indicate how arrays of one such type of cell may do the job in Figures 6–9 (see pages 217 and 218). We shall see that they have one great drawback from a biological modelling point of view, and then find to our pleasant surprise that the rectification of this deficiency yields a cellular behaviour which is more biological than that of the original.

We are to imagine an array which can conduct five types of pulse, an A pulse, a B pulse, a head pulse, a tail pulse, and a body pulse. An A pulse will propagate with unit velocity in a given direction until it hits the end of the string of cells, whereupon it is reflected and moves with unit velocity in the opposite direction. We decree that if there are ever two A pulses travelling in the array, then upon their meeting the pulse travelling to the left is annihilated whereas the pulse travelling to the right is propagated as if it had not encountered the other pulse. B pulses are propagated with half the speed of A pulses and are created at a boundary each time an A pulse is reflected there (see $t_0$ in Figure 6). If an A pulse overtakes a B pulse going in the same direction, then the two pulses do not interact but continue on their way at their specified velocities. If, however, the A pulse and the B pulse are going in opposite directions, then a fairly complex response is triggered.

If the A pulse is moving right and the B pulse is moving left, then the B pulse is annihilated whereas the A pulse will continue on its way. In addition to this effect, however, a tail pulse will be propagated with unit velocity towards the right-hand end of the string where it will be annihilated, and as it reaches each cell it will turn that cell into the tail state. At the same time a body pulse is triggered moving left at a velocity of 1/3 (which leads to some problems in discrete systems which I shall overlook, using a continuum in all my diagrams). In general, this pulse will not propagate all the way to the left-hand end of the string but will be annihilated on meeting an A pulse.

If an A pulse moving left meets a B pulse moving right, then the B pulse will again be annihilated, and this time a head pulse will be propagated left at unit velocity, turning each cell it encounters to the head state until it is annihilated upon reaching the left-hand end of the string.

To appreciate how these rules work look at Figure 6, in which the array of cells is graphed in space from left to right while its evolution in time is graphed with time progressing as we move down the page. We see that at time 0 an A pulse and a B pulse are both initiated at the left-hand end and they propagate until they meet at time $t_1$, whereupon the B pulse is annihilated, but we see that the tail pulses and the body pulses work to cause cells to differentiate into the appropriate states. Note, however, that at time $t_0$ the A pulse was reflected from a boundary, at which time a new B pulse was generated, and so we see that at time $t_2$ the new B pulse meets our A pulse, at which stage the cells in the front third of the array are told to convert to the head state. You will notice that the body pulse was annihilated by meeting an A pulse at just the same point in space and time at which the head pulses were triggered. It was to assure this coincidence that the strange propagation speed of 1/3 was chosen for the body pulses. We now see that as time goes by the A pulses and B pulses ricochet back and forth, interacting as they do so, but the messages they send out do not change the configuration and so in fact at time $t_3$ the stable state with properly differentiated head, body, and tail is attained.

Let us now see what happens to our array when pieces are cut out of it. We have marked on Figure 6 three sections which are to be removed. The evolution of Section I is indicated in Figure 7, that of Section II in Figure 8, and that of Section III in Figure 9. We add one more condition. When a portion is cut from an array, an A pulse is triggered at the point or points of damage. We shall see how this works by tracing through the fate of Section I in Figure 7 (the reader is invited to carry a similar step-by-step analysis of Section II and Section III). We see that at time $t_0$ not only is there an A pulse and a B pulse travelling in the section but also there is a new A pulse initiated as a result of injury, but this A pulse only lasts until time $t_1$, when it is annihilated on encountering the first A pulse. The B pulse and the first A pulse meet at time $t_2$ just as they would have in the uninjured specimen, and so the head pulses propagate forward from time $t_2$ so that by time $t_3$ we do have an array in which the first portion is head, the second portion is body, and the third portion is tail. However, the proportions of the three components are not equal and so a stable state has not been attained. As we follow the evolution we see that at time $t_4$ the adjustments are made which ensures that exactly one-third of the array is in the tail state, but it is not until time $t_6$ that another encounter between an A pulse and a B pulse sets up the final equilibrium, in which head, body, and tail each occupy one-third of the array. Note that the propagation forward of a body pulse at time $t_4$ has no apparent effect until time $t_5$, when it first emerges from the front end of the string of cells that were already in the body state at $t_4$.

The reader will note one unpleasant feature of this evolution. Just after time $t_4$, a piece of body is caught between two pieces of tail. Similarly, in Figure 8, we

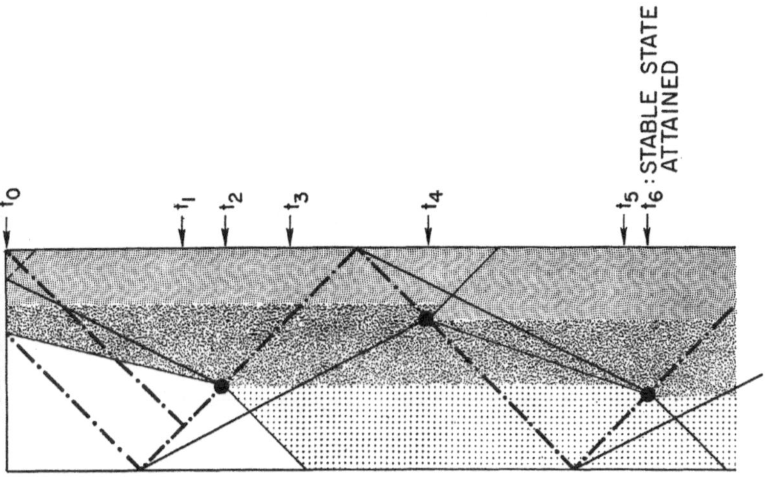


FIGURE 7

FIGURE 6 *(left)*

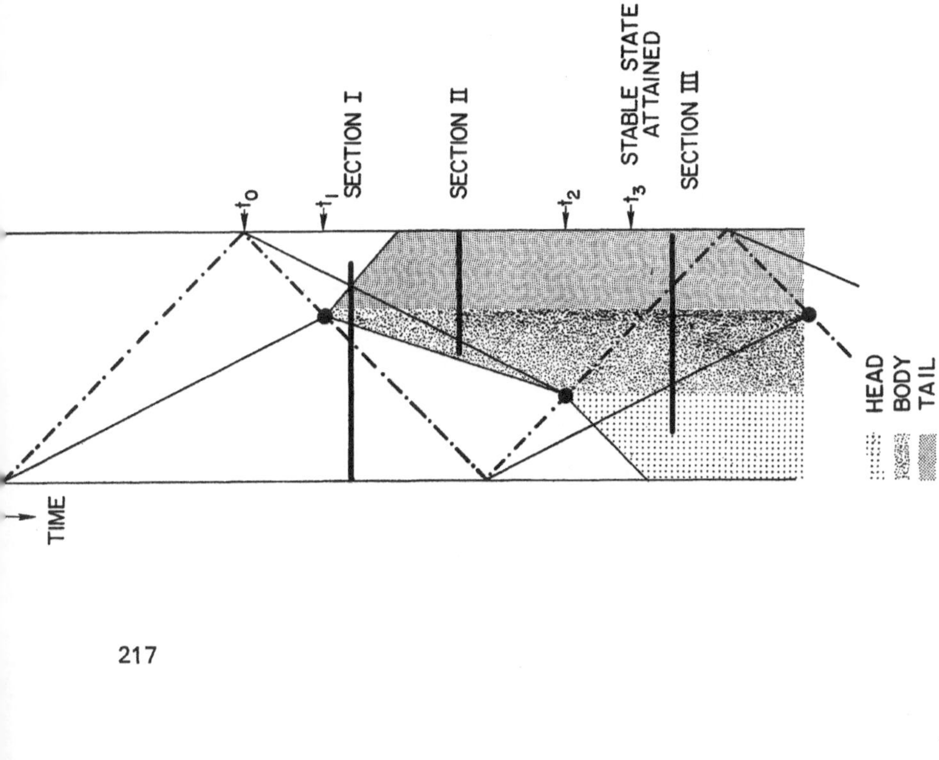

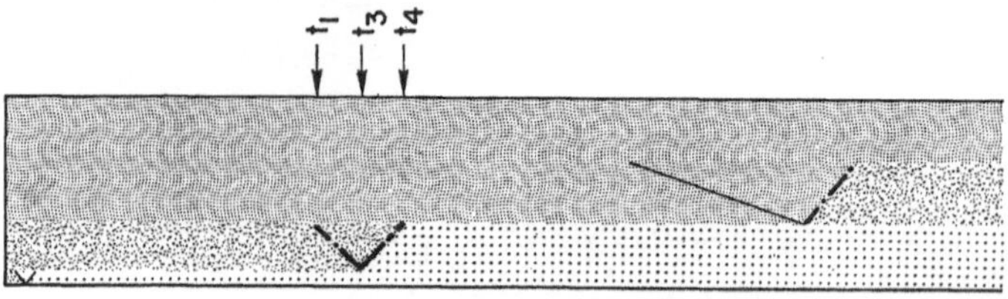


FIGURE 10

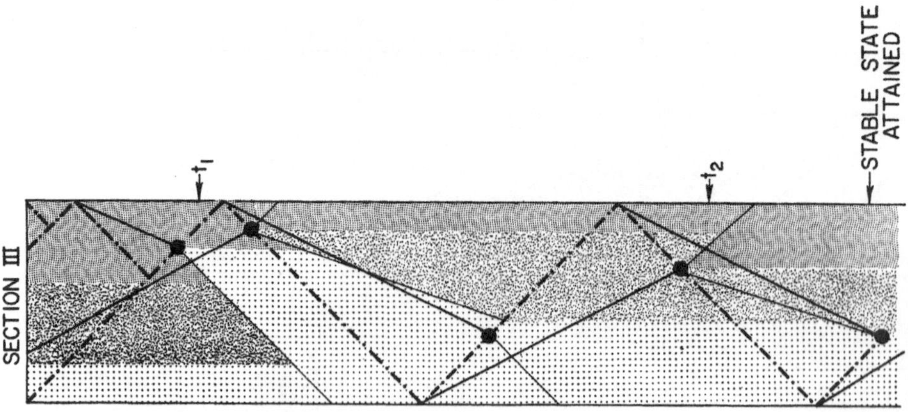


FIGURE 9

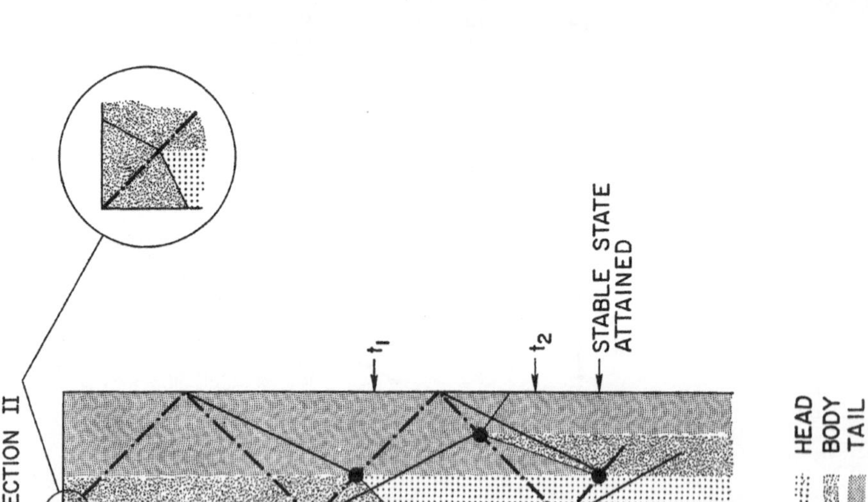


FIGURE 8

see head-body-head-tail at $t_1$ and head-tail-body-tail at $t_2$; while in Figure 9 we see head-body-tail-head-tail at $t_1$ !

I do not know whether embryologists have seen such effects, but it seemed to me worth while to look for a simple modification in my model which would avoid them.* We simply postulate that the states shown in my diagrams are not *actual* states, but *potential* states.

Let x be the location of a cell, $x-1$ the location of the cell to the left of it, $x+1$ the location of the cell to the right of it. Let $A_{xt}$ be the actual state of cell x at time t, $P_{xt}$ the potential state of cell x at time t.

We then have the following rules:

If $P_{xt}$= head, then $A_{x,t+1}$= head unless $A_{x-1,t}$=body or tail.
If $P_{xt}$= body, then $A_{x,t-1}$= body unless $A_{x-1,t}$= tail or $A_{x+1,t}$= head.
If $P_{xt}$= tail, then $A_{x,t+1}$= tail unless $A_{x+1,t}$=head or body.
In the other cases $A_{x,t+1} = A_{x,t}$.

Thus the cells are polarized, and a tail cell can repress the cell behind it from turning into a head cell or a body cell, etc.

In Figure 10 we have shown how the potential states of Section II, shown in Figure 8, give rise to actual states without any 'islands' when we make the repression assumption. For example, at time $t_1$ a pulse is sent forward changing cells to a *potential* head state, but until time $t_3$ each cell so changed finds a cell in front of it still in the actual body state. It is only at time $t_3$ that a cell can first change its actual state, and when it does so, the cell behind it can change one moment later, until at time $t_4$ all cells in *potential* body state have changed to *actual* body state.

I shall not explore this model further here. My purpose has simply been to show that the regenerative properties of the embryo are not incompatible with the automaton approach to development. It becomes, of course, a vital question to study how we may construct arrays which combine the properties of those in the first section with the regenerative properties we have presented here.

▸ *4. The bridge to evolution.* Among the many other questions which our discussion of self-reproducing automata raises are 'Whence come the components out of which our automata are made?' and 'Given that such automata exist, how might one imagine them to evolve?' It is not our purpose in this section to answer these questions—would that we could—but rather to suggest some interesting avenues towards their solution.

* Note, too, how in Section II, there is a period of time when no body cells are visible. Does such resorption occur in real preparations?

## Self-reproducing automata

Pattee [8] has suggested that the crucial problem of the origin of life may be reduced to the problem of finding a molecule which is both stable enough to survive but yet not so stable as not to evolve. In our automaton-theoretic approach we tend to neglect these crucial problems of reliability. However, we may note in passing that Cowan and Winograd have erected a theory of reliability in neural nets which proceeds by the judicious application of redundancy, and it seems to me quite conceivable that a similar redundancy scheme might be used to replace the function of one cell in our iterative array by the function of a number of cells in the new array, it further being the case that each cell in the new array shares the properties of several cells of the old array. However, the question still remains of what is the simplest component with which one may hope to build up interesting self-reproducing automata. It has been pointed out that if the only system one wishes to reproduce is a falling domino, then the problem of self-reproduction becomes very trivial. If, however, one's criterion of interesting self-reproduction is that the organism involved can have computational power akin to that of a universal Turing machine, then the problem becomes non-trivial. It is known that if one embeds one's automaton in a tessellation in which the state of each cell is affected by only the states of its four immediate neighbours, then eight states will suffice to yield computation universality [14]. Codd has further shown that two states are insufficient, but no minimal number of states for such an array has yet been shown. We should note that even an eight-state cell is rather complex in that each transition, depending as it does on the state of the cell and its four neighbours, requires the specification of actions for each of 32,768 possible circumstances. Unfortunately, it is not at all clear how this sort of discussion can give us any insight into the minimal complexity required of the macromolecule which can enter into a process of hereditary evolution. This is especially so since we cannot be sure that the present DNA system occurred at any stage at all near the origin of life. Just as a virus, which appears to be intermediate in complexity between simple molecular systems and the cell, could not have evolved until after cells had evolved to provide an environment for their reproduction, so it may well be possible that DNA could only evolve after various other subsystems, such as those now available in the cytoplasm, were available to help it in its replication. At the present stage it still requires cytoplasm or Kornberg for the replication of active DNA! While on this topic of possible precursors for present replicating systems we would draw the reader's attention to the papers of Pattee [8] and Cairns-Smith [23].

Turning now to the broader questions of evolution, once we have already obtained a suitable supply of existent automata from which to evolve, we see that a radically different point of view must be imposed on that of our basic automaton-theoretic approach. We have already mentioned, in Section 3, the transition to the Mark II module model in which the whole program of the organism is contained in each cell, differentiation serving to activate different portions of the program. To study evolution, one must proceed to the Mark III model, in which these internal programs are themselves modifiable (mutable) programs. It will be a matter of some interest to see how necessary stochastic factors are in modelling this situation. We must then study not only what is the simplest program for a self-reproducing organism but also how the program may be complicated to yield 'fitter' programs. Let me just briefly sketch the framework one might employ for such a study (the only attempt in the literature of which I know at erecting such a framework is in the work of John Holland; see, for example [24]). In such a study we move to a higher level at which we neglect the details of the dynamics of cell division and work instead, perhaps with some measures of mitosis intervals, and stable life-times of cells. In fact, it is not at all clear to me whether study at this level will depend upon its automaton-theoretic base or will in fact become indistinguishable from Fisher-type statistical studies of natural selection. Having set up such basic parameters we might then study free movement in the tessellation and consider for various aggregates what is their reproduction time and stable life-time. One would hope to prove that the initial aggregates arise by a transition from symbiosis to functional reproducing organism (compare [23]). One might expect that aggregates would be better suited to utilize material. One should also be able to prove theorems that parallel operation would be favoured, as would information and material transmission channels, to increase the use of materials encountered by the cell (Crick [4] has observed that if one removed all the DNA from a single human being and stretched it end to end, it would span the whole solar system—surely a very strong argument for parallel processing in any organism!) We might then ask for theorems which contrast the short generation time of individual cells with the long generation time of aggregates. Simon [25] presented a rather relevant parable of two watchmakers, one of whom had his watches arranged in a hierarchical organization so that whenever his work was interrupted the only damage done was that his work was put back to its component subsystems, whereas another watchmaker was in the unfortunate position that whenever he put down a watch it dismantled right down to the very smallest components. Clearly, the first watchmaker could get much, much more done than the second watchmaker. With such parables in mind we may hope to proceed to a really rigorous theory of evolution in which we can show the selective advantage of systems which are not only highly parallel but also hierarchical in structure.

Let us turn now to the relation between the origin and evolution of life on the one hand and the laws of thermodynamics on the other, and see wherein a statistical mechanics may prove useful.* Attempts so far to deduce thermodynamics from mechanics require an *additional hypothesis*, namely one of random phase, and I suspect that living systems are ones for which random phase hypotheses do not apply. There may be subsets of phase space, albeit of very small measure, where thermodynamics does not hold true, and such regions may contain the trajectories of populations of living systems. More strongly, we may even question whether thermodynamic parameters are well defined in such regions. Thus it is only as a sop to a limited statistical mechanics that in what follows I shall follow the standard line of reconciliation between evolution (increasing order) and the second law of thermodynamics (decreasing order) by saying that 'in the *total* system, the entropy must increase'.

I hope that developments in statistical mechanics, such as discussed by Prigogine [26], will eventually show how this simplistic statement may be refined.

Thermodynamic approaches to problems of organized complexity, integration, adaptation, or evolution fail to expose the mechanisms underlying these phenomena, and at best deal only with the direction in which a situation will develop and not with any quantitative aspects. In the same vein, we viewed evolution and progressive adaptation as directed processes. But, fine conceptually though this may be, it cannot tell us what will evolve, or what properties of the organism are going to emerge. A self-organizing system is characterized by change, whether it be growing or adapting. Thus we need to find a mechanism for the description of such a system which is essentially dynamic in nature. It must be further noted that the bulk of present thermodynamics and statistical mechanics is designed to deal with reversible systems which are essentially homogeneous. Self-organizing systems, on the contrary, are highly structured, and have a definite 'positive arrow in time', and so great care must be exercised in applying thermodynamic notions.

Be that as it may, let us now consider in more detail what we want our dynamic mechanism to be. It seems clear that whether we wish to analyse biological systems or synthesize adaptive machines, we are going to need a theoretical

* The rest of this section owes much to discussions with Warren McCulloch in May of 1962.

study of automata. We may think of automata as messages, and we may do so in two interconnected ways:

a. An automaton is a structure—as such it is essentially characterized by its form, and so contains information. In this sense of form (not of signal) we may consider it to be a message.

b. We may interpret as a message the regularity of the event the automaton realizes, i.e. its information-handling behaviour.

These messages are self-perpetuating and propagating in two senses:

i. They endure. (Endurance has three aspects: growth, learning, and homeostasis. Homeostasis must repair what would otherwise be permanent damage. The correction of purely temporary errors is a problem of information handling rather than self-perpetuation.)

ii. They multiply.

Although, in the first three sections, we gained some insight from the study of deterministic systems, it is clear that the eventual form of our theory of biological automata must be statistical, since they are continually subject to noise, both environmental and internal. The noise affects both aspects of the message:

a. It interferes with enduring and multiplying, and if uncompensated would cause the automaton to disintegrate rapidly and would prevent it from begetting its like.

b. It interferes with the information handling, and if uncompensated would cause drastic computational errors.

Clearly these aspects are not independent. We have already mentioned the work of Cowan and Winograd on combating computation errors in automata by applying information and coding theory to the problem. A major task for our theory is to learn how to apply coding theory to the problem of enduring and multiplying in the presence of noise.

Thermodynamically, the probability of making a living creature is very, very small, given a box of 'components'. But this is not the way it came about. The great trick of the living system is that it slows down the entropic degradation. The huge structural complexity of a human being is coded into the chromosomes selected for the job the organism is to do. The genetic instructions interact with the environment to produce the organism. (N.B. The environment of a cell includes adjacent cells, e.g. a zygote usually gives rise to an organism, but if the two halves are separated, two organisms will often be formed.) This organism is, as a result of the interaction, adaptive both in structure and in function. When something new hits the organism, be it a virus or an idea, it has to make new protein structures fitted to the job. There is a great difference between animals that learn much and those that learn little. A human baby is relatively immature when he leaves the womb, and a lot of his formation comes from subsequent learning and adaptation. This means that he can learn more, and adapt to a wider range of environments.

We may define *adaptation* as the property of that which has been kicked once which enables it to change in such a way that when kicked again it undergoes less internal change. Nature uses this process for adjustment to the environment. Note that this adaptation can only occur if the organism has the requisite variety from which to choose—this is the important point that Ashby has made time and again. Whether we build a machine or investigate a biological system, we must have components capable of entering into sufficient relationships to each other to gain flexibility and reliability. (We note that bisexual reproduction also serves to increase the variety!) Greater diversity in function might also handle these problems. But variety is useless unless it can proceed from a sufficiently rich structure. It is this combination of natural structure and requisite variety which yields adaptability.

The necessity of the basic structure can be illustrated by an example [McCulloch, personal communication] from neurosurgery. If certain parts of the brain are removed, the patient becomes 'stimulus-bound'—he only acts in the present. He has a poverty of associational structure—his speech becomes less ordered in the large, his sentences shorter. He lacks adaptability. Now, his brain is still very complex and abounds in variety—but a vital part of its structure is missing. The warning is clear: in building a self-organizing system we must ensure that we transfer enough information in forming the machine to allow it the complexity of structure requisite for the adaptive tasks we set it.

We should also reiterate at this point the importance of hierarchical structuring. The unity of an organism overrides the unities which make up the lesser parts. In large groups of organisms, the regularities are greater than for the individual (this is reminiscent of the behaviour of fluctuations in the statistical mechanics of gases). One would like a theory that is adequate to treat, and distinguish between, on the one hand, 'genesis, growth, adaptation, and learning', all of which are in terms of a single device, and, on the other hand, effects due to the interaction of organisms, which thus develop 'language' in the sense of Occam's conventional terms and Pavlov's second signalling system. A human derives energy and information from food, perception, and verbal communication. (It is interesting to note that the method of gaining information by verbal inputs is itself a product of learning—we have a regenerative loop of adaptation.) The food is composed of useful proteins, sugars, etc., and so contains much necessary energy and structure; perception and verbal inputs provide him with information. He grows and learns. His output is informed when he communicates and reproduces, but is otherwise degraded. Thus, his internal entropy is generally decreased, (despite environmental and internal noise), but, again, the net entropy of the universe may increase. In any case, we are reminded of the importance of a richly structured environment in securing the development of an organism.

What we recognize as order is something describable in our language in a small number of words. We strongly suspect that we have had the wrong way of looking at living systems. We are beginning to see proper order. If we exclude the case of a crystal, we find that, when we evolve something, there will be an apparent disorder simply because we have not figured out how to look at it. Just as the very concept of evolution itself brought a great deal of order to biology, so our further studies should reveal to us even further order in the resultant systems.

When we analyse biological systems, we are trying to uncover a rich order. In the optimum code for a noise structure each message may be encoded in many ways. All possible words will be used as often as possible—and the result will look like noise. Thus, *apparent* disorder may well be due to adaptation!

Our goal is to develop an automata theory which will be helpful wherever, as in embryology, we have that integration of diverse processes which yields the progressive emergence of increasing functional stability in complex systems.

## Notes and References

This paper owes much to the fruitful discussion at the Symposium on Theoretical Biology held at the Villa Serbelloni at Bellagio on Lake Como in August of 1967, and I should like to record my gratitude to my fellow symposiasts for the stimulation of their ideas.

Travel to the conference was made possible by Martin Garstens under grant Nonr 225(87) from the Office of Naval Research, while research was supported in part by the U.S. Air Force Office of Scientific Research, Information Sciences Directorate, under grant no. AF-AFOSR-1198-67.

1. Hotchkiss, R.D. in (Gerard, R.W., Ed.): *Concepts of Biology, Behavioral Science 3,* No. 2 (April, 1958) 129.

2. Elsasser, W.M., *Atom and Organism: A New Approach to Theoretical Biology* (Princeton University Press, Princeton, New Jersey 1966) 61.

3. The scheme we give for self-reproduction is clearly realizable with electronic circuitry which is subject to the laws of quantum mechanics (questions of reliability will be taken up in Section 4). It is thus disconcerting to find Wigner [5] claiming to show that

self-reproduction is virtually impossible in a quantum-mechanical system (as part of a desire to show that the explication of much of biology, as well as of consciousness, requires 'biotonic' laws supplementary to those of physics, cf. Elsasser [6]. Section 6 of Arbib [7] shows wherein we believe Wigner's argument fails. However, Pattee [8] has pointed out that Wigner's argument is well taken in that the problems of transition from genotype to phenotype are inextricably bound up with the yet unresolved problems of the quantum theory of measurement. However, this does not counter the claims of the reductionist–it merely shows that reduction will not solve all our problems!

4. Crick, F., *Of Molecules and Men* (University of Washington Press, Seattle, 1966).
5. Wigner, E. P., in *The Logic of Personal Knowledge: Essays Presented to Michael Polanyi* (The Free Press, Glencoe, Illinois, 1961) 231–8.
6. Elsasser, W. M., *The Physical Foundation of Biology* (Pergamon Press, New York and London, 1958).
7. Arbib, M.A. in (Hart and Takasu, Eds.) *Computer and Systems Science* (Univ. of Toronto Press, 1968).
8. Pattee, H. H., in (C. H. Waddington, Ed.) *Towards a Theoretical Biology: I, Prolegomena,* (Edinburgh University Press, 1968).
9. Von Neumann, J., in (L. A. Jeffress, Ed.) Cerebral Mechanisms in Behavior, *Proc. of the Hixon Symp.* (Wiley, 1951) 1–31.
10. Turing, A. M., *Proc. London Math. Soc.,* Ser. 2, *42* (1936) 230–65; *43,* 544–6.
11. Von Neumann, J. (A.W. Burks, Ed.) *The Theory of Self-Reproducing Automata* (University of Illinois Press, Urbana and London, 1966).
12. Arbib, M.A., *J. Theoret. Biol. 14* (1967) 131–56.
13. Thatcher, J.W., *Universality in the von Neumann Cellular Model,* to appear in a book edited by A.W. Burks on Cellular Automata.
14. Codd, E. F., *Propagation, Computation and Construction in 2-dimensional Cellular Spaces,* Technical Report (Univ. of Michigan, March 1965).
15. Arbib, M.A. *Information and Control, 9* (1966) 177–89.
16. Myhill, J., in (M. D. Mesarovic, Ed.) *Views on General Systems Theory* (Wiley 1964) 106–18.
17. Goodwin, B. C., *Temporal Organisation in Cells: A Dynamic Theory of Cellular Control Processes* (Academic Press, London and New York, 1963).
18. Michie, D. and Longuet-Higgins, C., *Nature 212* (1966) 10–12. [Reprinted in (C. H. Waddington, Ed.) *Towards a Theoretical Biology: I, Prolegomena* (Edinburgh University Press, 1968), with a comment by H. H. Pattee.]
19. Thatcher, J.W., in *Proc. Symp. Math. Theory of Automata* (Polytechnic Press, Brooklyn, 1963) (Vol. XII of the Microwave Research Institute Symposia Series) 165–71.
20. Lee, C.Y., in *Proc. Symp. Math. Theory of Automata* (Polytechnic Press, Brooklyn 1963) (Vol. XII of the Microwave Research Institute Symposia Series) 155–64.
21. Maynard Smith, J., in (C. H. Waddington, Ed.) *Towards a Theoretical Biology: I, Prolegomena* (Edinburgh University Press, 1968).
22. Wolpert, L., in (C. H. Waddington, Ed.) *Towards a Theoretical Biology: I, Prolegomena,* (Edinburgh University Press, 1968).
23. Cairns-Smith, A. G. in (C. H. Waddington, Ed.) *Towards a Theoretical Biology: I, Prolegomena* (Edinburgh University Press, 1968).
24. Holland, J. H., *J. Assoc. Comp. Mach., 9* (1962) 297–314.
25. Simon, H. A., *Proc. Amer. Phil. Soc., 106,* 6 (Dec. 1962) 467–82.
26. Prigogine, I., *Mededelingen van de Klasse der Wetenschapper, Koninklijke Academie van België* (*Bulletin de Classe des Sciences, Académie royale de Belgique*), Series 5, *53,* 4 (1967) 273–87.

# What Biology is About

Christopher Longuet-Higgins
University of Edinburgh

At the Villa Serbelloni we have been trying to make up our minds whether there ought to be a subject of Theoretical Biology, and if so what it should be like. Assuming that biology needs a general theory, should the theory be a mathematical one—a kind of dynamic topology in Thomist or Waddingtonian vein ? Should it, perhaps, be more like theoretical physics, or theoretical chemistry ? Or a sort of amalgam of evolutionary theory and population genetics, properly brought up to date under the title of neo-Darwinism ? Perhaps it should be an outgrowth of chemical engineering (Scriven), or of thermodynamics (Kornacker), or of metaphysics (Bohm), or of Fourier analysis (Iberall) or of classical dynamics (Lieber) ? To judge from our conversation, Theoretical Biology would seem to be all things to all men, no part of science being less relevant to the theory of life than any other. If this were so, I submit, we should have been wasting our time in this idyllic spot. But I do not believe this; I think that we must do some ruthless pruning on the periphery if we are going to reach a clear appreciation of what lies at the centre of this new subject.

One trouble with regarding theoretical biology as a branch of theoretical physics or theoretical chemistry is that biology is at the same time a narrower and a wider subject than physics or chemistry. Narrower, in that organisms are, by common consent, physical systems and hence presumably subject to the laws of quantum mechanics and statistical mechanics. (Bohm has stressed, here and elsewhere, the metaphysical weaknesses of present-day mechanics, but I am sure he would agree that these weaknesses matter no less for electrons than for snails, which are at least fairly large systems.) Wider, because physics and chemistry lack concepts adequate for distinguishing life from non-life. As a sign of the times, the idea that organic chemistry is specially helpful to a biologist has become indefensible; the name 'organic' applied to carbon chemistry rings very hollow nowadays, when we are beginning to realize that the interest of an organism lies, not in what it is made of, but in how it works. On a personal note, that is the reason why my consuming interest in living things has led me away from the physical sciences, to which concepts such as organization and function are entirely foreign. Kornacker might say at this point that biological organization is just a manifestation of thermodynamic coupling—and I would agree that this idea can be most illuminating when applied to specific phenomena such as membrane excitability—but surely there is more to biology than just that. An astronomer might quite correctly assert that quasars are just a manifestation of the conversion of mass into energy; but such an assertion would do little to satisfy our curiosity about quasars. I rather incline to follow Pattee's strategy of trying to see what must be added to physics—that is, to its language—before we can describe inheritance and evolution in physical terms. But I cannot think, with all respect to Lieber, that we shall make much progress with the construction of Theoretical Biology if we feel it necessary to go right back to the universal constants of nature for our primary concepts. Variability is a much more conspicuous feature of the living world than constancy, as Elsasser has emphasized; and anyway, Theoretical Biology ought presumably to begin at the point where living things start to differ in important respects from non-living matter. A theoretical biologist cannot escape the necessity of defining his field so as to exclude some phenomena as firmly as it includes others.

All this may seem rather negative, but one gets nowhere with difficult problems unless differences of opinion are as clearly stated as points of agreement. Continuing, then, with our catalogue, let us consider Fourier analysis, topology, chemical engineering and evolutionary genetics. Which of these subjects is of most importance to Theoretical Biology ? Set as a question in an intelligence test this would probably evoke the answer: evolutionary genetics. Plainly there *is* an honest discipline of evolutionary genetics (I wish Maynard Smith could have held forth at greater length about it) and it deals quietly and unpretentiously with—not metaphysical, astrophysical, or astrological problems—but genuine biological ones. All right, so evolutionary genetics must take an honoured place in the Theoretical Biology of the future. Also, as most of us seemed to agree, must chemical engineering. Not perhaps the old mix-it-and-hope variety, but the science which enables the engineer to design and operate complex processes with optimal efficiency, whether under steady-state or limit-cycle conditions. Nature is a chemical engineer of vast experience, and to understand the internal economy of the cell we may find it helpful to make models ourselves, either concrete or abstract. What about Fourier analysis ? Of course we must pay due attention to the oscillatory character of some biological processes. But a good idea becomes a positive menace if it is allowed to dominate one's whole thinking, and I had a feeling that Iberall's emphasis on frequencies and relaxation times was in danger of taking this course. One swallow does not make a summer.

How about dynamic topology—if Waddington and Thom will allow this name for their related ideas about the development of forms ? Here is an original and sophisticated approach to a wide class of problems, both evolutionary and morphogenetic. But is its descriptive power equalled by its explanatory power ? Most bacteria cannot live in a broth which is rich in sulphathiazole, *but some can* (and more are learning to do so, regrettably). One may say, with a wave of the hand, that the chreods of most bacteria peter out when they run into high concentrations of sulphathiazole, whereas the chreods of the sophisticated bacterium do not; but this is hardly an explanation. Actually, of course, *we know* in general terms why some bacteria are drug-resistant and others are not: it is because the resistant bacteria have special genes for coping with the drug, and the others do not or cannot use them. We must not make the mistake which might be made by a statistician studying the movement of cars through Central London who tried to account for everything in terms of a hydrodynamic sort of theory. He might do quite well until suddenly one day there was a complete absence of buses. Unless his theory of London traffic included the concept of a bus strike, he would be entirely unable to account for such an extraordinary fluctuation. Like the traffic in London, the macroscopically observable contortions of a cell are in the last analysis under the control of programs which are not directly apparent, and in a sense this is the most significant single fact about organisms as opposed to pretty, inanimate systems like Liesegang's rings or the patterns in a convecting fluid.

If they have read thus far, Arbib and Gregory will begin to understand why I have not so far referred to their recommendations about how we should think of constructing a coherent theory of biology. It is because basically I agree with them that the most fruitful way of thinking about biological problems is in terms of design, construction and function, which are the concrete problems of the engineer and the abstract logical problems of the automata theorist and the computer scientist. Indeed, it was on this basis that Gregory, Michie, and I decided to set up shop together in Edinburgh. But what justification can one offer for this view ?

To make a sweeping statement–but one which seems to be much more than half the truth–it seems to me that the problems of biology are all to do with *programs.* A program is a list of things to be done, with due regard to circumstances. (This is not a formal definition, but it conveys the spirit of the notion.) It may be a recipe in a cookery book, an algorithm for calculating the zeros of the zeta function, a set of rules for writing a fugue on a given subject, or a set of genetic instructions for making a mouse. At the first Serbelloni Conference some people were exercised about the question whether a man contains more *information* than the strands of DNA which his parents put together. Waddington felt that the problem was like the question whether a textbook of Euclidean geometry contains more information than the axioms of Euclid–a nice parallel–but nobody was prepared to say how the concept of information might be applied precisely to either problem, so that we got no further. In the end we felt ourselves moving towards the idea of a set of *instructions*, and Michie and I tried to demonstrate the relevance of this idea in our Party Game Model of Biological Replication, from which we ventured to draw some tentative biological conclusions. At the second Serbelloni Conference Arbib put forward some thoughts on the logic of automatic self-replication, and by that time everyone seemed to feel at home with the idea that the secret of an organism is the program which its life expresses. In the rest of this note I want to explore a few of the ramifications of this idea, expanding the remarks which I made at the conference itself.

Here, to begin with, is an irreverent little parable *a propos* of Kornberg's synthesis, *in vitro*, of an infective virus. Once upon a time there was an applicant for a post in computer programming. He was asked: 'Have you ever done any programming ?' 'Oh yes: here's one of my tapes.' 'How long did it take you to write ?' 'Only about ten minutes; I copied it from a tape I found in a drawer.' 'You *copied* it ?' 'Yes; I used the tape copier in the computer room.' The applicant was not appointed.

A thoroughly unfair gibe at a fine scientific achievement. But it may help to put our present biological knowledge into perspective. What we would dearly love to know is how existing biological programs originated (the origin of life, that is to say), how they have developed (the course and mechanism of evolution) and how they are implemented (the principles of morphogenesis). Considering the extreme youth of modern biology, a promising start has been made. We know the alphabet (adenine, thymine, guanine, cytosine) and the vocabulary (the set of triplet codons) and even something about the compilation (protein synthesis) of Nature's programming language. But we are a long, long way from being able to write our own DNA programs without cribbing from Nature. Not that this is particularly shameful, of course. The development of human computing would be an impossibly slow process unless computer scientists frequently and shamelessly lifted subroutines from one another's programs, and genetic evolution certainly has involved–and still does–a great deal of cribbing, particularly of new tricks such as how to survive the onslaught of man-made antibiotics. Nevertheless there is all the difference in the world between writing a program oneself, or understanding someone else's by reading it through, and discovering the meaning of a program by running it through a computer and seeing what comes out. Not until we can interpret the DNA of a new species without actually growing an individual from it will we be able to claim a full understanding of epigenesis.

Now perhaps we can see a little more clearly how the various components of theoretical biology ought to be fitted together. Take for example the problems connected with the origin of life. Pattee's work, and the abstract models being studied by Arbib, are concerned with complementary aspects of the matter: the physical limitations and the logical demands which must be satisfied by hereditary systems. To understand the origin of life it is not enough to know what chemical substances (proteins, nucleic acids, sugars) can be formed under pre-biological conditions, interesting as such information is. Indeed, Katchalsky has argued that our ability to produce such materials by sparking gas mixtures means that if we find them in objects arriving from extra-terrestrial regions we *cannot* conclude that they are of biological origin! Not until we find clear traces of functional programs shall we be able to infer the existence of life on other planets.

Or again, take the problems of cellular metabolism. Here the chemical engineer's role is to provide a set of principles which will enable us to see how far a complicated chemical system can be made to regulate itself without continual intervention from outside, and how far, in order to achieve a specified pattern of behaviour, it must be directly controlled by an operator or a *program* of operating instructions. At certain stages in the life of a cell—probably during the process of cell division, for example—it may be that things will not go according to plan without direct reference to the DNA; whereas at other times the cell may be able to carry on quite automatically using the enzymes which it has already synthesized.

Turning to morphogenesis, we may note a striking parallel between the concept of a repressor or de-repressor in the biological case and the idea of a 'conditional jump' in computer programming. In a high-level computing language there is a facility for labelling chosen points in the program and later returning to such points—or not—according to the outcome of other operations of the program. 'If such-and-such conditions are satisfied, then return to (or advance to) L' is a typical conditional jump. 'If such-and-such an enzyme or substrate is present, then start making this other one' would be the logically equivalent instruction for a cell. The biological case is, of course, extremely complicated in that control can be exercised at various levels–by reference back to the genetic material or in the cytoplasm; but I feel that the analogy with the man-made program is bound to be illuminating if one can explore it more thoroughly.

At the evolutionary level, in the Darwinian spirit, we can discern two kinds of problem: how variation occurs, and how selection operates upon it. We now know that the point mutation is by no means the only mechanism of genetic variation; phenomena such as lysogeny–a close analogue of 'subroutine borrowing'–and gene duplication–to which there is no obvious analogue in human invention (or is there?)–are beginning to resolve some of the most formidable problems about the rate of evolutionary change. When it comes to selection we are faced with issues which are already the subject of much useful work in classical evolutionary theory, and which are beginning to be tackled by the powerful statistical methods which Kerner outlined for us.

Finally, on a rather different level, one may try to see whether the theory of programs can be fruitfully applied to one of the most lively areas of present research–the functioning of the central nervous systems of animals. (The word 'function', incidentally, has a message for us: to the mathematician a function is a rule for generating one set of entities from another, and there is a very real sense in which biological function means, basically, the same thing. To a programmer a function is nothing more or less than a subroutine in his program; to the biologist the function of an organ is nothing more or less than the operational contribution which it makes to the life of the organism. The concept of biological function need not, therefore, be a loose and undisciplined concept, if the concept is employed always *within* a wider context. It may not mean much to ask what is the function of an elephant, but it does make sense to discuss the function of its trunk.) The difference between a genetic program and the program which an individual implements by his behaviour is that the former is constant for the individual over his life span whereas the latter is being modified all the time in response to his experience. Evolution is the development of genetic programs; learning (a faculty of much less general occurrence) is the development of an individual's own behavioural program. The task of the psychologist is to define the relation between the two. In both cases the program is modified by the data on which it operates–using the word 'data' to include not only sensations but also environmental pressures. And in both cases the overall result is that the programs 'rewrite themselves', a phenomenon which it would be very nice to be able to reproduce in the world of man-made computation.

To sum up this line of thought: the organic thing about organisms is that they organize themselves and their environment in relation to themselves. Organization (in this active sense) is the following out of a program. Just as a program consists in the evaluation of a series of mathematical functions, so an organism lives by the performance of its various biological functions. If—as I believe—physics and chemistry are conceptually inadequate as a theoretical framework for biology, it is because they lack the concept of function, and hence that of organization. In a sense the ideas of structure and of function are complementary in Bohr's sense of the word; a clock is not just a rather curious distribution of matter, it is a device for telling the time. This conceptual deficiency of physics and chemistry is not, however, shared by the engineering sciences; perhaps, therefore, we should give the engineers, and in particular the computer scientists, more of a say in the formulation of Theoretical Biology.

## Comments by C.H.Waddington

I should like to make three comments—

1. *'The problems of biology are all to do with program.'* So far as I can see I agree a hundred per cent, but is there perhaps something more concealed in this statement than was included in my precirculated memorandum (Prolegomena, p. 8) 'that in the transition from the zygote to the adult the "information" is not merely being transcribed and translated, but is operating as instructions—if you want to put it in fancy jargon, as "algorithms".' The Waddington-Thomist concept of a chreod is a structure in a vector field—but a field under the control of programs.

2. *'But is the descriptive power (of dynamic topology) equalled by its explanatory powers?'* The examples Christopher gives, to turn this question into a 'nasty' one, seem to me unfair. Of course if 'thinkers in chreods', like myself and René, come across situations in which some strains of bacteria can live in higher concentrations of sulphathiazole than others we can't at first say anything more than 'its chreod has terminated'—but nor could conventional biochemists say anything more than that 'the concentration of a deleterious drug has risen above what its resistance can deal with', which amounts to the same thing. The real point about the explanatory power of the notion of chreods is to ask 'Has this idea suggested a more plausible explanation for any category of phenomena than we had before?' I should argue that it has done so by suggesting the (experimentally confirmed) hypothesis of genetic assimilation—namely that selection of a population for the capacity for the modification of a chreod in relation to environmental stresses will lead to the establishment of a chreod

which leads to the selected-for end result more or less independently of the environment. This provides a convincing explanation of all those cases of adaptations which look as though they could have been produced by physiological responses to environmental circumstances but which turn out to be 'genetically assimilated' to a degree which renders them almost independent of the impinging environment. The Mendelian-Darwinist theory of evolution can then stretch out its hand to take in the whole body of phenomena in which Lamarckism has found its main support. And the process is–apart from some pre-Mendelian and today almost uninterpretable speculations attributed nationalistically in America to Baldwin and in Britain to Lloyd Morgan–a radically new one. This is, so far as I know, the only new evolutionary process which has been discovered since the cytologists learned about allo-polyploidy and other mechanisms of chromosomal evolution in the twenties and thirties. What more can you demand of a piece of theoretical science ?

3. *'The overall result is that the programs "rewrite themselves".'* This seems to me to be the pay-off line in Christopher's contribution–and see the remarks made by Arbib in his synopsis of the second meeting (p. 327). But here Christopher throws it away as though it were a mere aside. The point is, is it a characteristic of biological systems that they rewrite their own programs ? The chreod approach emphasizes only the point that they rewrite them to the extent that they improve their error-correcting ability. Longuet-Higgins is, I think, suggesting something much more radical and far reaching. To build up an error-correcting chreod is after all no more than to preserve the *status quo.* The system performs a function akin to memory. Is not Christopher perhaps hinting at the possibility that biological systems may be able to rewrite their programs, not merely to incorporate the past, but in such a way as to be not inappropriate for a future which is speculatively, but not irrationally, forecastable ? And, if this is his meaning, would this be so far from my suggestion (*Prolegomena,* p. 21) that a long-term future in evolution goes to those biological systems which have both accumulated for themselves a good Bridge hand and developed strategies which tell them when to 'pass'–drawing back into hibernation or some other mere subsistence ecological niche–or aim to keep up their sleeves some other defensive trump card, such as an efficient antibody-production mechanism ? The point of being able to rewrite your programme is not to keep on doing it all the time in relation to every changing breeze–a Lamarckian exercise–but to do it only as you graduate from the amateur to the competition, and finally to the world champion, class of games players.

## Reply by C. Longuet-Higgins

1. It was good to see Wad, in the *Prolegomena,* promoting the idea that the right way of thinking about morphogenesis was in terms, not so much of information as of *algorithms* (a fancy word, perhaps, but a very precise one!). But in my essay I said more: '... the problems of biology are *all* to do with programs.' Biogenesis, evolution, morphogenesis, cerebral function–the lot.*

2. No, I was trying hard to be fair, but at the same time to wave a red flag. Where Hinshelwood went off the rails (in the opinion of many of my biological friends) was that he thought of the bacterial cell *just* as a delicately balanced homeostatic physicochemical system, ignoring the fact that it is under the control of a mutable set of genetic instructions. The trouble with Hinshelwood's account of bacterial adaptation was–so it seems to me–that he tried to make do with chreods *alone,* without reference to the fact that what determines their observable forms in real biological systems is the controlling genotype.†

3. I'm glad Wad has put his finger on that sentence, because I certainly wouldn't like it to be regarded as an 'aside'. Plainly *some* biological programs are not only adaptive but *predictive;* certain types of learning undoubtedly involve inductive generalization, and one must not overlook the possibility that other sorts of biological development also involve a predictive element. The only problem is one of mechanism; how could such a notion be fitted in with orthodox evolutionary theory? Or was Darwin wrong in supposing that the variations on which natural selection operates are random–with no particular rhyme or reason?‡

*Footnotes by C.H.W.*

* *Touché*–I agree.

† The concept of a chreod is capable of very wide application, as René Thom has shown. But I introduced it (*Strategy of the Genes*) in the context of the epigenetic working out of a set of genetic programs.

‡ Is it perhaps just because the raw materials offered to natural selection are produced by random mutations that it is possible to select adaptive mechanisms, which are 'predictive' because they do more than is immediately necessary? (e.g. the vertebrate immunological system).

# On how so little information controls so much Behaviour

R. L. Gregory
University of Edinburgh

Perhaps the most fundamental question in the whole field of experimental psychology is: How far is behaviour controlled by currently available sensory information and how far by information already stored in the central nervous system ? Considering the origin of neurally stored information, we believe that this has only two origins: (1) ancestral disasters, changing neural structure according to the principles and processes of other phylogenetic changes occurring by natural selection; (2) previous sensory experience of the individual, stored as 'memory'. We may call these two ways of gaining stored information phylogenetic and ontogenetic learning respectively.

It is important to distinguish two quite different kinds of stored information. We learn *skills* and *events.* Some skills (e.g. walking, swimming, fighting) may be inherited (though often showing as behaviour only after sufficient maturation) and so are examples of gaining information phylogenetically; while learning or storing *particular* events is always ontogenetic. For examples of inherited skills, babies walk without special training at about fifteen months, and as Coghill [1] showed, salamanders kept from all movement by anaesthesia will nevertheless swim normally as soon as allowed, once the neural connections of the spinal cord (visible in the living animal) are complete. For examples of learned skills we may take games such as tennis, piano playing and chess. We may be able to recall the odd particular games or concerts, but as skills it is not individual past events which are stored, but rather appropriate behaviour and strategies which give more or less complete success in later similar situations. Evidently crucial generalized features of the original situation are stored and used when appropriate. But sometimes stored features are used when inappropriate: then we have an example of 'negative transfer of training'–for example playing table tennis with the straight arm movements appropriate to tennis. This serves as a handicap.

It is an open question just how far individual events are stored as such, and how far they have to be 'constructed' for recall (cf. [2]). What is certain is that information gained phylogenetically is always of the general 'skill' kind. We are not able to recall individual events experienced by our ancestors.

R. L. Gregory

We know quite a lot about the stages by which skills are learned by individuals. I would like to suggest that this can provide clues to the nature of how behaviour is controlled by sensory information. It suggests that control is not direct, except in the special cases of reflexes, but is via internal neural models of reality. These internal models are essential for skills—including perception of the external world.

▶ *The Learning of skills.* It has been clear ever since the experiments of Blodgett [3] that 'latent' learning occurs—that is, some information storage which does not at once show itself in behaviour nevertheless occurs during the early stages of developing a skill. We find two features of learning curves characteristic of ontogenetic skill learning: first, in learning discriminations—which seem vital to 'map the ground' in the first stages of learning—learning curves are positively accelerated; there being at first no progress, then later progress appears at an increasing rate. Experiments have shown that the animal (generally a rat) is responding to other, and it turns out irrelevant, features of the situation. Secondly, learning curves of skill show marked 'plateaux', during which no progress is observed but each plateau is followed by a sudden jump in performance, associated with a different strategy. In learning Morse code, typing, or the piano, increase in speed of performance occurs in steps as the input is handled in larger and larger units. Thus in typing, while each letter remains a unit, speed is limited to about two letters per second; but later, letter groups up to whole words and finally groups of words become the neural units. Speed is then far greater than is possible with the maximum decision rate of about 0·5 sec per decision possible for the human neural system. Lashley [4] has described the process for piano playing:

'The finger strokes of a musician may reach sixteen per second in passages which call for a definite and changing order of successive finger movements. The succession of movements is too quick even for visual reaction time. In rapid sight reading it is impossible to read the individual notes of an arpeggio. The notes must be seen in groups, and it is actually easier to read chords simultaneously and to translate them into temporal sequence than to read successive notes in an arpeggio as usually written.'

This grouping of what is at first discrete inputs is however done at the cost of complete flexibility. Unusual combinations of inputs may be missed, or accepted as though they were in a more usual order, with consequent errors. Random music is very difficult to play and random letters very difficult to type.

A system which makes use of the redundancy, in space and time, of the real world has the following advantages:

▶ *Advantages for a system utilizing input redundancy*

1 It can achieve high performance with limited information transmission rate. (It is estimated that human transmission rate is only about 12 bits/second.) The gain results because perception of objects–which are always redundant–requires identification of only certain key features of each object. Some kind of search strategy for these features would save a great deal of processing time for object recognition. (This is open to experimental investigation and has implications to pattern recognition, which is *not* the same as object recognition, which is perhaps an artificial concept.)

2 It is essentially predictive. In suitable circumstances it can cut reaction time to zero. (Experimental situations for demonstrating reaction time are somewhat artificial, seldom occurring during actual skills, such as driving, typing, piano playing, etc.)

3 It can continue to function in the temporary absence of any input, e.g. turning the music page, blinking, or sneezing while driving.

Loss of input is very different from loss of output control (e.g. the steering wheel coming off), and this difference seems important for investigating these internal selected groupings, or as we call them 'models', of reality.

4 It can continue to function when the input changes in kind. Thus in maze learning, rats can continue to run a maze once learned though each sensory input in turn is denied it–vision, smell, kineasthetics, etc.

(There is an important implication here for interpretations of brain ablation experiments, for so-called 'mass action' might appear though each sensory and corresponding learning system were precisely located in the brain, for the other specific systems might take over after destruction. The fact that rats can swim a flooded maze after learning to run it dry is particularly striking, for evidently it is *not* primarily patterns of motor movements which are learned. This is important evidence for cognitive learning at the level of the rat, and we believe that it gets even more important higher up the phylogenetic scale.)

5 It can extract signals from 'noise'. If the internal models are highly redundant, they can be called up with minimal sensory information. This means that the models can enormously improve the effective signal/noise ratio of sensory systems.

6 Provided a particular situation is similar to the situations for which a 'model' was developed, behaviour will generally be appropriate. This, in the language of experimental psychology, is 'positive transfer of training'.

We come now, however, to disadvantages of conceivable systems (including robots) in which behaviour is based on internal models.

▸ *Disadvantages of internal model systems*

1 When the current situation is sufficiently similar to past situations which have been selected and combined to give an internal model, but the current situation differs in crucial respects, then the system will *be systematically misled by its model.* This is 'negative transfer'.

2 Internal model systems will be essentially conservative–showing inertial drag to change–for internal models must reflect the past rather than the present. (This implies that rapid change of environment or social groups is biologically dangerous, and of course it favours young members of such groups.)

▸ *Further implications of internal models.* Since no model can be complete, and few if any are entirely accurate in what they represent, biological or computer systems employing internal models can always be fooled. They are fooled when characteristics which they accept for selecting a model occur in atypical situations. It is always possible that a wildly wrong model may be selected when this happens. It will happen most often when only a few selection characteristics are demanded or are available to the system. We know from many learning and perceptual experiments that there are great individual differences in what kinds of features are demanded. (In general 'brighter' individual animals, such as rats, demand where possible non-visual features while the dimmer brethren are largely content with visual features. This is curious in the case of the rat, which is generally regarded as rather a 'non-visual' animal.)

A model may be selected on purely visual data, but once selected it is generally used for non-visual predictions. Thus, in driving a car, the road surface is 'read off' the retinal image: what matters is whether the road is slippery. Slipperiness, though not a property of images, can be read from the retinal image.

In general, the eye's images are only biologically important in so far as non-optical features can be read from the internal models they select. Images are merely patches of light–which cannot be eaten or be dangerous–but they serve as symbols for selecting internal models which include non-visual features vital to survival. It is this reading of object characteristics from images that *is* visual perception.

Gross errors may occur when a wrong model is selected. Errors of scale can also occur; and these, I believe, are the familiar perceptual distortion illusions. These illusions are interesting because they can tell us something of how internal models are made to fit the precise state of affairs in the outside world (cf. [5]).

We cannot suppose that there are as many internal models as there are perceptible objects *of all sizes, distances, and positions in space.* But it is important for the models to represent the current sizes, distances, and positions of external objects if they are to mediate appropriate behaviour. To solve this problem we may suppose that the models are flexible. They can be adjusted to fit reality. They are adjusted by 'size scaling' visual features, such as perspective convergence of lines—though not always appropriately.

In the absence of any available scale-setting data, perception is determined by average sizes and distances. These are modified by 'scale-setting' sensory information when available. When scale-setting information is inappropriate to the prevailing reality, then perception is systematically distorted. On this view we can use distortion illusions as quite basic research tools. In the Muller-Lyer, Hering, or Orbison visual illusions, typical perspective depth features are presented on a flat plane. Features which would be distant if these figures were truly three-dimensional are expanded in the flat illusion figures. Thus expansion is normally appropriate—since it is object size and not retinal image size which is biologically important—but here the system is misled by the scaling information and systematic distortions occur. By studying these distortions we can discover experimentally just how flexible the internal models are; what sorts of information are used to give object scale, and also something of how internal models are built by perceptual learning.

Biologically important features of the world must be read from available sensory information. To be useful, visual features must be related to the weight, hardness, and chemical properties of objects which have to be handled or eaten. Now it is well known that a small object of the same weight as a larger object feels up to fifty per cent heavier. This is the 'size-weight' illusion. Vision selects a model calling up appropriate muscle power for lifting the weight, but when the internal model is inappropriate the power called up is inappropriate—and we suffer an illusion corresponding to the error.

The weight setting adopted by the nervous system in the absence of information of the size of the weight corresponds to a density of one—about the average density of common objects.

It is interesting that scale distortion illusions are (a) similar in different individuals from the same culture, but differ somewhat in different cultures when the available characteristic features are different, and (b) are very slow to change in adults. (On the other hand, systematic changing of *all* inputs, with e.g. distortion glasses, does produce rapid appropriate adaptation in adult humans.)

In a case of adult recovery from infant blindness we found [6] that the newly available inputs were only accepted when they could be directly related to previous touch experience. In our present terms, vision was only possible after the corneal grafts when visual data could select *already available* internal models, based on earlier touch experience. Building new models was very slow, taking a year or more. The use of vision for size-scaling occurred within a few months, the initial distortions being very great in situations where touch or other information had not previously been brought to bear—as when looking at the ground from a high window, when the ground appeared almost within touch range though actually forty feet below. The normal systematic distortion illusions did not occur: I suppose that there was no 'negative transfer' of perceptual learning where there had been no opportunity for learning of the normal size-scaling features, such as perspective.

▶ *Sensory discrimination and the appropriateness of models.* I have distinguished between (a) selecting models according to sensory information, and (b) size-scaling models to fit the orientation, size, and distance of external objects.

Now let us consider an experimental situation which may tell us something about the 'engineering' nature of the models in the brain. The experimental question is: What happens to sensory *discrimination* when there is a scale distortion?

Consider the following paradigm experiment. We have two sets of weights, such as tins filled with lead shot. Each set consists of say seven tins all of a certain size, while the other set has seven tins each of which is, say, twice the volume of the first set. Each set has a tin of weight, in grams, 85, 90, 95, 100, 105, 110, 115. The 100 gram weight in each set is the standard, and the task is to compare the other weights in the same set with this standard and try to distinguish them as heavier or lighter. The tins are fitted with the same size handles for lifting to keep the touch inputs constant except for weight. Is the discrimination the same for the set of *apparently* heavier weights but which are in fact the same weights? The answer is that discrimination is *worse* for weights either apparently *heavier* or *lighter* than weights having a specific gravity of about one [7]. Why should this be so?

Suppose that sensory data are compared with the current internal model—as they must be to be useful. Now if it is not only *compared* with it, but *balanced against it,* then we derive further advantages of employing internal models. We then have systems like Wheatstone bridges, and these have useful properties. Bridge circuits are especially good (a) over a very large input intensity range and (b) with components subject to drift. Now it is striking how large an intensity range sensory systems cover ($1:10^5$ or even $1:10^6$), and the biological components are subject to far more drift than would be tolerated by engineers in our technology confronted with similar problems. So balanced bridge circuits seem a good engineering choice in the biological situation.

Consider a Wheatstone bridge in which the input signals provide one arm, and the prevailing internal model the opposed arm against which the input is balanced. Now the internal arm is part of the model–and will be set wrongly in a scale distortion illusion. In the size/weight illusion, visual information has set the weight arm wrongly. This means that the bridge will not balance. The illusion is the misbalance of the bridge. Now an engineer's bridge which is not balanced suffers in its ability to discriminate changes in its input, for it is no longer a null system but relies on scale readings of the galvanometer or other misbalance detector. Thus the supposed biological system gives just what a practical engineer's bridge would give–loss of intensity discrimination associated with an error in balancing the bridge. This is some evidence that internal models form arms of bridge circuits in the brain.

▸ *Speculations on mental events–normal and abnormal.* On this general view, perception is not directly of sensory information but rather of the internal models selected by sensory information. Indeed, the current perception *is* the prevailing set of models.

There are well-known situations in which the sensory information calls up two or more incompatible internal models with equal probability. The best-known example is the spontaneously reversing Necker cube. The available information is insufficient to decide between rival internal models, and each comes to the fore in turn. It is interesting that in this case the addition of tactile information–provided by holding in the hand a luminous cube viewed in darkness–does not serve to abolish visual reversals, though it does reduce their rate of occurrence [8]. Evidently the visual internal model system is largely autonomous, though it is partly under the control of other senses. Visual size and distance can be set by other senses, especially touch. It is also worth noting that size scaling follows not only currently available sensory information, but also changes in the internal model. Thus, a luminous cube appears as a cube when seen correctly–though the further face is smaller at the retina–but as a truncated pyramid when depth-reversed. Here there is no change at all in the sensory input, only in the internal model, so the scale changes *with the model,* though the sensory information remains constant.

Generally, the internal model is reasonably complete and appropriate, but a wrong model may always be selected, and even if appropriate it may be wrongly scaled. We know from perceptual experiments in situations where only minimal information is available that both selection and scaling can be quite wrong. So it is a small step to say that in the absence of any sensory information entirely wild models might be called up. This could be the case in dreaming, and in drug or fatigue-induced hallucinations. Hallucinogenic drugs might call up internal models either by increasing cortical noise or by reducing the threshold criteria for acceptance of the stored models.

Abnormal conditions such as schizophrenia might be caused by inappropriate models being built in the first place, or by wrong selection criteria being employed. Greater knowledge of the processes and conditions for perceptual learning might have implications for psychiatry. If the models *are* our internal world, we should find out more about them.

▸ *Implications for the design of robots.* Devices which respond to sources of information are commonplace. There is no difficulty in arranging for a door to open itself when someone breaks a beam of light to a photocell. But such devices do not 'see' or 'perceive' in the sense that we do. Similarly, our reflex blink to a sudden bright light is not 'seeing', 'perceiving', or 'observing'.

Theories of perception (especially the Gestalt theory) lay far too much stress on sensory characteristics, giving insufficient weight to the vital point about perception: perception is geared to *objects*, for it is objects which are biologically important. Objects are dangerous or useful, food or disaster; but retinal images, and vibrations of the tympanum, are of no importance except to indicate the identity of external objects. The patterns of sensory activity are but symbols from which reality may be read. This involves far more than the recognition of patterns. Pattern recognition is only an early stage of perception, for objects are more than patterns, and it is objects that matter. Objects have all manner of vitally important properties which are seldom sensed, so current sensory information cannot be adequate for dealing with objects.

On this theory, perception allows behaviour to be appropriate to the hidden properties of objects, when the internal models sufficiently reflect their properties. This is very like the notion of a medical syndrome—a few spots may indicate the past, present, and future course of a disease, together with an appropriate strategy for dealing with it. Once recognized, the syndrome—or perceived objects—may be accepted for guiding the most complex behaviour with but little current information.

The special feature of perception is that it does not mediate behaviour directly from current sensory information, but always via internal models of reality–which themselves reflect the redundancy in space and in time of the external world. This is where perception differs from devices such as photocells actuating doors, or biological reflexes, for these give control directly from the inputs. They do not use the current information to call up appropriate models, giving information drawn from the past of the hidden features of the present situation. The past is usually a reliable guide, and our memory contains vastly more information than can be transmitted in reasonable time by the sensory channels even when the relevant information is available–which is rarely the case.

One might be tempted to think that objects, as perceived, are no more than statistical groupings of sensed events–syndromes of sensation. But to say this is to miss a vital point. Sensed events are categorized also in terms of the use made of them. A book, for example, is seen as a single object. This is because we handle the collection of pages as one object. Sensory inputs are grouped according to the repertoire of behavioural skills of the owner of the perceptual system.

One man's object may be another's pattern–or be nothing but randomness.

This brings out the kind of difficulty we have in imagining the perceptual world of animals, or even people whose interests are very different from our own. It also has implications for designers of robots–machines to see and act on what they see. If they are to respond to objects via internal models–and all the biological advantages will apply to the machine–then its models must be appropriate to *its* sensory inputs and to *its* repertoire of actions. These will differ greatly from ours. But could we communicate with a robot having internal models very different from our own ? We should expect the same extreme difficulty that we have in trying to communicate with other animals or with schizophrenics. Even though we design and build our own robot, and know exactly how its circuits function communication could be impossible when its internal models are not ours.

▸ *The status of perceptual brain models.* We suppose that perceptual models are aggregates of data about objects, and about how objects behave and interact in various circumstances. Perceptual models bear a resemblance to hypotheses in science. We may think of sensory data suggesting, testing, and sometimes modifying perceptual models in much the same way that scientific data suggest, test, and modify theory and hypothesis in science. A precise comparison of perceptual processes with the logic and method of scientific inquiry could be highly rewarding.*

* This project was planned by Norwood Russell Hanson with the present writer, but tragically Russ Hanson was killed in his private plane.

We are concerned here with not only the logical but also the biological and the engineering status of brain models. Whatever they are, one thing is quite clear—they are not isomorphic pictures of external shapes. The Gestalt theory misses the point here, for all sorts of information about objects must be stored but pictures can only represent specific shapes and colours. Shape and colour have only indirect significance: what matters is whether the object is useful, a threat, or food. It is non-optical properties that are important. When we look at a picture, we can read all kinds of significance beyond mere shape and colour. The picture serves to evoke our internal models, which have been developed by handling objects, so that non-optical features have become associated. Similarly the pictures in the eye, the retinal images, only have significance when related to non-optical properties of objects. Without such correlations all pictures, including retinal images, would be meaningless—mere patterns. The artist by presenting selected visual features plays games with our internal brain models, and may quite drastically change them by evoking new associations. It is clear that the brain models cannot be logically at all like pictures, for though pictures can evoke models, their appropriateness is in terms of objects, not pictures, which in themselves are utterly trivial.

The computer engineer will ask: are these supposed brain models digital or analogue ? This distinction is important to the engineer, because analogue and digital systems have very different design features and advantages and disadvantages for various purposes. Indeed, it is possible to make an informed guess as to which system is adopted by the brain in terms of speed of operation, types of errors, and other characteristics typical of analogue or digital engineering systems cf., [9]. The engineering distinction arises from the fact that in practice analogue systems work continuously but digital systems work in precisely defined discrete steps. This difference is immensely important to the kinds of circuits or mechanical systems used, and vital practical implications follow. Discontinuous systems can have much higher reliability in the presence of 'noise' disturbance. Analogue devices can have much faster data transmission rates, but their precision is limited to around 0·1–1·0%. There is no limit in principle to the number of significant figures obtainable from a digital computer if it has space enough and time.

Because of the clear engineering distinction between continuous and discontinuous systems, there is a temptation to define analogue in terms of continuous, and digital in terms of discontinuous. But this will not do. We can imagine click stops fitted to a slide rule: this would make it discontinuous, but it would still be an analogue device. We must seek some deeper distinction.

The point, surely, is that analogue and digital systems both represent things, and so in both cases their internal states represent something else. The essential difference between them is not in their engineering, but rather that they represent logically different kinds of things. The distinction is between *actual events in the world,* which occur continuously, and *symbolic representations of events,* which are always discontinuous. (Even the continuous functions of differential calculus have to be handled as though they were discrete steps.)

A continuous computing device can work without going through the steps of an analytical or mathematical procedure. A digital device, on the other hand, has to work through the steps of an appropriate mathematical or logical system. This means that continuous computers functioning directly from input variables necessarily lack power of analysis, but they can work as fast as the changes in their inputs—and so are ideal for real-time computing systems provided high accuracy is not required. The perceptual brain must work in real time, and it does not need the accuracy or the analytical power of a digital system following the symbolic steps of a mathematical treatment of the situation. Perceptual motor performance only has an accuracy of around one per cent. It seems that a continuous analogue system is appropriate for perceptual data processing. This holds both for actual brains and future robots.

Perceptual learning involves not the development of software programmes for programming a digital system according to mathematical analyses of the behaviour of objects, but rather by developing quite crude continuous analogues of the organism's input-output functions, in the presence of recognized objects. From the point of view of the perceptual computer, objects represent transfer functions between the organism's input and output in various situations. Behaviour is given by selecting the appropriate transfer functions, which are stored in the perceptual model elicited by the recognized object.

To build a seeing machine, we must provide more than an 'eye' and a computer. It must have limbs, or the equivalent, to discover non-optical properties of objects for its eyes' images to take on significance in terms of objects and not merely patterns. The computer must work in real time. It need not work according to analytical symbolic descriptions of the physical world—all it requires are quite crude analogues of input-output functions selected by distinguishing features of objects. These collections of transfer functions give appropriate behaviour through predictions, made possible by the redundancy of the world of objects. Ultimately the perceptual brain models reflect the redundancy of the external world—when they do so correctly we see aspects of reality without illusion.

References

1. Coghill, G.E. *Anatomy and the Problem of Behaviour* (C.U.P., 1929).
2. Bartlett, F.C. *Remembering* (C.U.P., 1932).
3. Blodgett, H.C. *Univ. Calif. Publ. Psychol. 4* (1929), 113–34.
4. Lashley, K.S., in (Jeffress, L.A. ed.), *Cerebral Mechanisms in Behaviour* (Wiley, N.Y., 1951).
5. Gregory, R.L. *Nature, 199* (1963), 678–80.
6. Gregory, R.L. and Wallace, Jean G. *Recovery from Early Blindness: A Case Study* (Heffers, Cambridge, 1963).
7. Gregory, R.L. and Ross, Helen E. *Percept. and Motor Skills, 24* (1967) 1127–30.
8. Shopland, C. and Gregory, R.L. *Quart. J. Exp. Psychol., 26* (1964), 66–70.
9. Gregory, R.L. *Brit. J. Phil. Science, 4* (1953), 15, 192–7.

## Comments by C. H. Waddington

Richard Gregory asks the question—a very good one—how so little information controls so much behaviour, and answers it by arguing that the information triggers off pre-existing models of objects in the perceivable world which have already been formed within the brain. This is an attractive thesis, but I find the word 'model', with its suggestion of miniature clockwork railway locomotives and other object-like hardware, rather inhibiting to an imaginative grasp of what is involved. I should prefer to think of these 'models' as chreods. If X denotes a set of (informationally inadequate) external stimuli, these are mapped into the brain-states Y by some function $f:X \rightarrow Y$ dependent on the neural connections between brain and sense organ. Now we have to suppose that the space of brain-states is divided into a number of domains, each characterized by a vector field which, at any given time, is dominated by a particular attractor (cf. René Thom's discussion of words as chreods in his forthcoming *Stabilité Structurelle et Biologie*). The information contained in the brain-state y which corresponds to the external stimulus x then has added to it the information embodied in the chreod (vector field plus attractor) into which it falls. In such a picture it is easy to realize the provisional character of the 'internal information', i.e. the boundaries of the various domains, the characteristics of the vector fields and attractors, and to appreciate that this must be subject to continual change and adaptation through processes of learning.

# Cognitive Processes in Physics and Physiology

Karl Kornacker
Massachusetts Institute of Technology

It at first seems trivial to note that cognitive processes are involved whenever we perform experiments. Perhaps the clearest non-trivial example from physics is the measurement of heat, where the observed process stands in direct contradiction to the universal work-energy theorem of mechanics, thus forcing the formulation of statistical mechanical theories of macroscopic observation. The recording of a single quantum mechanical event is another example from physics where the experimenter's cognitive process seems to be explicitly involved. By mentioning these two examples I do not mean to imply that either is understood.

It appears to me that a striking aspect of physiological systems is that cognitive processes occur inside the system. The recognition of patterns by neurons in the central nervous system is an obvious case in point. Less obvious cases are enzyme-substrate specificity and the active transport of ions across membranes. In this paper I will outline a general theory of the cognitive process and show how it applies to heat, active transport, and neuronal pattern recognition.

In brief I argue that cognition, or the recognition of form, always comes down to the calculation of correlations. The usual form of a correlation calculation for the variables A and B is $(\langle AB \rangle - \langle A \rangle \langle B \rangle)$, where $\langle \; \rangle$ denotes an averaging operation

$$\langle A \rangle = \int A(x)p(x)dx \qquad \ldots. (1)$$

$$\langle B \rangle = \int B(y)p(y)dy \qquad \ldots. (2)$$

$$\langle AB \rangle = \int\int A(x)B(y)p(x,y)dxdy \qquad \ldots. (3)$$

and p denotes a density function. Randomness is not required. Note that a non-zero value of $(\langle AB \rangle - \langle A \rangle \langle B \rangle)$ implies that $(p(x,y) - p(x)p(y))$ is non-zero for some x and y, but that the converse may fail, allowing some correlations to pass undetected.

The quantity $(\langle AB \rangle - \langle A \rangle \langle B \rangle)$ is inherently macroscopic, since it vanishes if A, B, and AB are measured directly. Furthermore the calculation of this quantity requires non-linear interactions (multiplication) within the cognitive device. Let us now consider the relevance of these considerations to heat; active transport, and neuronal pattern recognition.

Despite the tradition in statistical mechanics which follows the ensemble theory of Maxwell and Gibbs, a tradition which emphasizes the non-reproducibility of molecular states in the repetition of thermodynamic experiments, the fact remains that spatio-temporal averaging completely defines the thermodynamic (macroscopic) point of view. For example, pressure is the finite spatio-temporal average of mechanical forces exerted by molecules on the transducer surface of a pressure gauge. The need for spatio-temporal averaging is basically the need to increase the signal-to-noise ratio of the measurement. Spatial and temporal averaging are complementary in the sense that, for a given acceptable noise level, the smaller the transducer area the longer must be its response time. If for example one wishes to use local equilibrium assumptions in a general theory of irreversible processes, then clearly the implied local measuring devices must have a long response time.

Now suppose that the time-averaging operation performed by a local thermodynamic measuring device is invariant under translations in time, that is

$$\langle A \rangle = \frac{\int_{-\infty}^{t} f(t'-t)A(t')dt'}{\int_{-\infty}^{t} f(t'-t)dt'} \qquad \ldots (4)$$

For example, in viscous mechanical damping, or capacitative electrical smoothing,

$$f = e^{(t'-t)/\tau} \qquad \ldots (5)$$

where $\tau$ is the time constant (response time). For any such invariant averaging operator which in addition has 'finite memory', meaning that $f(-\infty)A(-\infty)$ is zero, it is easy to show by direct calculation that

$$\frac{d\langle A \rangle}{dt} = \left\langle \frac{dA}{dt'} \right\rangle \qquad \ldots (6)$$

Now consider the effects of time-averaging on the work-energy relation. The relation itself is

$$\frac{dE}{dt} = F \cdot v \qquad \ldots (7)$$

where the force vector F and the velocity vector v are considered in the system phase space, so that the product $F \cdot v$ includes summation over all coordinates of all particles. For invariant finite memory time averaging we have, combining equations 6 and 7:

$$\frac{d\langle E \rangle}{dt} = \langle F \cdot v \rangle \qquad \ldots (8)$$

On the other hand, we calculate the macroscopic rate of work done on the system as

$$\frac{dW}{dt} = \langle F \rangle \cdot \langle v \rangle \qquad \ldots (9)$$

so that we recognize the dynamic cross correlation $(\langle F\cdot v\rangle - \langle F\rangle\cdot\langle v\rangle)$ as the rate of heat flow dQ/dt into the system, based on the first law definition

$$\frac{dQ}{dt}=\frac{d\langle E\rangle}{dt}-\frac{dW}{dt} \quad \text{....(10)}$$

In this case the time-averaging is performed by the measuring devices and the multiplication is performed by the observer.

Active transport, like heat, has no microscopic counterpart. By definition active transport is a net transmembrane flow which is not caused by any macroscopic energy gradient. The energy for active transport comes from chemical reactions which have no macroscopic spatial direction, and the coupling of this energy to the macroscopically directed flow takes place in an anisotropic membrane structure. I will now show that such coupling requires a dynamic cross correlation calculation in the membrane.

The electrical conductance of a passive membrane, with respect to the $\gamma$ charged species, is defined as

$$g_\gamma=\frac{I_\gamma}{\mu_\gamma} \quad \text{....(11)}$$

where $I_\gamma$ is the transmembrane current carried by the species and $\mu_\gamma$ is the electrochemical potential difference for the species. If there is active transport of the $\gamma$ species, then, treating active transport as if it were a current generator, the modified conductance equation is

$$g_\gamma=\frac{I_\gamma-J_\gamma}{\mu_\gamma} \quad \text{....(12)}$$

where $J_\gamma$ is the active transport current.

Let us now consider explicitly the electrical time-averaging performed by biological membranes. The averaging is due to the membrane capacitance (about one microfarad/cm$^2$) which lies across the passive membrane resistance (about one thousand ohms/cm$^2$), giving a time constant of about one millisecond. Introducing the fluctuation potential

$$e_\gamma=\mu_\gamma-\langle\mu_\gamma\rangle \quad \text{....(13)}$$

Equation 11 may be rewritten as

$$I_\gamma=g_\gamma e_\gamma+g_\gamma\langle\mu_\gamma\rangle \quad \text{....(14)}$$

Averaging equation 14 gives

$$\langle I_\gamma\rangle=\langle g_\gamma e_\gamma\rangle+\langle g_\gamma\rangle\langle\mu_\gamma\rangle \quad \text{....(15)}$$

which, rearranged, becomes

$$\langle g_\gamma\rangle=\frac{\langle I_\gamma\rangle-\langle g_\gamma e_\gamma\rangle}{\langle\mu_\gamma\rangle} \quad \text{....(16)}$$

Comparing equation 16 with equation 12 allows us to identify

$$J_\gamma=\langle e_\gamma g_\gamma\rangle=\langle\mu_\gamma g_\gamma\rangle-\langle\mu_\gamma\rangle\langle g_\gamma\rangle \quad \text{....(17)}$$

so that the active transport current is a calculated cross correlation between the fluctuations of conductance and electrochemical potential. In this case the calculation is done by the membrane. The molecular mechanisms which could generate the required correlated fluctuations are considered in detail elsewhere [1].

Let us finally consider how it would be possible for neurons to calculate cross correlations between various synaptic inputs. From the physiology of synapses we suspect that excitatory actions generally add with each other, inhibitory actions generally add with each other, and inhibition generally divides excitation. The nervous system must therefore find a special way to obtain subtraction and multiplication if it is to calculate any correlations. The following method is not unique, but its physiological implication is so striking that further close examination seems warranted.

We note first that an exponentially weighted averager, as in equation 5, has the property that

$$\left\langle \frac{\partial A}{\partial s} \right\rangle \quad \frac{A-\langle A\rangle}{\sigma} \qquad \ldots . (18)$$

If s is time, then $\sigma$ is the temporal averaging time constant $\gamma$; if s is space, then $\sigma$ is the spatial averaging space constant $\lambda$. Equation 18 shows that the required subtraction can be obtained if the gradients or velocities of certain physiologically important quantities, rather than the quantities themselves, are fed into the neuronal correlation calculator. A correlation calculation of the form $\langle (A-\langle A\rangle) B\rangle$ could then be performed without explicit subtraction.

As regards the required neuronal generation of multiplication, two successive divisive interactions would accomplish this. In other words, disinhibition is a multiplicative form of excitation. I propose therefore that the central nervous system could be organized as an analogue computer for calculating spatio-temporal correlations between various neuronal activities. The above arguments might then point to the fundamental functional significance of disinhibition and mathematical differentiation in the nervous system. The cerebellum, for example, seems to function as an organ of disinhibition and might therefore be expected to connect to major correlation calculators (cognitive centres).

## References

1. Kornacker, K. in (Douben, R. M., ed.) *Biological Membranes* (Little, Brown Co, in press.)

# A General Property of Hierarchies

Ted Bastin
Cambridge Language Research Unit

At the 1967 meeting at Bellagio the idea of a hierarchy frequently entered the discussions. This was to be expected. If one thinks at all about biological systems, it is natural to simplify the problem of getting a comprehensible picture of what is going on in them, by drawing boundaries around blocks of activities which have a fairly obvious coherence and autonomy; and having one such block exercise control over others, while being controlled by yet another or others. When we have done this, we have already imposed the concept of a hierarchy. Sometimes it is clear that the imposition of hierarchical structure is natural, because the 'blocks' are very apparent in their action. Other times it is extremely hard to find them at all, yet the hierarchical idea has such a grip on us that we still find biological systems of this second kind being thought of in terms of these blocks. This latter situation is particularly common in theories of brain physiology.

My task in this note is to think of hierarchical structure as abstractly as possible, in order to get a simple picture of the logical and mathematical relations that must exist between the 'blocks' if they are to constitute a hierarchy. The picture that I shall put forward will be remote from detailed biological experimentation, but this requires no apology, for, if it is really the case that biologists are in the habit of using hierarchical ideas as a conceptual framework for biological systems, then it is important to be as clear as we can what consequences this habit really imposes on us.

To pursue this task I shall use mathematical ideas from a hierarchy theory that was developed by Amson, Bastin, Parker-Rhodes, and Kilmister [1] for application to foundational questions in quantum theory, and I go back to the root from which the hierarchy idea sprang—namely the priestly and then political hierarchy—to provide a simple picture to which these mathematical ideas apply. I do not of course claim that this simple picture of a hierarchy formalized in this way is the only one that can be formalized, or even that the formalization in question is the only possible formalization of that picture. It is, however [2], usual in science, if one has a model or a theory—particularly one with some degree of mathematical articulation—which has been applied to the explanation of phenomena in some field, to try to apply it in other fields. Partly, doubtless, this is because it is natural to expect to be able to generalize; partly, again, the discovery of a piece of mathematics which fits the world in a new way is a rare event (a fact that is emphasized in the paradigm philosophy of Kuhn and that was disastrously overlooked in the old hypothetico-deductive view of the use of mathematics in science).

I shall call my picture the 'similar units model' (the 'units' being the discernibly separate blocks to which I have already referred). In operating with these units there is one question which hits us at the outset. We may be able to discern coherent entities, or 'blocks', or units, which control each other in a hierarchy, when we look at some biological–or other–system; but what is it, experimentally speaking, that we discern ? Bohm, at the Bellagio conference, answered that when we strip our observations to bare essentials what we discern is 'order', and he has written at length on the importance of this concept. He also maintains that 'order' is the colligating concept to which we should first look when we wish to assimilate ideas from biology to those from the foundations of physics. I think his position would be that order (defined within finite sets of recognizably similar entities) is a concept that really comes into its own as soon as we think in terms of hierarchical control, so that 'order' and 'hierarchy' are mutually supporting concepts. I am sure that this association is a correct one, and that it would prove a rewarding study to justify it from the general standpoint of the logical foundations of mathematics, but I cannot attempt this investigation here, and for the present I shall assume the validity of the association.*

A brief account of the stages which led to formalization of the similar unit model, though not of the mathematics, is given in appendix I. For my present purpose I need only the central idea which underlay the work and which became more definite as the formalization process progressed, which is this: If units in a hierarchy of control are similar in their intrinsic properties and differ only in their function, then the only possible distinction that can serve to classify them into levels (and hence delineate the hierarchy) arise through their behaving with different characteristic reaction times. Hence *levels of control must be distinguished by different time constants.*

There were indications of a similar idea in the remarks made by several speakers at Bellagio (notably Goodwin and Iberall), which fact possibly supports my belief that this characteristic may be a general property of control hierarchies. In

*At the Bellagio Conference I illustrated the concept of order in the context of Bohm's remarks by describing one particular *'ordering relation'*–namely that which we employ when we ascribe 3-dimensionality to the space continuum. This illustration is given as appendix II.

order to illustrate it I shall go back to the primitive case of a hierarchy from which the idea comes and to which we refer when we use the concept–namely to a political (originally more specifically a priestly) hierarchy of control or authority. ▶ The feudal system of medieval England provides an extremely simple idea of a hierarchy of command with its different levels of authority each consisting of a set of similarly privileged and similarly powered individuals (king, barons, knights, squires, yeomen, villeins) in descending order of authority. It is unlikely that this simple picture would ever have been accurate: some barons would matter more than other barons, and some would have more rapid or easy access to the ear of the king, and so on, so that the simple type of order specified by the hierarchy with the set of individuals at a given level having equivalent places in the structure (and hence unordered in respect of command) would be true, at best, only as a first approximation. Nevertheless the mere fact that ranks were devised and allotted to individuals demonstrates that the ideally simple pattern of control presented by the simple hierarchy was necessary, at any rate to enable those individuals who themselves formed the hierarchy to have a basic picture of its function, even if that picture had then to be modified in detailed ways.

This then is the picture that is to be generalized into a hierarchy of orders, with higher levels consisting of abstractions from the structure at lower levels, in order that the picture may serve as a useful theoretical model of aspects of nature. For such a model to give an essentially new way of thinking, moreover, which is not reducible again to a simple set of mechanical relations or orders at one level* the hierarchical structure must be a dynamical one, with changes at each level being dictated from other levels, and themselves dictating changes at other levels. This dynamical character is quite evident of course in the case of the feudal structure where there is constant conflict, or at any rate interaction, between the ways the individuals at a given level would act if left to themselves and the dictates of higher levels of command. (This conflict of interaction is not

*Bohm emphasized the 'non-mechanical' character of mechanically interdependent ordering relations, which follows indeed from the fact that we certainly have no mathematics at present that is capable of specifying such a scheme in its general form. (My similar units hierarchy model is a special case which will–I hope–give clues to treatments of greater generality.) Bohm argues that in so far as a mathematical specification can be found, explanation is to that extent reduced to one level. This view would mean that the phenomena of life always elude exact specification. This I find too extreme. More likely there is a class of mathematical constructions, a case of which is familiar to us as the mathematics of classical dynamics, which could not afford explanation at more than one level. However, I do not think that all mathematics has to be of this type, and in particular I think recursion theory is not.

necessarily one way, but I do not want at present to discuss the need–or lack of need–for a one-way restriction.) What is not evident–is indeed strange–is the idea that an individual at a given level–say a baron–could be an abstraction from the structure existing at lower levels. However, when we look into it this idea, too, is reasonable. The baron, as an individual, is no different essentially from individuals at all other levels. As a baron–a member of the level of barons; on the other hand–he has his particular status in recognition of the fact that he acts on behalf of, and with authority over, the whole segment of the total hierarchy that is under his command. Thus the concept 'baron' itself is an abstraction, it refers to the power of acting on behalf of a sub-hierarchy.

The situation can be clarified further if we imagine a being from another planet–a Martian, say–to visit a feudally structured society. We shall suppose that he is so different himself from the individuals composing the hierarchy that he sees no essential differences between them that accord either with the status they have in their own eyes or in those of the other individuals that compose the hierarchy. If he notices differences at all between the individuals (in circumstances where all seem to him completely remote from himself) he will be as likely as not to pick up differences that have nothing remotely to do with the hierarchy structure. Thus he may notice the height of the individuals, or their weight, or whether they wear blue or red clothes. Let us therefore idealize the position in which the Martian finds himself, by considering all individuals to be identical. This Martian will then have to tell them apart by their *function*. We will suppose that the Martian is familiar with the logic of control, so that he is all set to discover a hierarchical structure, with individuals at different levels. Also, we will suppose that he can impose limited constraints in the behaviour of any given individual to observe the effects of the constraint on the other individuals.

This latter power does not enable him directly to infer control. For example, in a tightly knit society, a constraint on one member at a given level will have repercussions on some or all of the other individuals at that level, as well as on the individual that it controls or that controls it.

So where does the Martian look for his *first evidence* on which to classify individuals ? I argue that he can find it at all, only in cases where there is a great difference in the *time constant* (or relaxation time) associated with individuals at different levels. This argument seems to accord with our everyday experiences, for we know that if we are members of a hierarchy of command, then we cannot operate at all if commands come to us with a frequency comparable with the rate at which we can implement them in detail. Since we are similar individuals to the individual who sends (or the individuals who send) us commands, we must perform each detail at approximately the same speed as that at which the overall command–implying many details–was delivered. We need far more time.

The Martian, accordingly, will be able to make a consistent start in discerning the levels of the hierarchy if and only if he can classify individuals by the rate at which they interact with their own kind. Or–to put it another way–he will be able to discover which are the same kind as a given individual by their all having a distinctive time-constant for their mutual interactions.

I conjecture that the intuitive argument that I have sketched is in fact capable of general application to cybernetic systems where control (or command, as I have hitherto called it) is hierarchical in character. If this is so then we have a non-formalized (but probably formalizable) theorem underlying hierarchical control which asserts that two levels are specifiable if and only if the ratio of the time constants characterizing the two levels is large (numerically). The *ratio* rather than either of the separate time constants is the important quantity here, for it is clearly only the ratio which has any experimental meaning in the absence of an independent time reference.

I have given a simple illustration of the thinking underlying a model of hierarchical structure by abstracting certain essential features from the hierarchy of the political type. It was such thinking, using these abstractions, that permitted mathematical development of this very difficult concept (the hierarchy), and this development proved to be possible only in the case of one simplifying assumption (namely that the time constants characteristic of different levels were very disparate numerically). There is a *prima facie* case therefore for trying to use the concept of a hierarchy with levels differentiated by their possessing different time constants, in biology, and I shall now look more closely at this possibility.

▸ A biological system usually shows some clear differences from our idealized feudal hierarchy. Firstly, the complex biological structure is not obviously an association of essentially similar individuals whose differentiation into levels is a matter of function rather than of design, as it is in the case of the feudal men. The latter we think of as men first, and as elements of a feudal hierarchy second. Thus there is a contrast; true, we immediately find ourselves looking for reasons to establish similarity between the two cases. A feudal man, we may say, who is trained as a baron is really very different from a villein, and, so we can argue, the real entity–a man–is not the biologically separable mechanical entity, but is that mechanical entity together with all its learned functions in the society of other individual men. However, that we find ourselves arguing in ways like this really only re-emphasizes our dependence on the hierarchy concept, for in macroscopic biology there is little which is conspicuously set before us to emphasize the build-up of diverse structure out of assemblages of essentially similar elements.

When we get to the cell level the situation is different; there is direct evidence for hierarchical organization in the very existence of cells, and the cells are the feudal men in our pattern of control. To make this comparison would be pathetically simpliste (indeed, that is the reaction we immediately feel) if it were not that current ideas about patterns of control among biologists at the cell level and below often seem to be even more lacking in subtlety.

Indeed, I think there is a danger that biologists may be inhibited from formulating their underlying assumptions about control relationships just because these ideas would seem too crude if brought to light. Certainly a great deal of detailed knowledge of control relationships—particularly of the combinatorial sort that is provided by modern molecular biology—does exist, but little of it is systematized in such a way as to provide general guides to thinking. In the absence of such systematization it seems that what the scientist must do is to work on hunches to guide him in his estimates of the plausibility of hypotheses, and presumably these hunches incorporate implicit judgments about control. Nevertheless, however crude these judgments would turn out to be if formalized, it would surely be much better to have them out in the open. For example, if the analogy with feudal villeins has something to do with our feeling that there ought to be a basic, undifferentiated unit of control in the shape of a cell, and thus the existence of such a thing is something that one would expect, whereas anything to the contrary would need explanation (and I guess that this is indeed what people mostly think) then I should very much like to have this strand in the biologists' thinking thoroughly out in the open and explicit.

The mere existence of biological structures built out of large numbers of roughly similar cells hardly constitutes a call for the 'similar unit hierarchy' model. The structure is usually simply too evident, and there is no need for elegant mathematical methods to be applied to answer questions about the functions of cells in—say—a bird's wing. We already know the answers. That is not to deny that an enormous mystery remains, but the mystery of control has been pushed back into the microscopic and even into the molecular conditions of development of the cells. By the time we can look at a functioning bird's wing there remains very little of the flexibility of function to individual cells that would constitute good ground for application of the 'similar units model'.

We might find what we were looking for in something like the slime fungus where the cells can wander about in conditions of slow, highly viscous, mass flow and yet the organism retains some sort of primitive structure and identity.

However, the case we are most likely to find worth looking at is the brain. There is obviously a great deal of structure of some sort to the brain, as is obvious from how our memories can work. But this is not structure in the same sense as we apply the term to the bird's wing. It is indeed more like the structure of a feudal society as our Martian sees it, where the hierarchical status of an individual has to be inferred from his function. A good deal of brain-physiological research does take the form of investigation at the cell level into the effects of stimulation of individual cells by different means, either on the organism as a whole or upon other cell complexes in the brain. This research is rather like the efforts of my Martian to map out the control patterns by putting spanners into the works at particular points to stimulate or inhibit particular individuals, and the general conclusion to be drawn from the mathematics of the similar unit model would be that, in conducting such research, one should look first for any evidence of a division of individuals into classes according to their reaction rates.

In making these suggestions I have spoken rather loosely of 'cells' and of 'units' without definitely asserting that cells are to be regarded as the units that compose the hierarchy. In fact I want to keep open the possibility that the correct identification for the hierarchy individuals are small complexes of cells. There are a variety of cells known to exist in the cerebral cortex, so that some structure is visually evident which shows that it would certainly be inappropriate to treat members of different varieties as similar individuals. On the other hand, I think it is widely assumed that some repeated structure of cell complexes must exist at some level in such a large mass of cells as constitutes the cerebral cortex. The alternative – that no repeated structure exists – is logically possible but hardly conceivable. And wherever repetition sets in, we find the units for our hierarchy.

I have suggested how models like my 'similar unit hierarchy' may provide concepts which will assist in the understanding of control relationships in cerebral tissue. A critic, however, may argue that a model is not likely to be of much use if its applicability depends upon our *lack* of knowledge of structure. This model – it seems – is going always to be in retreat from the progressive discovery of what the detailed nerve net shape actually is. I think this criticism contains a serious error. Obviously one should be prepared to trace control chains as far back as one can. On the other hand, if one supposes that in doing so a stage will come at which one suddenly meets a kind of uncaused cause, or beginning of the control chain, one is obviously in a muddle. One may even find oneself asking the nonsense question 'at what point does consciousness enter the control process ?' Concepts appropriate to hierarchical organization and to the centralization of control are needed for their own sake. I think moreover that it is a fact strongly in support of my position that biological tissue gives every appearance of becoming more homogeneous and less mechanically differentiated as we approach the centre of control.

This article is at best only the prolegomenon to a piece of work. It is not the work: that would be on a larger scale. It would go on the assumption that there was scope for a real application of the similar unit model in biology (for example to the cerebral cortex) and assume a sorting of units into levels by time constants. So far we have already discussed the problem. But then the work would go on to deduce from the mathematical properties any similar unit hierarchy must have, what form detailed structures within the hierarchy must have. This problem would be important as soon—for example—as one wished to ask how particular memories are stored. Clearly in such an enquiry there is a conflict between the requirement for similarity of the units and the requirement for diversification of structure. This conflict has to be resolved by considering the dynamical characteristics that are allowed by the hierarchy mathematics within the similar unit model, and using these to create diversity of structure.

## Appendix I

*Note on the stages in the development of the 'similar unit model' for hierarchies.*

*Stage I* An attempt to replace the intuitively defined continua of space and time by patterns of ordering or *ordering relations* [3].

*Stage II* Realization that the ordering relations could never be mapped on to the physical continua. The latter had to be constructed in terms of operationally defined interpolation of points starting from finite (discrete) sets. See also [4].

*Stage III* Hierarchical construction seemed to be the only way to extend the sets of points

*Stage IV* In absence of any rigorous way of understanding such hierarchical relationships, and with help from Gordon Pask, a three-level continuously variable feedback system with binary switches acting at thresholds, was constructed [5, 6].

*Stage V* These models suggested to Parker-Rhodes [1, 6] a mathematical method which made it possible to represent the relations between levels in a hierarchy by using matrix algebra over the binary field with symbols 0, 1 as elements of matrices and vectors. An information preserving mapping of $p \times p$ matrices on to $p^2 \times p^2$ matrices was the key step [1, 6].

*Stage VI* Amson, Bastin, Kilmister [1] showed that this calculus required for its basic logical operation a process whereby one entity could discriminate the other entities in a set by elementary decisions of the type, like/unlike. Its history of such decisions then defined its place in the hierarchy. Intrinsically it had no place. In this way the concept of a similar unit hierarchy was given a definite shape.

*Stage VII* It was then found that for a hierarchy of a given depth of level (number of levels starting from the simplest) there existed an upper bound to the number of discriminable entities. A way was found to relate this upper bound to the quantum limit of measurement in the physical world, with numerical success.

*Stage VIII* It is now possible, with the conclusion of Stage VII, to look back to the control model of the hierarchy and fit into it the insights supplied by the formalization. We then see that the different reaction times which characterize the levels as a matter of mechanical expediency in a hardware hierarchy of the Bastin-Pask type, must exist as a matter of logical necessity if we apply the constraint that units must be intrinsically similar. This gives us the main argument that we used in the body of this paper.

*Note.* It is of course very difficult to construct a similar unit hierarchy in practice and one has to use imagination and guesswork in interpreting its behaviour. A part of the motivation of this whole line of work, indeed, so far as the present writer was concerned, was an idea that quantum uncertainty must be due (if one were assuming a hierarchical construction of the physical continuum) to the fact that one could never with complete certainty settle the allocation of hierarchy units into levels because complete certainty could only be attained at the expense of infinite ratio of time constants of the levels. This reasoning was based upon a sort of engineering guesswork.

## Appendix II

*An example of an ordering relation in a hierarchy*

Consider the *quadratic group S* which consists of the elements *a, b, c, e,* and in which all the operations in the group can be read off from the following 'multiplication table': thus the result of the operation $i \cdot j$, where $i, j$ are elements and where the dot between them represents the group operation, can be found as the content of the square in the table where the *i*th row and the *j*th column intersect.

| | a | b | c | e |
|---|---|---|---|---|
| a | e | c | b | a |
| b | c | e | a | b |
| c | b | a | e | c |
| e | a | b | c | e |

The table gives a set of relations, therefore, between all possible triples of group elements, which together define the group. This set of triples is usually presented as equations: $i \cdot j = k$, or as transformations: $i \cdot j \rightarrow k$, but for my purpose it is more convenient to write them in the form $R(ijk)$ which we can read as a triadic relation, defined by an ordered set of 3 elements. Looking at the table with these relations, *Rijk*, in mind, we see immediately that one of the elements–namely *e* – differs from the other three in its behaviour. In fact it is, in group theoretic terms, the unit element. However, we can find no distinction of this sort or any other among the other elements. Their relations seem to be completely symmetrical. In fact we should like to say that the sub-set *a, b, c* of the elements has the particular *ordering relation* that no order can be defined among *a, b, c* by the group operation. This fact can be restated by saying that all the relations *Rijk* which can be formed by giving *i, j, k* values taken from the set *a, b, c,* and taking each once only, are true. We call this the *S*-ordering relation (*S* for simultaneous).

How much importance, if any, we may give to this statement about the symmetry of *a, b, c* is unclear. In a paper [3] by Kilmister and the writer an attempt was made to relate the 3-dimensionality of physical space to the *S*-ordering relation of *a, b, c* in this algebraic structure. I think that no difficulty really arises from the fact that the *S*- ordering relation specifies the *absence* of any *preferred* order rather than anything positive; the statement of the *S*-ordering relation can be put in positive form by saying that everything that is true under any one ordering is true under every other. This last way of putting the matter, indeed, leads straight to the intuitively strongest case for relating the algebraic structure of the quadratic group to the dimensionality of space. We do know that any equation from physics has to be precisely equally valid in every way if the coordinates are interchanged– a fact which strongly suggests the *S*-ordering relation.

However, none of the remarks I have so far made about the significance of the *S*-ordering relation are precise, and a brief account of the efforts made to get them so (p. 253) and how these efforts forced us to look for a new mathematical tool–namely ordering in hierarchies–may be useful here. These efforts were the first attempt to formalize the order concept in anything like the sense that was being used at Bellagio, though, particularly in relation to space-time, they owed a lot to Whitehead's theory [7] of the logical construction of space-time from ordered events.

Let us discuss some criticisms that were made of the position that the ordering relation ($S$) of $a, b, c$ in $S$ underlie the 3-dimensionability of space, to show the stages by which the attempt to state an ordering relation led to a hierarchical point of view.

*Criticism* (*1*). The specification given in the $S$-ordering relation to take elements of the group $S$ other than the unit element must be a remark in some metalanguage. (It cannot be part of the group theoretic calculus. If it could, then the deduction of the ordering relation $S$ might indeed be a deduction about groups instead of merely about the number 3.) The importance of this criticism appears when one analyses the intuitive argument that was used to relate the ordering relation $S$ to physical space—namely that physical equations must be invariant under interchange of coordinates. This argument gets its appeal from the (unstated) assumption that the $S$ group will be mapped on to a number field to give a 3-dimensional Cartesian space with $a, b, c$ as unit vectors. In fact, however, the group itself does not provide the ordering relation $S$, and so no such development could be undertaken.

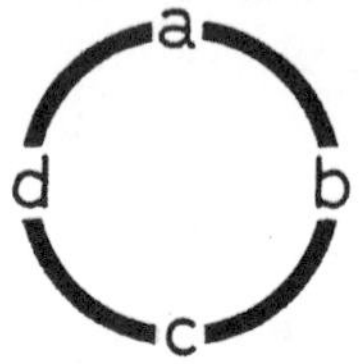


*Criticism* (*2*). The alleged ordering relation $S$ is simply a property of the number 3 wrapped up in a suggestive way, for $Rijk$ cannot be given values from $a, b, c$ in just *any* way. Each must be chosen only once. (For example, $Raab$ is not true, since $Raae$.) Moreover, in view of criticism (*1*) this restriction is arbitrary. Hence, all that is being said is that if three things are arranged on a ring then nothing we can say about the relative positions of the things will be rendered untrue by interchanging them. This—so the criticism admits—would not be true of four things arranged *abcd* on a ring, for $a$ (for example) has a different relationship to $b$ and $d$ (which are neighbours) from what it has to $c$ (which is not). The algebraic equivalent of this demonstration is simple: if $Rabc$ and $Rabd$, then $c=d$, and there are not four things but three, and if it is not the case that both $Rabc$ and

*Rabd*, then the rules distinguish *c* and *d*, contrary to hypothesis. Hence the *S*-ordering relation cannot be extended to four, or therefore to higher numbers of things. It is just a property of the set of 3 things. But it is no more.

*Criticism* (*3*). Physical reality has to be attributed first to space-time and only subsequently to 3-space. This could never be the sequence of development within the proposed basis for the space-time structure.

*Criticism* (*4*). The proposed basis for the 3-dimensionality of physical space is most implausible. The 3-dimensionality really specifies the number of *independent* assertions that can be made about the position of an object. The nature of the symmetry or otherwise of these assertions can therefore not determine their number which is just a fact about the world.

I include criticisms (3) and (4) because physicists will usually make them, but discussion of (4) is not relevant to my treatment of *S* as an ordering relation (though it opens up fundamental questions of physical interest). Of (3) it will have to suffice to say that if we examine with care the processes we actually might use to ascertain the number of spatial dimensions by means of actual operations, we see that we in fact reject many that are actually necessary to give complete specification of the kinematic state of a test-particle at a point (such as the electric potential) and it is actually plausible to suggest that we do so just because these are quantities of diverse sorts.

The substantive criticisms are therefore the related ones (1) and (2), and the *point d'appui* in finding a way around the difficulty that these criticisms raise—namely that what we wish intuitively to say seems to be forbidden mathematically—is likely to be the distinction upon which our imaginary critic has made these criticisms completely hinge: namely that between formal language and metalanguage. Clearly, simply to insist that statements from the metalanguage be allowed in the formal language will do no good, since we should then have no criterion of what to allow in and what not. We may not take that course even if we are doubtful of the validity of the distinction as an ontological necessity.

To avoid the impasse in which we seem to be, let us look at the mathematician as he manipulates the symbols of the structure that we have been discussing. How does he know that he is meant to try *every* operation that the calculus allows him, and not, for example, to stop short at those operations that for our purposes we should like to forbid? There is no answer to this, logically, except that he has been trained as a mathematician, and that is what the other mathematicians do. So we seem to have demonstrated that two can play at the metalanguage game, for we have caught the mathematician allowing himself to be guided by a rule that can only come from the metalanguage.

What sort of assumption would the mathematician have to make in order to avoid appealing to the way in which he happens to have been trained ?* The mathematician might well argue that he could have been replaced by a machine which was programmed or otherwise constrained to obey the rules of the calculus but by nothing else, and in which the starting positions are subject to random variations. Then he could be sure that the machine would, in the course of sufficient time, run through all the cases without there being any question of appeal either to the mathematician's insight or to his upbringing.

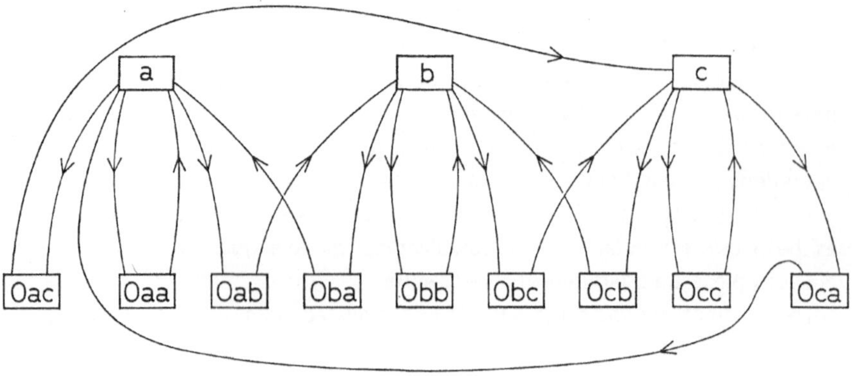


Now, however, an interesting change has come over the total picture. Because we have introduced a machine to supplement the mathematician we have raised a quite new sort of question. What status now have the rules of the calculus which before seemed in no need of investigation ? The mathematician was forced to justify his conduct by an appeal which, like the magician's apprentice, he has no power to call a halt to. In particular, the choice out of the original rules of operation which we desired to make in order to represent principles of ordering turns out to be more natural for a wide range of mechanical systems than does the conventional set of group operations, 'complete'. Let us therefore be guided by the logical

*And the validity of my argument depends upon the mathematician's being compelled to make some such appeal to a mechanism as the one I have outlined.

implications of this train of thought and try to think of the structure of that machine which will embody our ordering ideas with the fewest *ad hoc* assumptions. It will incorporate the metalanguage spectre quite explicitly as a level of operators upon the original ordered entities–one for each pair of the latter, as shown in the figure.

To get a self-functioning system, of course, there has to be a second level of operators that operate upon the first level of operators, and so the total system rapidly goes off into indefinite complexity which we are unable to follow and have to represent by a random input at some level not too far distant.

The fact that we draw a diagram to illustrate certain features of a hierarchy of orders by no means implies that we understand in detail how it could work (and in those circumstances it is doubtful whether we should describe the construction as a machine at all). Nevertheless our immediate task is accomplished if we have mapped the stages in the argument which lead from a certain simple mathematical structure to a crude realization of the hierarchical organization of orders of which, we claim, it is an abstraction. (Even if the immediate reaction of the mathematician, confronted with the horrors thus exposed, is to set the whole argument in reverse as fast as he can.)

### References

1. Amson, J.C., Bastin, E.W., Parker-Rhodes, A.F., Kilmister, C.W. Work summarized in Bastin, E.W., *Studia Phil. Gandensia, 77* (1966) 4.
2. Kuhn, T. S. *The Structure of Scientific Revolutions*. (Chicago, 1962).
3. Bastin, E.W. and Kilmister, C.W. *Proc. Camb. Phil. Soc., 50* (1953) 2.
4. D. Bohm. *Proc. Int. Conf. Elementary Particles.* (Kyoto, 1965) 252.
5. 'Self-Organization and the Notion of Level'. U.S.A.F. Office of Scientific Research Technical Note. (Cambridge Language Research Unit, Information Structures Project 1960.)
6. *U.S.A.F. O.S.R. Annual Reports* 1961–5. Information Structures Project.
7. Whitehead, A.N. *Phil. Trans. Roy. Soc.*, 1905.

## Comments by C.H.Waddington

I think it is worth drawing attention to the way in which several contributors to these papers have, from different points of view, approached one another in attributing great significance to the association within biological systems of a number of time-scales differing markedly in magnitude. From a purely biological standpoint, I have myself been remarking on this for some years. For instance, in 'The Strategy of the Genes', 1957, pp. 5, 6: 'Perhaps the main respect in which the biological picture is more complex than the physical one, is the way in which time is involved in it . . . to provide anything like an adequate picture of a living thing, one has to consider it as affected by at least three different types of temporal change, all going on simultaneously and continuously. These three time-elements in the biological picture differ in scale: . . . evolution . . . life history . . . turnover of energy or chemical change.' And the same point was made in my summary of the results of the first meeting (*Prolegomena*, p. 218). These statements were made as descriptions of what we in fact come across in biology, where we are confronted by systems affected by controls at the levels of evolution, development, and functioning, with very different time scales.

It is more significant to find the same, or at least a very similar, argument being advanced by several of the physicists on grounds of logical necessity. Bastin, in the article above, argues that it is possible to analyse a complex control system into a hierarchy of subsystems only if 'the time constants of the different levels are very disparate numerically'. At the first symposium Pattee (*Prolegomena*, pp. 69–93) came to the conclusion that such a fundamental biological process as heredity could only operate with adequate reliability if it involved non-holonomic constraints, i.e. 'metastable configurations with relatively long relaxation times compared to our time of observation'—an example being a DNA polymerase enzyme molecule which persists for a time long in comparison with the time required to catalyse the bonding of one pair of bases. In the same vein, surely, but at a more abstract level, would seem to be David Bohm's insistence that an order of orders ultimately demands a reference to a 'timeless order'—which perhaps need be timeless only in the sense of having a time scale much greater than that of the lower orders, since David explicitly contemplates the possibility of its changing (p. 57). Finally, there would seem to me to be at least an analogy, and perhaps something more, between non-holonomic constraints and Christopher Longuet-Higgins' programs, which have relaxation times much greater than the time required for performing each item in the program.

It is true that there have been other contributions, dealing with time, which have not laid such stress on the importance of discontinuities between time scales. This was, for instance, in the main true of the discussions of limit cycles by Kerner, Iberall, Goodwin, and Cowan. But one may point out that these authors were mainly concerned with some one level of biological operation, e.g. Iberall found his continuous spectrum of limit cycle frequencies all within the realm of metabolism.

It is surely significant when so many of those who approach the problems of general theoretical biology from the standpoint of professional physicists advance such closely allied arguments attributing major importance to changes of time scale. It seems probable that we may here be approaching a major theorem of theoretical biology, such as it was hoped these symposia would produce.

# Physical Problems of Heredity and Evolution

H. H. Pattee
Stanford University

My purpose in this paper is to present questions about the nature of life and evolution which can still generate deep curiosity among physicists in spite of the assurances of molecular biologists that the physical basis of life is now clearly understood. I have learned that this is not a popular purpose, to say the least; for these days to admit uncertainty about the basic nature of heredity, enzymes, or evolution is often regarded only as evidence that you have lost contact with modern biological knowledge.

It was not always this way. In fact, credit for some of the origins of molecular biology must go to physicists who were puzzled by the peculiarities of living matter [1]. The attitude of physicists twenty years ago towards biology was recently characterized in the following words [2]: 'Thus it was the romantic idea that "other laws of physics" (Schrödinger) might be discovered by studying the gene that really fascinated the physicists. This search for the physical paradox, this quixotic hope that genetics would prove incomprehensible within the framework of conventional physical knowledge, remained an important element in the psychological infrastructure of the creators of molecular biology.'

One example of this type of so-called 'romantic idea' was the speculation of Delbrück [3]: 'It may turn out that certain features of the living cell, including perhaps even replication, stand in a mutually exclusive relationship to the strict application of quantum mechanics, and that a new conceptual language has to be developed to embrace this situation.'

But today, twenty years later, these questions do not often arise. It is now commonly asserted, not only that the secret of heredity is understood in terms of physics, but that 'No paradoxes had been encountered, no "other laws of physics" had turned up' [4], and that '. . . up to the present time conventional, normal laws of physics and chemistry have been sufficient. . . .' [5].

Now while these evaluations may be true in one sense, I believe it is correct to say that no 'strict application of quantum mechanics' to the living cell, including replication and specific catalysis, has even been attempted by molecular biologists. I am not even aware of any 'loose' treatments of the physics of heredity by molecular biologists. Therefore I do not see any evidence that the physical basis of heredity has been worked out.

Of course there is no question that the molecular biological revolution has uncovered facts of life in incredible detail. There is no doubt, either, that all this knowledge will be increasingly useful to the medical sciences and helpful in unifying the many areas of biology which have not had, until now, a common foundation. My discussion, on the other hand, has to do with whether these new complex facts of life really have, as yet, any relationship to the discipline of physics.

What is it then about the fundamental facts of living matter which are still not easily understood in terms of physical theory ? To answer this question we must have some idea of what it means to a physicist to understand his observations in terms of a theory. Biology and physics have long had different traditions with respect to the relation between observations and theories, so that it is not unexpected that this difference in style often causes misunderstanding. Biologists have always had to struggle with an enormous variety of facts and with great complexity and lability in the simplest living units. Consequently molecular biology appears from this perspective as a set of relatively simple unifying observations about all life, and the ability to reproduce many of the basic functions of cell components in the test tube is often interpreted by the biologist to be a 'reduction of life to physics'. Indeed, this certainly is one type of significant reduction in the complexity of even the simplest living cell.

But the physicist sees a different problem. He has already learned a set of unifying foundations or theories which quite accurately describe an enormous range of his experience. The physicist also collects facts, but it is the overwhelming concern of the physicist to find out whether or not these facts can be predicted by or 'reduced' to theory. Moreover, his rules for 'reduction' require relatively high experimental and formal precision compared to that expected in biology [6]. Historical experience has also taught physicists that elementary concepts are clarified not by applying them to more and more complicated systems, but rather by asking deeper questions about the elementary concepts themselves. For example, the laws of classical dynamics would not have evolved into the more profound relativistic or quantum theories simply by applying these classical laws to systems with more and more degrees of freedom. For this reason it does not seem reasonable to expect the current interpretation of heredity in terms of the Watson-Crick template model and the formal description of the coding and synthesis process to develop into a more profound theory of life simply by applying it to more and more complex organisms or by exhaustive description of, say, one strain of bacterium.

▸ *What is the central question ?* If living matter differs from non-living matter because of its 'hereditary' property, which allows for evolution by natural selection, then the central question for the physicist is to explain what 'heredity' means in the language of physics, and to explain how hereditary structures would arise and persistently propagate and evolve in a relatively disorderly environment. In other words, if the nature of heredity is 'the secret of life' then surely we must know something of the physical significance of this concept if we are to claim we understand this secret.

Let me make clear at this point that I shall assume that the current laws of physics are the most universal laws we have for describing the behaviour of all matter, dead or alive. I am not searching for 'other laws of physics'; but neither am I claiming that life is now 'explained in terms of physics' just because we know the average structure of DNA or can make enzymes work in a test tube. I am not questioning whether matter in the living state obeys the laws of physics, but rather how the laws of physics can actually explain the *difference* between living matter and non-living matter. Karl Pearson [7] put the question long ago, 'How, therefore, we must ask, is it possible for us to distinguish the living from the lifeless if we can describe both conceptually by the motion of inorganic corpuscles ?'

We are all familiar with the traditional biological generalizations about life, e.g. living matter depends on enzyme-controlled reactions, living matter self-re-produces, living matter is cellular, living matter evolves, etc. We can also find more general distinctions, e.g. life is distinguished by adaptability, order, design, purpose, etc. These distinctions between the living and the lifeless forms of matter not only suffer from ambiguity at the most elementary levels of living processes, but they are not in the language of physics. Thus it is difficult to say exactly where protein enzymes and simple catalysts differ, or where self-reproduction and crystal growth differ in terms of physical observables [8].

▸ *Two approaches to the central question.* Molecular biology has evaded this general question by using only explicit descriptions of life; and this has satisfied many biologists. No longer do textbooks begin with discussion of the general 'characteristics of life' but rather with the specific 'building blocks' of life. Text-books emphasize that it has been demonstrated that life is built out of normal atoms and molecules, and since most cellular reactions can now be demonstrated *in vitro* there is the implication that life is just a certain collection of DNA, RNA, enzymes, and other molecules in the proper spatial configuration. So if there can be said that there is any difference between living and non-living matter, then according to the molecular point of view, this difference lies only in how these molecular parts are organized.

This is a deceptively simple and convincing attitude, for we do understand the essential behaviour of many types of molecules in terms of physical laws. What is deceptive is the idea of 'how these molecules are organized' because the laws of physics do not account for special types of structure or organizations of matter called 'living' as distinct from the organization of non-living matter. So our question of how we distinguish life and non-life has simply been displaced by the question of what is the physical difference between organizations which live and organizations which do not.

In the language of physics, matter is 'organized' only by the 'energy of interaction' or equivalently by the 'forces' which act on it [9]. For example, the structure or organization of the solar system is largely the result of gravitational interaction or forces, the structure of molecules is partly the result of electric forces, and the structure of nuclei depends on nuclear forces. A few other types of forces are known, and all types may be combined in a great variety of ways, but as yet there is no clearly recognized special force or special combination of interactions or forces which can be uniquely associated with living organizations.

We find, then, that there are two basic attitudes toward the nature of life. I will simplify these attitudes somewhat to contrast them more clearly. By far the predominant attitude today is that of molecular biologists who do not worry about the obvious physical distinction between the behaviour of living and lifeless matter. Their point is simply that they have looked at the parts of cells down to their essential molecules without being able to perceive any special physical distinctions. Since all the parts of living systems work in a test tube, it is asserted that life has been reduced to physics'[10].

The physicists, along with many older biologists, focus on the obvious differences between the behaviour of living and non-living matter, principally the evolutionary fact that living matter becomes more and more orderly, and non-living matter, less and less. Since they presently see no basis in physical law for this difference, it is asserted that some principle is missing from physics, or if no such principle exists, that life is irreducible [11].

Both of these assertions may have some truth in them, but I do not believe that either of them provides a satisfactory answer to the question which was asked. The first position says, in effect, we discern no basic physical difference in detail between living and lifeless matter, therefore life can be reduced to physics. The second position says, in effect, we discern a clear difference between living and lifeless matter but there is no physical basis for this difference, therefore life cannot be reduced to physics. Stated so bluntly, there are obvious logical fallacies in both these statements which are often obscured or softened by more elaborate formulations. But I do not think the lack of logic is the main difficulty. The fallacies in both cases, I believe, are more in the nature of seeing too narrow a view. Strangely enough, the biologist often takes a narrow view of the conditions for life to exist, while the physicist often takes a narrow view of the conditions for physics to exist. In particular, molecular biologists tend to ignore the nature of evolution and underestimate the exceptional physical requirements for persistent hereditary processes, whereas the physicists tend to ignore the nature of observation and underestimate the exceptional hereditary requirements for measurement processes.

▶ *Summary of a third approach: Three problems.* My approach will be to assume that the potential for *hereditary evolution* is the primary characteristic of life which distinguishes it.from other collections of matter. The central question will then reduce to formulating *first* a physical description of an elementary hereditary process; *second,* the physical conditions which allow hereditary machinery to persist in the face of the inevitable disordering interactions with its environment; and, *third,* the nature of descriptive symbolism in hereditary systems which have the potential for evolving the enormous degree of 'self-determination' that we observe in living matter.

The first problem, the formulation of elementary hereditary events in terms of physics, will lead to the conclusion that hereditary transmission depends entirely on specific rate-controlling mechanisms. Macroscopic hereditary machines (automata) require special flexible constraints (non-holonomic constraints) which are not directly derivable from equations of motion, but can be interpreted by non-equilibrium statistical mechanical descriptions. Molecular hereditary machines must satisfy similar conditions, but the quantum mechanical analogue of non-holonomic constraint is a more difficult conceptual and formal problem that has not been adequately solved. The second problem, the formulation of conditions for persistence of hereditary machines in a disordered environment, will lead to the thesis that classical hereditary machines cannot have exceptional reliability, and that only quantum mechanical hereditary machines, without thermal dissipation, can operate more reliably. The third problem, the conditions for evolving increasing self-determination, will lead to the requirement for a separate *description* or *symbolic representation* of the hereditary machinery, along with the actual hereditary machinery itself which can read (decode) the descriptive symbols. This is closely related to the condition for universality in abstract automata [12]. It is represented biologically by the genetic description, the code, and the synthesis machinery found in all cells we now observe and, at a higher level of symbolism, in the brain.

▸ *The first problem: Physical conditions for hereditary processes.* Hereditary processes require the existence of a set of relatively fixed objects or traits any one of which can be transferred in a recognizable form from parent to offspring in the course of time. The first essential condition for hereditary propagation is the possibility of a set of more than one trait. For example, even though the energy or velocity of an undisturbed mass or the number of degrees of freedom of a system will be 'propagated' unchanged in the course of time, we do not speak of these constants of motion as hereditary traits since there is no possible alternative allowed by the laws of motion. The biological idea of a hereditary 'trait' also implies the possibility of an alternative trait, which in turn implies a record for identifying which one of the more-than-one alternatives is to actually be propagated. This record for identifying traits may be called the description of the trait or the genome.

In order to see more clearly what the physics of heredity requires, consider a very simple hereditary tactic copolymerization in which two types of monomer are added to a chain in alternating sequence, *ababab* . . . The fact that such an alternating copolymer exists is, in itself, no assurance that a hereditary reaction has occurred, for it is quite possible that this alternating sequence is the *only* stable sequence. In other words, the homopolymers *aaaa* . . . and *bbbb* . . . , or even dimers *aa* and *bb,* might be energetically unstable so that no alternative sequence is energetically allowed. In this case the sequence is uniquely determined directly by the equilibrium configuration and hence no hereditary propagation can be said to occur.

Therefore one condition for heredity propagation in such a copolymer is that more than one sequence is more or less equivalent energetically. Let us say then that all bonds *aa, ab, ba, bb* have approximately the same ground state energy so that the probability of each distinguishable sequence is approximately the same at equilibrium. But now for heredity propagation we must have a second condition that the probability of the growth of one (or a subset) of the possible sequences is increased relative to the others because of the previous existence of one of these sequences—the genetic sequence. In other words, although the initial sequences were required to be initially equiprobable (degenerate) so as to have more than one alternative, we then require that the existence of these sequences leads to unequal addition probabilities in the growing sequences. How can this be ? Since the bonds being formed are the same between all monomers, we must assume that no difference can occur in the equilibrium energies of different sequences and therefore only the relative *rates* of monomer addition can be changed. In other words, in such a copolymer, *hereditary propagation must depend on specific catalytic control of the rates of monomer addition* [13]. This also implies that to the extent that sequences are determined by the equilibrium energies, they are not properly hereditary. It is also important to realize that in so far as some other *inherent* physical property of the bonds joining the monomers controls relative rates of monomer addition, i.e. different activation energies or affinities, then the sequence has no alternatives and is not hereditary. In other words, only when bond formation is *correlated* to specific, rate-controlling elements *other than the bond itself* can we imagine the physical alternatives necessary to call it a hereditary process. The same distinction holds if the monomer addition step requires energy. If there is an energy coupling reaction that joins all monomers with equal facility then this is not a hereditary process. It is the specificity interaction correlated with the catalytic reaction which effects the reduction of alternatives necessary for the hereditary process.

A more general physical description of a hereditary process can be given in terms of the initial conditions and equations of motion. A physical system is defined in terms of a number of degrees of freedom which are represented as variables in the equations of motion. Once the initial conditions are specified for a given time, the equations of motion give a deterministic procedure for finding the state of the system at any other time. Since there is no room for alternatives in this description, there is apparently no room for hereditary processes. Alternatively, since the specification of the state of a system at one time serves to completely determine the course of the system at past and future times, there is no consistent way to introduce the additional concept of a record or memory of past events within this deterministic system. The only useful description of memory or heredity in a physical system requires introducing the possibility of alternative pathways or trajectories for the system, along with a genetic' mechanism for causing the system to follow one or another of these possible alternatives depending upon the state of the genetic mechanism. This implies that the genetic mechanism must be capable of describing or representing *all* of the alternative pathways even though only *one* pathway is actually followed in time. In other words *there must be more degrees of freedom available for the description of the total system than for following its actual motion in the course of time.* This can be accomplished classically only by introducing dynamical constraints which are additional equations relating coordinates to the trajectories in a way which is not derivable from the ordinary equations of motion or initial conditions. Such constraints are called *non-holonomic* following Sommerfeld [14], although other definitions are found.

Almost all man-made machines are interesting and useful because they introduce non-holonomic or hereditary constraints. From the simplest clocks to the largest computers, the essential dynamical elements such as ratchets, relays, switches, and escapements, which make the machine function according to some human design, depend upon non-holonomic constraints. Of course, it is seldom questioned if these machines can be explained by the laws of physics. As a matter of fact, non-holonomic constraints are seldom discussed in physics texts except for very simple cases, because of the serious formal mathematical problems which arise and because of the difficulty in making generalizations. All that it is necessary to say here is that since a classical non-holonomic constraint has more static than dynamic degrees of freedom it cannot be described by formal systems with a fixed number of specified degrees of freedom. Therefore, in order to effectively reduce the number of dynamic degrees of freedom we must introduce the *statistical* idea of time-averages or time-dependent *correlations* between degrees of freedom. But then the dynamical system becomes essentially irreversible and dissipative. It is for this reason that any classical device which operates as a logical element by reducing the number of alternatives must dissipate energy (approximately kT per bit) for each reduction. This must be true for the simplest reset mechanism [15] as well as for a complex Maxwell demon [16], otherwise the Second Law of Thermodynamics could be violated. In general a statistical mechanical treatment of 'discrete' classical non-holonomic or hereditary constraints will require highly non-equilibrium and non-linear equations with at least two distinct time scales, so that it is always easier to design the logical behaviour of machines and treat the physical reliability as a separate problem.

The first essential point I want to emphasize here is the inherently statistical nature of non-holonomic constraints and hence the inherently noisy nature of hereditary or rate control processes. In the copolymer example recall that all sequences had to be energetically possible in the absence of the specific catalytic control in order to satisfy the hereditary requirement for physical alternatives. Therefore the overall accuracy with which a given sequence is propagated depends on the precision of the monomer recognition steps and the correlated catalytic rate difference in the addition step. For example, if the catalytic mechanism increases the addition rate of *a* on to a terminal *b*, or vice versa, by a factor of $10^4$ over the uncatalysed rate for all monomer additions, then there will necessarily be a mutation rate of $10^{-4}$; i.e. there will be on the average a non-alternating pair per $10^4$ additions. The second point, which is also fundamental, is that although the *reliability* of a hereditary mechanism can be very high or very low in a given situation, our formal mode of describing the situation is not a matter of degree. Either it has or it has not the hereditary property. That is, in the alternating copolymer example, the mutation rate could be much lower, as in DNA replication, or much higher, say one error for every two additions, on the average, but both reactions would be hereditary provided we are describing *a reduction in the set of possible alternatives in the course of time.* It is significant that this formal distinction between non-hereditary and hereditary descriptions is the same type of distinction as that between reversible and irreversible descriptions, or between strictly deterministic or probabilistic descriptions.

The biologist should be aware that these distinctions have appeared irreducible to many physicists [17] and that hereditary and non-hereditary behaviour is therefore a serious problem even in simple physical systems, quite apart from the question of whether the hereditary reliability of living systems is exceptional or not. For practical calculations it is sometimes possible to pass over the gulf by 'approximations' without any physical bridge, but it is important not to ignore the depth of this gulf. Biologists should also realize that once we have decided to pass this gulf and operate, isolated, without any physical bridge, on the hereditary side, then we are in the unreal world of deterministic automata in which we can largely design what amounts to our own error-free laws of motion (though we still will have unsolvable problems) [18]. Yet some matter has not only crossed this gulf, but also literally lives on the other side with exceptional persistence. To explain by physical laws the bridge over which matter passes into a hereditary world, I would call the reduction of life to physical laws. Nothing less could be called the secret of life.

▸ *The second problem: Physical conditions for hereditary reliability.* We have argued that in order to introduce the idea of hereditary behaviour into physics we must allow an internal (genetic) selection of alternative trajectories which amounts to a dynamical reduction in the number of degrees of freedom in the system, i.e. a non-holonomic constraint. But any process in which we lose track of degrees of freedom has passed from the dynamical world into the statistical world, and therefore in spite of the formal precision used to symbolize abstract hereditary rules, no real physical system can follow these rules without statistical error.

Let us look at the physical nature of error in more detail. What is the simplest classical system in which we would recognize error ? If we choose to describe our observations by the equations of motion of classical mechanics we know that there is no room for error, no matter how complicated the system. We do not consider that the motion of the stars or the galaxies could be in error. If the earth is struck by a meteor it might be called a catastrophe, but not an error.

In order to speak of a system making an error, it is necessary to have alternative trajectories possible within the system in the same sense that we find alternatives necessary to speak of hereditary systems. For a simple example of error, let us return to our copolymer, where we defined the monomer sequence as hereditary if from a set of energetically possible sequences there was a sequence-dependent, rate-controlling rule of growth of new sequences. This rule of hereditary propagation we saw could be executed with high or low reliability without jeopardizing the existence of the abstract hereditary rule itself. In this context I believe it is useful to speak of one type of error as simply the failure of a physical system to follow a formal hereditary rule. This type of error occurs in the dynamic motion of the hereditary machinery and therefore interferes with a time-dependent course of events. A second type of error is said to occur in time-dependent patterns or memory storage structures, such as DNA, in which an unpredicted alteration in the relatively static memory pattern occurs. Most error-correction schemes operate on this second type of error, since there exist general methods of coding or checking against error. However, the concept of memory storage or message has no meaning without specifying the code or the hereditary rules for reading-in and reading-out the memory or message patterns. Furthermore, any error-correcting code or checking scheme for messages must use the hereditary rules of coding to determine what is correct. Therefore I believe that the concept of error as *the failure of a dynamical rule of hereditary transmission* [19] is the most fundamental for this discussion.

Now we are ready to turn again to the fundamental question of whether classical non-holonomic machinery is reliable enough to assure the persistent hereditary evolution in a disordered environment, which is our basic criterion for life. As we said earlier, it is difficult to generalize about hereditary devices or non-holonomic constraints, because once you admit their simplest forms they can be concatenated to produce the most ingenious dynamical behaviour that man can conceive. On the other hand, man is by now reasonably familiar with the practical characteristics of classical machines, such as automobiles, clocks, and computers, and in most cases we know that high speed, low error rate and small size are incompatible, and that eventually they all wear out. The limitations on the speed and accuracy of man-made machines can of course be stated more elegantly [20], but I can foresee that even with a more formal treatment of error someone would raise the objection that although it is true that our man-made machines do not have the high speed, high precision, small size, and overall reliability of biological machines, this could still be only a matter of better design on the part of nature. It could be claimed that if we knew more about the error-correcting and self-repair properties of cells we could then match their behaviour with our macroscopic machines. Now this appears very unlikely to me, for the following reason: The elementary rate-controlling or logical machinery of living systems are not macroscopic organs but individual molecules–the enzymes. So the only relevant question is whether the reliability and speed (or, in chemical language, the specificity and catalytic power) of enzyme molecules can be predicted quantitatively by classical models.

To briefly summarize the argument, we began by agreeing that persistent hereditary evolution is the essential characteristic of life, but that the abstract hereditary process itself is clearly not the distinction between living and non-living matter, since many man-made machines exhibit elaborate hereditary behaviour. However, we find that no hereditary rules can be represented by physical systems without introducing statistical description and hence error in hereditary propagation. Furthermore, small size, high speed, and low energy dissipation is not consistent with low error rate in hereditary transmission in classical machines. On the other hand, the living cell executes all its hereditary rules with incredible speed and reliability using single molecules.

Therefore I conjecture that one fundamental physical distinction between living and lifeless matter is the exceptional *reliability* of hereditary transmission in living systems. Specifically, this implies that the dynamical behaviour of enzymes (and possibly other single, non-holonomic molecules) cannot be quantitatively described by classical models, that enzymes do not dissipate kT per bit of hereditary information transmitted, and that their hereditary logic is in some sense isolated from the thermal motions in the classical environment [21].

Let me make it clear that I have no doubt about the continued role of classical enzyme models both for pedagogical and technological advances. But that is not the fundamental question. We have asked if there is a clear physical distinction between living and lifeless matter. If there is none, and if the present classical descriptions of the hereditary rules of replication, transcription, coding, and synthesis tell the essential story, then the molecular biologists do indeed approach the truth in saying that we understand the secret of heredity and that life can now be explained in terms of physical principles.

But, on the other hand, if there is a crucial distinction between living and lifeless hereditary machinery, and if it cannot be explained classically, then the physicist will not only have a very strong point but also a profound problem. The point is that the secret of life is neither simple nor understandable by classical models; the problem is to express hereditary rules in the language of quantum mechanics, that is, to describe how an exceptionally reliable non-holonomic constraint can arise in a single molecule. This is not primarily a problem arising from the complexity of enzymes or our inability to calculate solutions to certain equations. It is of the same nature as the problem of interpreting the measurement process in quantum mechanics where we must use both reversible (deterministic) and irreversible (statistical) descriptions for a single physical situation. But in the case of molecular hereditary processes it is even less clear where to apply each type of description [22].

▸ *The third problem: Physical conditions for descriptions.* So far we have paid attention only to the most elementary physical conditions for hereditary machinery. Such machinery associates the specification of some action with the action itself, the description of the trait with the trait itself, or ultimately, the genome with the phenome. All hereditary machinery controls the rate or path of a system towards equilibrium, but in living matter this hereditary transmission is performed by specific catalytic molecules which associate the shape (or other interactions) of a molecule with a change of rate of reaction of a particular bond. We have conjectured that one fundamental physical distinction between living and artificial hereditary systems is in the exceptional reliability of the non-classical correlations between the specificity and catalytic power of enzymes. But clearly the course of biological evolution requires more than this. No hereditary machinery can operate entirely isolated, free from perturbations outside the system.

We are therefore led to my second condition for life, which is closely related to the traditional necessary condition of self-reproduction. I believe it is just as sensible to say that for persistent evolution of hereditary machines they must have an inherent capability for self-repair. But it is not primarily the question of whether self-reproduction or self-repair is the better concept which I think is fundamental. Rather it is the physical significance of the concept of 'self'. Now while it is difficult to find anyone who will dispute the necessity of self-replication for biological evolution, it is almost as difficult to find anyone who will consider the nature of 'self' in this context as a serious fundamental problem. The problem has been made most apparent to me in thinking about the origin of life and the physics of heredity. When can we say, with some physical justification, that a certain collection of molecules has a 'self' to reproduce ? For example, does ordinary crystal growth or the occurrence of a repeating copolymer sequence demonstrate some aspect of self-replication ? In fact, one should even ask exactly what aspects of the Watson-Crick model of DNA template fitting may be usefully called self-replication.

To answer these questions I would like to return to the simple alternating copolymer example. There we noted that the mere existence of a chain with alternating monomers, *ababab . . .*, is not sufficient to establish a process of hereditary propagation. A necessary condition for a hereditary sequence is that a different sequence could exist, and that the particular sequence which occurs is the result of a specified rate control process. For a hereditary step there must be both a 'specification' and a 'catalysis' with the hereditary machinery correlating the two. The same type of condition must hold for self-replication if it is to be considered as a type of hereditary process. The idea of 'self' here must refer to the specifications of the catalytic reaction and not to the reaction itself. Therefore the idea of 'self' in self-repair and self-replication can only make sense as a subclass of hereditary descriptions in which *the hereditary machinery is a part of what is symbolically described.* This concept of self-replication is considerably more stringent than the simple idea of copying or repeating or growing the same thing again. The 'self' prefix is used here to mean not primarily the specification of the incidental traits being copied, but the description of the hereditary mechanism which executes the reading of the description itself (i.e. the coding mechanism). This concept of self-replication is very nearly equivalent to the concept of the condition that the automaton be capable of producing a more complicated automaton [23].

To return to the physical meaning of symbolic 'descriptive' function in molecules, let us assume that our first and second problems are solved and that we have a quantitative physical description of the specificity, reliability, and catalytic power of a molecular hereditary machine (e.g. an enzyme). This is difficult enough, but our third problem is to explain what a *molecular description* of this machine could mean in physical language.

Now most biologists would say that a particular base sequence in DNA is a specification of the amino acid sequence in the enzyme, and in this sense the

D N A molecule functions as a symbolic description of the enzyme. This is a useful logical statement as long as the concepts of 'description' and the 'object described' (the D N A and the enzyme) are kept logically and physically separated, that is, as long as the real physical dynamics is not followed too closely. The cell actually physically separates the descriptive, coding, and synthesis functions quite clearly, so that in a sense the cell helps obscure the physical laws which in simpler systems appear to determine all behaviour. But of course that is exactly what any automaton is designed to do. It is designed to 'follow' its own symbolic laws with as little physical interference as possible.

However, there is no escaping that the symbolic description molecules must be read-out by interacting with hereditary molecules, and in fact the entire genetic meaning of the description depends only on how the hereditary machines are constructed. Yet it is just this construction which is being described in a self-reproducing system. Von Neumann was the first to recognize and indicate one solution of this logical problem, but it is far from clear that his or subsequent formal discussions [24] have much to do with any real physical system, certainly not a quantum mechanical system [25].

Now we may state the third problem again: How do the laws of physics account for the assembly of molecules into a hereditary machine which includes a *description* of this machine so that it can be a self-reproducing hereditary system ?

This is hardly more than the chicken-egg problem, and hardly less than the matter-symbol problem reduced to a primeval level. I would call the physical account of such a molecular system from reasonable initial conditions the fundamental problem of the origin of life.

## CONCLUSION

Living matter is an exceptionally reliable form of molecular hereditary machinery. The most general and exact formal language for describing molecular behaviour is quantum mechanics, whereas the most highly developed formal language for describing symbolic hereditary behaviour is automata theory. One might expect then that a satisfactory account of the essential material and symbolic aspects of living matter could use both quantum mechanics and automata theory. But this would now require a dualistic theory of life, for in their present forms quantum mechanics and automata theory have very little in common with each other. Furthermore, where they do share a common language there arise suspicious inconsistencies.

For example, the intuitive foundation for the idea of an automaton is an idealized classical machine, as in Turing's machine [26]. Gödel has gone even further to define any 'formal' symbolic system of logic or mathematics as a 'mechanical procedure' for producing formulas called 'provable formulas'. He also equates the idea of 'finite' procedure with 'mechanical' procedure [27]. But as we have pointed out, any real mechanical device which is designed to execute logical or hereditary steps is non-holonomic and hence dissipative in some sense. In other words, the laws of physics say that logical or hereditary operations are necessarily statistical in nature and hence inherently subject to error. This physical situation should disturb logicians who rely so heavily on idealized models of macroscopic machines to guide their intuition. Modern molecular biologists should be even more acutely aware of the problem of hereditary reliability, since the hereditary machinery they observe in cells is truly microscopic and can be even less accurately idealized by large computing machinery.

It is also by no means clear that quantum mechanical description will easily solve the hereditary reliability problem; first, because of the completeness principle which says that there are no 'hidden variables' to statistically correlate in forming hereditary dynamics, and second, because of the indeterminacy principle which says that complete deterministic precision is impossible. However, there exists a suggestive complementary relationship between the quantum mechanical picture of the material world and the automaton picture of the formal symbolic world. In the quantum mechanical picture it is assumed that there exists a complete formal description of the state of a system with the consequence that attempts at exhaustive observations on the system are necessarily indeterminate. On the other hand, in the automaton picture it is assumed that all operations and observations are strictly deterministic with the consequence that attempts at exhaustive description are necessarily incomplete. These two pictures were developed in the first place to describe entirely different sets of observations, and yet in living systems there appear at a fundamental level both the material and symbolic aspects of each of these pictures.

Although the idealized models of living machinery will undoubtedly improve in detail and utility, any profound reduction of life to physical laws will require understanding the quantum mechanics of elementary hereditary processes. This must include not only an account of how molecular codes and descriptions can originate, but also how they can continue to operate so reliably in a disorderly environment. To the physicist this still appears as a deep enigma.

This work was supported by the Physics Branch of the Office of Naval Research Contract Nonr 225 (90) and the Environmental Biology Division of the National Science Foundation Grant GB 6932.

## Notes and References

1. One fruitful area of molecular biology is the study of the structure of macromolecules by X-ray diffraction. This area was pioneered by physicists, beginning with W.T. Astbury. Another well-known example is M. Delbrück (see J. Cairns, G. S. Stent, and J. D. Watson, eds., *Phage and the Origins of Molecular Biology*, Cold Springs Harbor Laboratory on Quantitative Biology, 1966).
2. G. S. Stent in *Phage and the Origins of Molecular Biology*, 4.
3. M. Delbrück *The Connecticut Acad. of Arts and Sciences 38* (1949) 173. (Reprinted in ref. of note 1, 20.)
4. G. S. Stent in *Phage and the Origins of Molecular Biology*, 6.
5. J. C. Kendrew *Scientific American 216* (1967) 142 (reviewing *Phage and the Origins of Molecular Biology*).
6. The physical explanation of the chemical bond is an appropriate example. A predictively useful concept of valence was developed in the 1850s (E. Frankland, A. Kekulé), and by the 1870s the idea of the chemical bond permitted the development of three-dimensional structural chemistry (Van 't Hoff). But, however useful such models are, the concepts of valence, saturation, bond angles, isomerism, etc., were not explained or reduced to physical laws until Heitler and London (1927) showed how these concepts could be calculated from quantum theory.
7. K. Pearson *The Grammar of Science* (J. M. Dent and Sons: London 1937) 287 (first published 1892).
8. For some examples of earlier discussions of the physical nature of life see Lottka *Elements of Physical Biology* (Dover Publications: New York 1956) ch. 1 (first published 1924); A. E. Oparin *Life: Its Nature, Origin, and Development* (Academic Press: New York 1964) ch. 1.
9. We exclude as a scientifically verifiable hypothesis the fortuitous occurrence of initial conditions which are equivalent to a state of living matter simply as the result of rare fluctuations from a more probable configuration. This would amount to a 'special creation'.
10. Examples of some molecular biologists' attitudes may be found in J. D. Watson *The Molecular Biology of the Gene* (W. A. Benjamin: New York 1965), esp. p. 67; F. Crick *On Molecules and Men* (Univ. Washington Press 1966), esp. pp. 56–7.
11. Examples of some physicists' attitudes may be found in N. Bohr *Atomic Physics and Human Knowledge* (John Wiley and Sons: New York 1958) papers 1 and 2; W. M. Elsasser *Atom and Organism* (Princeton Univ. Press: 1966).
12. For an elementary discussion of *universal* computing machine see B. A. Trakhtenbrot *Algorithms and Automatic Computing Machines* (D. C. Heath: Boston 1963). Also M. Minsky *Computation: Finite and Infinite Machines* (Prentice-Hall: N.J. 1967) Part II.
13. This idea is expanded in H. Pattee (A. D.

Ketley, ed.) *The Stereochemistry of Macromolecules* Vol. III (Marcel Dekker: New York 1968) 305.

14. A. Sommerfeld *Mechanics* (Academic Press: New York 1952) 80. Also E.T. Whittaker *A Treatise on the Analytical Dynamics of Particles and Rigid Bodies* 4th ed. (Dover Publications: New York 1944) ch. 8.

15. R. Landauer *IBM J. Res. and Development 5* (1961) 183.

16. L. Szilard *Z. Physik 53* (1929) 840. This paper is discussed in L. Brillouin *Science and Information Theory* 2nd ed. (Academic Press: New York 1962) 176 *et seq.*

17. For example, M. Planck from his *Survey of Physical Theory* (Dover Publications: New York 1960) 64: 'For it is clear to everybody that there must be an unfathomable gulf between a probability, however small, and an absolute impossibility'. And on p. 66: 'Thus dynamics and statistics cannot be regarded as interrelated'.

Also H. Weyl in *Philosophy of Mathematics and Natural Science* (Princeton Univ. Press 1949) 203: '... we cannot help recognizing the statistical concepts, besides those appertaining to strict laws, as truly original'.

And of course we should include von Neumann in his discussion of measurement in quantum mechanics: 'In other words, we admit: Probability logics cannot be reduced to strict logics, but constitute an esssentially wider system than the latter, and statements of the form $P(a,b) = \theta (0 < \theta < 1)$ are perfectly new and *sui generis* aspects of physical reality.' [By '$P(a,b) = \theta$' von N. means: If a measurement of $a$ on a system has shown $a$ to be true, then the probability of an immediate subsequent measurement of $b$ showing $b$ to be true is equal to $\theta$.]

18. I am referring to Gödel type proofs of unsolvability and incompleteness theorems. See M. Davis, ed. *The Undecidable* (Rowen Press: New York 1965).

19. For more discussion on the physical basis of error in hereditary molecules see H. Pattee in (C. H. Waddington, ed.) *Towards a Theoretical Biology, I* (Edinburgh Univ. Press 1968) p. 86 *et seq.*

20. A collection of fundamental theorems on fluctuations and noise can be found in C. Kittel *Elementary Statistical Physics* (John Wiley and Sons: New York 1958) Part 2.

21. The idea of dissipationless change of shape conceivably playing some biological role was suggested by F. London *Superfluids* Vol. I (Dover Publications: New York) 8.

22. The relation of the quantum measurement process to molecular heredity is discussed further in H. Pattee *J. Theoret. Biol. 17* (1967) 410.

23. J. von Neumann in (A. W. Burks, ed.) *The Theory of Self-reproducing Automata* (Univ. of Illinois Press: Urbana 1966), esp. Lec. V.

24. M. Arbib *Information and Control 9* (1966) 177; C. V. Lee in *Mathematical Theory of Automata* (Polytechnic Press: New York) 155; J. W. Thatcher, ibid., 165; J. Myhill in (M. D. Mesarovic, ed.) *Views on General Systems Theory* (John Wiley and Sons: New York 1964) 106.

25. E. P. Wigner in *The Logic of Personal Knowledge* (Routledge and Kegan Paul: London 1961) 231, remarks on the '... "tailoring" of what substitutes for equations of motion ...' being only the result of hard, macroscopic systems with discrete variables. He also notes that the 'reliability' of classical models of self-replication needs to be evaluated and compared with experience.

26. See reference, note 12.

27. See reference, note 18, p. 72.

# Statistical Mechanics and Theoretical Biology

Martin A. Garstens

University of Maryland

It is hardly an understatement to say that nothing in the history of biological research has appeared and been found acceptable as a formulation of theoretical biology. A variety of disciplines have evolved in recent times contending for this title. Included among these are cybernetics, information theory, theory of automata, systems analysis, mathematical models, analyses based on computer technology and adaptations of physical models. Current philosophic discussion on a possible formulation of theoretical biology is endless but with little consensus. A large school of thought works on the assumption that biology is in a state similar to that of the early days of physics and that, given time, a theory will develop naturally, as it did in physics. The analogy is not good, since physics has always had more substantial relative strength of theory to empirical content than biology ever had. There is a feeling among some that biology is inherently incapable of yielding any theory; that it is the field of the unique and unpredictable; that its uniformities exist only in its physical aspects.

Not only has no acceptable general working theory appeared, in spite of the many attempts to generate one, but even more frustrating, there have not been indications as to which theoretical or mathematical techniques show promise of leading to a theory acceptable to biologists. It is also not clear how the above-mentioned disciplines (i.e. cybernetics, automata theory, etc.) relate to each other or to the establishment of a theory for the field of biology. It is the object of these remarks to clarify the issue and suggest a possible direction for more hopeful coordination of current activities in theoretical biology.

It is the thesis of this approach to the problem that the field of statistical mechanics, properly developed and expanded, contains the clue for a theory of biology. Statistical mechanics can be a coordinating area for all the disciplines which have attempted a theory of biology and supplies a general outlook suitable to settle many philosophical controversies such as those which arise between the reductionists and the non-reductionists. In addition, statistical mechanics is believed to show promise of infiltrating operationally into professional biological research, allowing a deep probing into the nature of biological measurement, a necessary activity if biological theory is to emerge.

## Statistical mechanics and theoretical biology

There are several obvious reasons why statistical mechanics should be considered as a clue to theory in biology. Statistical mechanics is the only area where a large degree of success has been attained in dealing with complex systems. The problems which remain to be solved in the field are reminiscent of those to be solved in complex biological systems. A most important reason for remaining 'close' to statistical mechanics is that it diminishes the tendency, particularly in philosophic discussions of biology, to wander too far afield from the realities of the physical world, without necessarily succumbing to a reductionist point of view.

The primary objective of statistical mechanics is to determine the connection, both in equilibrium and non-equilibrium, between the microscopic properties of ensembles of objects, such as atoms or molecules, and the macroscopic variables and laws observed in these ensembles. Such objectives are clearly manifested in thermodynamic studies where there also arises the important question as to the relationship of macroscopic measurement and microscopic events. How does a microscopic event result in a determinate macroscopic quantity during measurement ? (This problem is concerned with what is known as the reduction of the wave packet in quantum theory, a topic extensively discussed in connection with the theory of measurement in physics.) One of the main problems in irreversible thermodynamics is determination of the conditions for the existence of macroscopic quantities. It is also important to enumerate the complete set of macroscopic variables of a many-body system, given the Hamiltonian of that system.

The fact that some of the macroscopic quantities in statistical mechanics are impossible to observe, although they show their existence by the presence of a relaxation time, is of interest to biology. In some cases in statistical mechanics there are indications of the presence of several levels of macroscopic states manifesting themselves in different relaxation times. Both situations are suggestive of possibilities in biological phenomena where interactions between microscopic and macroscopic processes also give rise to such levels.

If the comparison of ensembles of biological systems with those occurring in statistical mechanics is valid, an expanded statistical mechanics would be expected to predict reliably those biological phenomena which are reproducible, in the sense of being common to all individuals. An expanded statistical mechanics should, in principle, be able to determine those aspects of behaviour which are common to systems of a given class. It is clear how this happens in inanimate systems. An examination of the foundations of statistical mechanics makes plausible the extrapolation of its methods into biology, constituting the sought-for foundation for theoretical biology.

In physics we deal with collections of identical microscopic particles, each of which is described by the laws of quantum theory. These particles, however, manifest their existence in terms of macroscopic measurements. Thus a particle passing through a cloud chamber leaves a macroscopic trail of cloud behind it. To deal with the macroscopic aspects of matter, through which one learns all that is known of nature, including its atomic properties, one proceeds in two ways. First the macroscopic properties of the physical world around us are observed, including the relationships among them. This is the subject of such branches of classical physics as thermodynamics, hydrodynamics, mechanics, etc. For example, in thermodynamics one deals with the macroscopic properties of heat, temperature, entropy, and pressure, and one finds them related through the first and second laws of thermodynamics. These properties and laws, historically, were inductive and empirical findings. Secondly, the microscopic or atomic properties of matter are observed and the macroscopic properties are reinterpreted in terms of them.

The discovery of atomic phenomena meant that certain macroscopic events were indicative of, and could be used to measure, microscopic events. The theory of measurement which attempts to trace this connection between macroscopic and microscopic events is still unresolved in modern physics. When clarified, it should play a fundamental role in theoretical biology.

Statistical mechanics has the task of relating the atomic domain with the macroscopic, determining how macroscopic measurements can lead to microscopic information and how microscopic events are recorded by macroscopic ones. It must show the connection between observed macroscopic variables and averaged microscopic properties. In order to do so, two types of assumptions are made. One consists of all the assumptions underlying quantum theory. The other is a set of auxiliary assumptions required to ease the difficulty in dealing with complex systems. To suit the needs of physical systems, various standard auxiliary assumptions are made. However, for a long time there has been an ongoing activity, called ergodic theory, to reduce all auxiliary assumptions to quantum theory or mechanics. This is now known to be impossible [1]. It is now seen that there cannot be any rigorous mathematical derivation of macroscopic equations from microscopic ones. Additional postulates or principles must be assumed to make this transition. This is precisely what must be done to establish the theory of complex systems in biology. Among the postulates which have been in common use in physics are: (*a*) the postulate of random phases, (*b*) the ergodic hypothesis, (*c*) the master equation assumption, (*d*) the assumption of equal *a priori* probability, (*e*) the metric transitivity assumption, and (*f*) the hypothesis of maximum entropy.

The hypothesis of maximum entropy is found capable of replacing most of the above assumptions. This replacement, first suggested by Elsasser [2] and recently extensively amplified by Jaynes [3], is receiving a great deal of current attention.

It is important to note that the whole set of auxiliary assumptions required to go from quantum theory to statistical mechanics are no less basic than quantum theory itself and are not reducible to the latter. However, while quantum theory, as applied to atoms and molecules (excluding high-energy phenomena) seems very well substantiated, the same is not true of the auxiliary hypotheses needed in practice to deduce the macroscopic variables from microscopic foundations. In fact the 'art' of obtaining good statistical mechanical analyses consists in knowing how to choose suitable auxiliary hypotheses (not reducible to quantum theory). The same art, much extended, needs to be applied to biology.

Accepting this account of the relation between the macroscopic and the microscopic in non-living systems, how can we profit by it in attempting an account of the living? There seems to be no reason to believe that quantum theory, the theory of the microscopic, does not carry over without change into biology. There has never been any evidence that individual particles violate Schrödinger's equation, or its equivalent, when applied to the atoms or molecules making up biological systems. However, in trying to describe such systems auxiliary assumptions are needed, as in the statistical mechanics of inorganic systems. An important question which arises is: Are unique auxiliary assumptions required to lead to observed biological macroscopic variables and laws?

If unique assumptions are required, they cannot violate presently accepted physical principles, nor can they be reduced to them. Evidently no amount of searching among current concepts in physics will fill the bill if biology is indeed irreducible. Just as in present-day fundamental particle research there is wide-spread 'feeling' that new principles are needed to explain the newly observed particles, so there is reason to 'feel' that more general ideas are needed in biology.

One method of approach to statistical mechanics, which shows promise of generating such general ideas, is that of E.T. Jaynes. In a series of articles [4] he has been able to relate information theory to statistical mechanics. By injecting the methods of statistics more directly into statistical mechanics it appears to be possible to obtain the required auxiliary principles and macroscopic variables by inductive methods. Assuming only that the entropy

$$S = -\kappa \sum_i P_i \ln P_i$$

is maximized in any system, observed physical quantities are equated to the average values of the operators representing them (see Appendix). Subject to these constraints, the density matrix $\rho$ of the system can be obtained. Knowing $\rho$, additional macroscopic observables characterizing the system can be calculated from the inductively obtained partition function and checked by experiment. This procedure can and has been applied to physical systems in a straightforward manner. For biological systems one is faced with difficulties of complexity, although the method seems sufficiently general to be applicable. One must expect of course that, at first, prediction must be confined to common aspects of species behaviour rather than to individual behaviour. But before a prediction can be made or a biological species characterized the knowledge of an excessive number of essential parameters seems to be required. So many as to make the problem seem hopelessly complex unless a means is devised for selecting only parameters which are most essential and thus greatly reducing their number.

There would appear to be some ways of accomplishing such reductions. One method is to eliminate from consideration those aspects of behavioural systems which do not seem essential to continued existence. It is known that certain parts of living systems can be removed without ensuing death. Secondly, by suitable restrictive hypotheses, gained from experience, one could diminish the generality of Jaynes' approach and make it more adaptable for biological considerations. Such restrictive hypotheses might involve topological constraints on the system rather than quantitative ones. These restrictive hypotheses combined with quantum theory could yield laws of developments describable in topological language. Hypotheses of this type would correspond to the auxiliary principles mentioned above, which are also required in physical systems in order to extract the familiar macroscopic variables.

G. Ludwig [5] has pointed to the need for specific additional hypotheses for distinguishing macroscopic variables in microscopic systems. The macroscopic variables in his analysis are characterized as continuous observables with respect to a certain defined macroscopic measure of discernibility $d(W_1, W_2)$, which is of the commensurable type and for which groups of continuous, weakly continuous, and strongly continuous operators coincide. Different measures of discernibility are then defined with respect to the ensembles $W_1$ and $W_2$.

Microscopic discernibility, for example, is defined as the distance:

$$d(W_1,W_2) \equiv \frac{1}{\sqrt{2}} \| \sqrt{W_1} - \sqrt{W_2} \| = \frac{1}{\sqrt{2}} (\sqrt{W_1} - \sqrt{W_2} ; \sqrt{W_1} - \sqrt{W_2})^{\frac{1}{2}}$$

This is followed by a similar but special definition of macroscopic discernibility. The axiomatic method is not the only approach to statistical mechanics nor even the most desirable. In practice one may use an empirical procedure as was done in thermodynamics and arrive at the variables inductively. In biology all methods will be needed and will play a rôle, since each presents a special picture and a unique point of view.

The importance of the above techniques for biological theory is that they maintain an intimate and necessary connection with the quantum theory of matter and still do not demand reducibility to the latter. This should produce great flexibility of method without separation from physics.

The auxiliary hypotheses under consideration should not be reducible to known principles, since they would then be redundant. They also should not be contrary to established principles. Finally, they must be capable of synthesis with quantum theory in the sense in which the above-mentioned principle of maximum entropy is used in conjunction with quantum theory. This means that they should not constitute an independent doctrine making no conceptual contact with quantum theory.

Special definitions and postulates, capable of being grafted on to basic physical foundations, are needed for biological theory. It is likely that all quantitative regularities in biology can be explained in terms of established physical principles. The required auxiliary hypotheses must therefore be qualitative in character, i.e. topological or statistical. Among the regularities in organisms which seem unique and demand explanation are their stability, unity, development, reproduction, memory, and evolution. All of these phenomena and many others have in common the incredible difficulty of explanation in terms of current physics.

There is a particular problem in trying to set up auxiliary postulates that could lead to or explain regularities of the above type. Whereas in physics it is generally easy to spot the macroscopic variables, usually small in number, this does not seem to be the case in biology. The concepts here are fuzzier, overlapping, and of far greater number. To postulate conditions giving rise to such variables and laws will therefore be difficult. Ultimately a large number will be necessary. At all times the questions of their interdependence will be a concern.

An example of the modelling of development or morphogenesis in biology can be obtained by setting up the conditions in irreversible thermodynamics under which a macroscopic variable grows, fades away, and then is transformed into another variable. It would be of interest to study the conditions under which several macroscopic variables interact so as to enhance some, weaken others, and eliminate the rest, or the conditions under which a given macroscopic variable is transformed into two, or the conditions under which two variables join to give rise to a third (reproduction ?). Many variations of these situations resembling processes in living organisms can be imagined. The importance of these considerations are that they are not too divorced from accepted, successful, and fruitful procedures in physics and that they supply ready-made formalisms. The mode of analysis involved also indicates the directions to be taken in search of needed mathematical techniques.

Since statistical mechanics is the only well-established physical science of the complex, it should have bearing on biological systems. At the same time statistical mechanics does not rule out empirical or inductive findings which must ultimately be incorporated within it. An amplified statistical mechanics can allow for pluralistic (or not completely integrated) functioning of systems, if such exist. It avoids the narrowness of an approach like the ergodic, which attempts to reduce all phenomena to quantum mechanics.

Through the proper use of auxiliary hypotheses, statistical mechanics can become sufficiently flexible to act as the much-needed area of integration of the various new disciplines which attempt to set up theoretical foundations for biology. The need is not only to integrate such disciplines with each other but to show their connection with the basic discipline of physics. The approach through statistical mechanics shows much promise of accomplishing these aims.

## Appendix

The method used is called maximum entropy inference, using the Shannon expression for entropy:

$$S_i = \kappa \Sigma P_i \ln P_i$$

or

$$S_i = -\kappa T_R(\rho \log \rho)$$

where $T_R$ stands for trace and $\kappa$ is an arbitrary constant. $S_i$ gives a measure of the amount of uncertainty in the probability distribution $P_i$ or the density matrix $\rho$. A density matrix or probability distribution is chosen which maximizes $S_i$ subject to the constraints supplied by the available information. Thus the most uniform assignment of weights to the states in the ensemble representation consistent with the given information is obtained. Any other assignment of probability over states would be equivalent to assuming some arbitrary additional information not warranted by the observed macroscopic measurements. The constraining equations are:

$$\sum_i P_i = I$$

and $$\sum_i P_i\, g_r\,(X_i) = \langle g_r \rangle, \qquad r = 1, 2, 3, \ldots$$

where the $\langle g_r \rangle$, the average or expectation values of $g_r$, are considered known and constitute the given data. The Lagrange method of undetermined multipliers maximizes $S_i$ subject to the above constraints, yielding:

$$P_i = e^{-\lambda_o - \lambda_1 g_1 (x_i) - \lambda_2 g_2 (x_i)}$$

which is the solution, with the multipliers $\lambda_i$ all determined, $\lambda_o$ being the log of the partition function. In fact

$$\frac{\delta \lambda_i}{\delta \lambda_r} = -\sum_i P_i\, g_r\,(X_i) = \langle g_r \rangle$$

Thus, knowing the partition function, the macroscopic observables can be determined.

### References


1. I. E. Farquer *Ergodic Theory in Statistical Mechanics* (1964).
2. W. M. Elsasser *Physical Review 52* (1937) 987.
3. E.T. Jaynes *Information Theory in Statistical Physics* (Benjamin: New York 1963).
4. E.T. Jaynes *Physical Review 106* (1957) 620; *Physical Review 108* (1957) 171.
5. G. Ludwig 'Axiomatic Quantum Statistics of Macroscopic Systems' in *Ergodic Theories* (Academic Press: New York 1961).

# Aspects of Evolution and a Principle of Maximum Uniformity

**Paul Lieber**
University of California

*General introduction and theoretical background.* This essay is concerned with the identification of various aspects of evolution and their connection with a universal evolutionary process, which emanates from irreducible and universal processes identified here with the Dimensional Universal Constants of Nature. These Aspects of Evolution are specifically concerned with Constancy, Force, Process, Equilibrium, Stability, Adaptation, and the Nature of Information. This first paper is specifically concerned with the nature of force, equilibrium, non-uniformity, and stability, envisaged here as particular aspects of evolution, conceived as a universal process which reconciles everywhere in nature, constancy, and change.

In this paper an outline is given of ideas and reasoning which led to the conception of a proposition that may prove to be a general law of nature, fundamentally endowed with aspects of evolution. This proposition embraces the laws of classical mechanics, a general stability law, historical thrust and commitment, and information relevant to the formulation of a theory conditioning strongly non-equilibrium thermodynamical processes. The stability law so obtained bears the same kind of relation to stability, envisaged here as a general aspect of the performance of classical mechanical systems, as do the laws of classical mechanics to another equally general aspect of their performance, namely equilibrium.

The conception of the general proposition which embraces this stability law is inextricably linked with a conception of the nature of force. By this conception, force is the universal and fundamental global aspect of non-uniformity posited in nature to sense perception and sense awareness, and from which all sensation, experience, information, and consequently knowledge, ultimately emerge. In the particular case of classical mechanics, force is here conceived as the universal manifestation in sensation of global non-uniformity in nature, that is, as the resultant of all non-uniform connections that exist between an inertial body instantaneously situated at a particular location, and the universe in which it is contained. From these considerations it follows that the dynamical aspect of classical mechanics (more specifically the kinematical aspect) which is based on a conception and description of process ascribed to immutable bodies in motion, is significantly more restricted and consequently less fundamental than is the aspect of nature symbolically designated by $\bar{F}$ in Newton's propositions. I use the word designated, rather than represented, in order to emphasize that this symbol, as it is used in classical mechanics, is not brought into correspondence with the anatomy and structure of nature's space-time manifold. Indeed, a critical examination of Newton's formulation and use of the known laws of classical mechanics does in fact suggest that he may have also tacitly conceived of force as an ultimate and global aspect of nature, and of his law of motion as a relationship between this ultimate aspect of nature and the motion of a body endowed with inertia. This point of view differs essentially from and denies the one taken by most of his followers, as well as from one expressed as a consensus among contemporary scientists, who choose to interpret his law of motion as a definition of force.

According to the ideas of this paper, force, as designated by the symbol $\bar{F}$ in Newton's propositions, in fact dominates the established laws of classical mechanics which are here understood to express only some of its fundamental aspects in nature. According to this paper, $\bar{F}$ in fact assumes a fundamental and dominant role in Newton's propositions. It dominates the dynamical term appearing in Newton's law of motion, which expresses only one of its particular manifestations within the domain of classical mechanics, and consequently does not define it. Instead, by this symbol Newton implicitly designated the resultant and thus total connection between a body endowed with inertia and the universe in which it exists. In so doing he implicitly assumed that this connection is independent of the frame of reference in which the motion of the body is described and calculated. This is tantamount to postulating, by implication, that the global aspect of nature symbolically designated by $\bar{F}$, and the connection it represents between a body and the universe, is covariant under all coordinate transformations. By treating force in this way Newton evidently displayed humility and wisdom. Humility, because he instinctively realized that the nature of the global connection between a body and the universe in which it exists is the most fundamental and least understood aspect of mechanics; and wisdom, by treating force as primitive, and thereby not imposing arbitrary restrictions on what is not understood. The present interpretation given to force as it appears in the propositions of classical mechanics is not explicitly represented in Newton's writings, but rather is inferred here from its usage and the way the symbol $F$ is formally treated in his propositions. From the above considerations it follows that the global connection designated by $F$, and called force, has the same stature in classical mechanics as does inertia, interpreted according to Mach's principle, according to which inertia is also a manifestation of a global connection between a body and the universe—a connection which, however, is characterizable by a scalar, and is therefore intrinsically endowed with high uniformity.

What is strictly local in Newton's propositions refers to their kinematical content on which their dynamical aspect is based. It is this dynamical and local aspect which restricts their covariance to inertial frames and consequently limits their generality. From this it follows also that it is naïve to interpret Newton's law of motion as a definition of force, as it is nonsense to define a fundamental aspect of nature which has unrestricted covariance in terms of an aspect whose covariance is limited to inertial frames. We see here again, from this point of view, that force does in fact dominate the laws of classical mechanics.

These considerations show that, in classical mechanics, the presence of a resultant force impressed by the universe on an inertial body, which is consequently not free, implies a non-symmetrical and thus non-uniform connection between the body and the universe. When the connection between an inertial body and the universe is symmetrical, and thus uniform, in the particular sense that individually impressed forces cancel vectorially—the body is then said to be free, according to the established laws of classical mechanics, and consequently moves according to Galileo's principle. We shall show in the section subtitled 'Hierarchies of Uniformity' that there in fact exists a hierarchy of free bodies, that is, bodies which can with meaning be distinguished as being more or less free, but all of which are equivalent and therefore not distinguishable by the established laws of classical mechanics.

These and other considerations concerning the nature of force, made within the edifice of classical mechanics, are sufficient to demonstrate that *all forces* in nature may be conceived as manifestations of the existence, in nature's space-time manifold, of non-uniform connections between inertial bodies and the universe. According to this thinking, forces that are revealed in the domain of classical mechanics emerge from the same ultimate and universal processes in nature as do all other forces. Force, thus conceived as the universal manifestation of non-uniformity in the space-time manifold, posited to sense-perception and to inertial bodies embedded in this manifold, brings into universal correspondence the various domains of physical theory which we have by convention learned to distinguish as classical and modern. These ideas and considerations are particularly designed to point out the fundamental connection between non-uniformity in nature and force, and to establish the thesis that force is the universal manifestation of these non-uniformities invoked in sensation.

## Evolution and maximum uniformity

I shall now introduce the observations and ideas which led to the conception of the principle of maximum uniformity, in which force is there conceived as the fundamental physical aspect of non-uniformity. The identification of a natural law is not an exercise in formal logic ; nor is what its propositions assert, provable. General propositions about nature can be tested only by experience – by what they predict and explain of it. In the present case the principle of maximum uniformity was discerned by fully generalizing explicit global information, which was obtained as a theorem for a class of dynamical systems, by suitably modifying and using Gauss's and Hertz's formulations of the principles of classical mechanics [1]. This information pertains to a global, positive, definite, scalar measure of the internal forces generated at each instant within such a system. The modifications of the Gauss-Hertz Variational Principles of Mechanics which render this general information explicit and without quadrature, consist of ascribing to force *the* dominant role in mechanics, and of identifying all forces in nature with an onto-logical-geometrical basis for the production of stringent geometrical constraints, which were [1–4] originally conceived to emerge from the impenetrability of matter understood as a property of position. This information, which bears directly on the fundamental problem of continuum mechanics, has not been made explicit, and as far as I see *cannot* be made explicit by Newtonian Mechanics, in which the only representation, given to force in its propositions, is vectorial. This means that in the significant sense of information rendering, the various formulations of the principles of mechanics are only conditionally equivalent. This development led me inexorably to the concept of 'Categories of Information', in terms of which questions concerning the equivalence and non-equivalence of various formulations of the principles of mechanics can be rationally examined and resolved. This led to identification of eleven distinct, yet related, Categories of Information by examples derived from familiar as well as more sophisticated aspects of experience. Once cited, these examples invoke consensus [5].

The global information, so explicitly obtained as a theorem on the distribution of internal forces, asserts that a positive, definite, *scalar* measure of all the internal forces is *instantaneously* less for the actual motion than it is for any other motion which satisfies the initial conditions and the geometrical constraints (as well as the external forces) which are instantaneously impressed upon the dynamical system. This theorem was established for a particular (non-trivial) class of dynamical systems. For this class, the scalar measure of the internal forces can be directly interpreted as a global measure of non-uniformity in momentum space.

The Principle of Maximum Uniformity, as it pertains to classical mechanics and classical continuum mechanics, was obtained by (a) interpreting the information obtained from the above theorem as a particular aspect of a general law which holds in all mechanical systems, and (b) introducing the concept of conditionally stringent geometrical constraints and relating these to material properties through which they are implemented in nature. This brings the Principle of Maximum Uniformity into correspondence with thermodynamical aspects of the equations of constitution of various materials and relates the idea of conditionally stringent geometrical constraints to uncertainties in the initial conditions from which historical commitment, causality, and a general stability principle naturally emerge.

The phenomenological description of the performance of classical mechanical systems reveals two general and mutually independent characteristics, namely equilibrium and stability. The known propositions of classical mechanics refer strictly to equilibrium, by invoking the condition that forces be instantaneously in equilibrium everywhere and for all time in the system. This is their information content. They report nothing of stability, which is an equally general and fundamental aspect of the behaviour of classical mechanical systems. The laws of mechanics give but limited expression to the Principle of Maximum Uniformity by asserting that the forces acting everywhere in a system sum vectorially to zero in all directions. This restriction allows a multiplicity of directional and spatial distributions in the magnitude of the forces impressed upon a body, without therefore exercising a condition on preferred distributions which the stability principle presented here in fact does.

*1.1. Concerning the nature of evolutionary adaptation*

The considerations noted above help demonstrate that force, equilibrium, and stability are particular manifestations of an overriding tendency in nature to increase a global measure of uniformity identified with the global structure of the space-time manifold. This process is envisaged here as universal and conditioned by the principle of maximum uniformity, namely (a) that force is the instrument for increasing uniformity in nature, or what is equivalent, the instrument for effecting reduction of global non-uniformity existing in the space-time manifold, (b) that all forces in nature emerge from these global non-uniformities, and constantly act to reduce them, and (c) that forces are the universal manifestations of non-uniformities in nature as they are directly posited to sensation.

Evolutionary adaptation is envisaged here as a universal aspect of all process in nature, an aspect which reconciles constancy and change in all of their ramifications in natural phenomena. The thrust of evolutionary adaptation, so conceived, derives from the ultimate processes, embedded in the space-time manifold, which drive and construct the manifold by irreversible connections which must necessarily exist between these ultimate processes and the manifold. The irreversible connections are implied by the immutability of these ultimate processes, called here the universals, as they are reflected in, and revealed by, the Dimensional Universal Constants of Nature with which they are here identified.

The universal adaptive process described above has been conceptually identified with, and emerged from, a conceptual model of nature's space-time manifold which is endowed with certain essentially ontological features inferred from the Dimensional Universal Constants [6]. These (essentially) ontological characteristics were independently discerned in concurrent studies that initiated in 1947, which is based on C. F. Gauss's [7] and Heinrich Hertz's [8] formulations, of the principles of classical mechanics which in each case was motivated by a quest to grasp the nature of force by attempting to establish force on a strictly geometrical foundation. This endeavour was initiated by Gauss in 1829 and culminated at the turn of the century by Hertz's last and monumental work entitled 'Principles of Mechanics'. In this profound and beautiful work Hertz formally constructs a $6N$ dimensional Euclidean Manifold in which the motion and state of a classical mechanical system consisting of $N$ bodies free of prescribed forces are described and represented. Hertz restricts the admissible motions and states of the mechanical system by formally subjecting the coordinates of its bodies to constraints which in the most general case are considered as non-integrable and therefore non-holonomic. As the application of these geometrical constraints to a body restricts its freedom geometrically, these restrictions must emerge in the Newtonian scheme as forces.

The study based on the Gauss-Hertz formulations* has produced two results which bear on the conception of the Principle of Maximum Uniformity and on the identification of a physical, that is, of an ontological-geometrical basis for the production of actual stringent holonomic as well as non-holonomic constraints in nature's space-time manifold. This ontological-geometrical basis gives physical support and justification for the existence in nature of the geometrical restrictions which Hertz used to effect a formal reduction of force to geometry, and serves to identify the formal representations he gave to non-holonomic constraints, with experience and thus with nature.

The same study revealed that Hertz's edifice, in which he formulated a general law governing the motion of forceless mechanical systems subjected to non-holonomic geometrical constraints, and which renders obsolete all previously

* Some results of this study are presented in [2].

known formulations of the laws of classical mechanics, also accommodates the formulation of a new and general stability law cited above. This stability law, which bears the same kind of relation to stability as do the established laws of classical mechanics to equilibrium, is found to be independent of the known laws of mechanics and to embrace fundamental and general information not included in these laws. This information bears on historical thrust and commitment and derives from an adaptive-evolutionary process ascribed directly to the geometrical restrictions which impress non-uniformities on the space-time manifold, and from which all forces are understood here to emerge. This entails the identification and classification of holonomic and non-holonomic ontological-geometrical constraints into the following types : ( 1 ) Active Stringent Constraints, ( 2 ) Passive Stringent Constraints, ( 3 ) Conditionally Stringent Passive Constraints. This classification led naturally to the idea that the annihilation of conditionally stringent passive constraints which are ascribed here to universal congruence restrictions impressed on the space-time manifold by the irreducible universals identified by the Dimensional Universal Constants constitutes a fundamental and general instrument of adaptation in the space-time manifold. It is this crucial instrument which allows one to posit a general stability law for classical mechanical systems and which affords according to the Principle of Maximum Uniformity the mechanism which is essential for physically producing *the required many to all mappings evident* in biological systems.

The annihilation of conditionally stringent constraints is accompanied by consequent modifications of the forces emanating from the non-uniformities induced by them in nature's space-time structure. According to the observations and reasoning of this paper, the annihilation of conditionally stringent constraints is envisaged as an essential feature and instrument of adaptation, conceived here as a general and universal aspect of all process in nature seated in the space-time manifold. This universal process of adaptation in nature, and consequently the process of annihilation of conditionally stringent constraints upon which it incisively depends follow by the thesis of this paper the Principle of Maximum Uniformity.

### *1.2. Concerning aspects of uniformity and non-uniformity*

Some significant aspects of uniformity and non-uniformity revealed in human experience are cited in this section. This is done in order to point up their universal role in natural phenomena, and as a consequence the strong implications of the principle of maximum uniformity noted here and which will be further examined in depth in subsequent papers under the present title. These aspects of uniformity include :

1. Symmetry
2. Equilibrium : local, global, spatial, and temporal
3. Stability : local, global, spatial, and temporal
4. Isotropy : a local aspect
5. Homogeneity : a global aspect
6. Constancy
7. Invariance
8. Covariance
9. Law
10. Correspondence
11. Element
12. Order
13. Reproducibility : the ultimate criterion and requirements of scientific investigation

The Corresponding Aspects of Non-Uniformity include :

1. Force : the most fundamental and universal aspect of non-uniformity posited to sense perception and sense awareness
2. Asymmetry
3. Information
4. Curvature
5. Symbol
6. Language
7. Anisotropy : a local aspect
8. Inhomogeneity : a global aspect
9. Gradient
10. Structure
11. Shear
12. Constraints
13. Uncertainty
14. Fluctuations
15. Disorder

To each set of conditionally stringent constraints there corresponds a positive definite scalar measure of non-uniformity as manifested in experience by the internal forces. The relaxation of such constraints increases uniformity, and the selection among a possible set of conditionally stringent constraints is made to maximize global uniformity in adherence with the Principle of Maximum Uniformity amplified in the following section 2. The universal constants embrace constancy and process and thus both uniformity and non-uniformity. This is the synthesis in the elementary processes which they reveal.

The process of reducing the non-uniformities in nature's space-time manifold is here envisaged to be the ultimate aspect of all adaptive phenomena in nature. Evolution becomes then a word labelling this universal adaptive process. An aspect of evolution that is both essential and universal is *force,* and its nature we evidently do not grasp more in physics than in biology.

*1.3. Concerning the nature of non-holonomic constraints*

Non-holonomic geometrical constraints are constraints that emerge physically from contact. These constraints are therefore directly tied to the printing process, and consequently, necessarily to local impenetrability, conceived here as the ontological-geometrical basis and support for all printing and sensation in nature. Indeed, non-holonomic constraints are significant aspects of local impenetrability. The question concerning how non-holonomic constraints are made in nature is necessarily linked to the fundamental question which emerged from the study of Hertz's Mechanics, cited above, namely: What is the ontological-geometrical basis in nature of the stringent geometrical constraints formally invoked by Hertz to geometrize force? These considerations, which bear on Pattee's [9] interesting discussion of the role of non-holonomic constraints in relating the genotypes and phenotypes of organisms, will be developed in a separate essay.

The process of randomization in gas flows has been elucidated in terms of a dissipation mechanism [3]. This mechanism is based on the propositions of classical mechanics and unilateral non-holonomic constraints, which come directly from the ontological-geometrical property of matter described here as local impenetrability. This dissipation mechanism bridges conceptually unilateral non-holonomic constraints, oblique collisions, asymmetries in momentum-configuration space and randomization, with a process of intrinsic irreversibility which emerges from them and the laws of classical mechanics. The evolution of shock waves in gas flows and their associated zones of high temperature and dissipation have been explained by this dissipation mechanism. As randomization is under certain conditions an aspect of uniformization, we are able to establish a correspondence between the Principle of Maximum Uniformity and this mechanism of randomization.

*2. Concerning the nature of force, equilibrium and stability*

Among all the symbols which appear in the formal statements of the propositions of classical mechanics, the symbol $\bar{F}$, used to abbreviate the *name* Force, desginates what is recognized here as the universal and the most fundamental aspect of nature posited to experience within the domain of classical mechanics. Although, as previously noted, Newton did not explicitly associate with this symbol the universal aspects of nature which are revealed in physics by the Dimensional Universal Constants, his instinct for its dominant role in the propositions of classical mechanics is nevertheless suggested by the constancy he implicitly ascribed to it, under transformations connecting non-inertial as well as inertial coordinate systems. Since the symbol $\bar{F}$ is simply a designator, that is, a name, and not a representation of a well-defined aspect of nature relating to the space-time manifold, it is in fact inconceivable how Newton could have, in principle, formally prescribed to it geometrical conditions which refer to coordinate systems and to their transformation. To do so he would first have to describe geometrically the architecture in the space-time manifold, from which *all* forces derive. I emphasize *all* forces because this is the meaning that we must with logical necessity ascribe to the symbol $\bar{F}$, since it must necessarily designate the sum of all connections which exist between a non-free body and the universe in which it is contained. This way of thinking is amplified below, and is sufficient to draw the inference, on strictly logical grounds, that the aspect of nature designated by the symbol $\bar{F}$ in the propositions of classical mechanics must necessarily be endowed with information content that refers to the Universals of Nature and which is therefore directly related to Dimensional Universal Constants of Nature. Although this logic, which by itself is sufficient to infer that the symbol $\bar{F}$ necessarily labels universal aspects of nature, is clearly not included in the familiar body of knowledge reported under the aegis of classical mechanics, it nevertheless evolved from a critique of the concepts which led to the emergence of this symbol in the history of classical mechanics.

In this regard it is important to emphasize that the unrestricted covariance Newton tacitly ascribed to what the symbol $\bar{F}$ designates in the space-time manifold allows the interpretation I have given to it here and therefore the universal information content which is here ascribed to it. In other words, Newton's intuition led him to a formulation which allows us to assign to the symbol $\bar{F}$ appearing in his propositions, the role of designating the connections between the most fundamental and universal aspects of nature and all experience manifested within the domain of classical mechanics. This inference, which reports that in classical mechanics the symbol $\bar{F}$ necessarily designates universal aspects of nature, was derived by implication from a critical examination of the concepts which led to its emergence in the development of the known laws of classical mechanics.

Historically, according to Aristotle, the motion of a body calls for the constant application of an externally applied impetus to sustain it. This deeply entrenched view was subsequently denied by Galileo on experimental grounds which led him to the conception of a free body in motion, that is, a body which can sustain a uniform velocity indefinitely, free of an externally applied agent. Newton, as did Galileo before him, conceived of inertia as an innate property of a body, and consequently as a local property of its location in space. Mach, however, conceived of the inertia of a free body as the manifestation of its global connection with the whole universe in which it is embedded. According to Mach's Principle, a so-called free body is nevertheless joined to the universe, but in a very special way, as revealed by a scalar connection manifested by its inertia. The connection between a free body and the universe is not necessarily strictly isotropic, that is, completely and thus maximumly uniform in all directions. The particular and indeed very special case of a maximumly uniform connection may be envisaged as equivalent to a strictly isotropic force field. Although for this very special case the resultant of all forces connecting a free body with the universe vanishes, this condition may be satisfied under much less stringent conditions of uniformity; all of which equivalently satisfy the definition of a free body as presented by the established laws of classical mechanics. This observation led me naturally to the concept of 'Hierarchies of Freedom' which correspond to differences in the degree of uniformity or order of the connections existing as forces between bodies and the universe, all of which equivalently meet the condition of a free body as required by the laws of mechanics.

By Newton's view of inertia, a free body must necessarily be understood as disjoined, that is, as insulated from the universe in which it exists. This on reflection violates intuition, unless the body is envisaged as existing in a strict vacuum, that is, in a strictly empty space which insulates the body from all matter and radiation in the universe. With this view in mind, Newton extended Galileo's principle by in effect posing the question: what is the case when the body is not free; that is, not strictly insulated? In so doing he vectorially *designated* the sum of all the individual connections that exist between a body and the whole universe by the symbol $\bar{F}$, and evidently understood these individual connections as emanating from sites strictly external to the body-sites which are individually manifested by individual agents called forces, externally applied and which follow the law of linear superposition.

*2.1. The principle of universal correspondence*

The symbol $\bar{F}$ is strictly a name, that is, a label which is devoid of representational content of the space-time architecture and anatomy from which the connections named forces emerge in the space-time manifold. In classical mechanics we must therefore necessarily interpret $\bar{F}$ as a dummy label designating the resultant of all connections existing between a body and the whole universe. Accordingly we must associate with the symbol $\bar{F}$ a universal and completely global aspect of nature, endowed with the highest information content* manifested within the domain of classical mechanical experience—a content which it however does not explicate in the sense of a representation.

We shall see that the interpretation given here to the symbol $\bar{F}$, which is indeed compatible with its usage in Newton's formally stated propositions of classical mechanics, allows the accommodation and satisfaction within the domain of classical mechanics, of a fundamental principle which was conceived in a study based upon the Universal Constants of Nature [6]. This principle stems directly from a philosophical axiom which, if denied, leads to a profound philosophical catastrophe, that is, to a conclusion which is conceptually absurd, and which denies common sense as well as philosophical intuition. This axiom asserts that the universal and elementary aspects of nature are prominent and indeed imminent in *every* natural phenomenon, every process, and every experience, and in particular therefore in experiences which transpire within the domain of classical mechanics. In reference [1] the universal and irreducible aspects of nature are identified with the universal constants of nature which are revealed in the domain of experience, conventionally called physics, by the Dimensional Universal Constants. The Dimensional Universal Constants of Physics are accordingly conceived as being manifestations in the realm of what we by convention call physical reality, of the truly elementary processes in nature which universally and immutably transpire in the space-time manifold. This realization is, I believe, tantamount to a partial fulfilment of man's search and sustained quest for the identification of the elements of nature, which of course derives from his faith and conviction that they indeed exist. The axiom cited may be paraphrased to read that what is universal and fundamental in nature can in no sense be incidental, either phenomenologically or in the laws that condition the phenomena incurred in various realms of scientific experience. To claim the contrary is considered here absurd and to deny common sense and philosophical intuition.

* The notion of information content is further considered in the section 3, in which 'Hierarchies of Information' was conceived as aspects of 'Hierarchies of Uniformity'.

Paul Lieber

The extension of experience immediately accessible and manageable by the human hand to domains that are remote from everyday experience and that require particular technological skills to engage has revealed natural phenomena which cannot be accommodated by the laws discovered earlier in physical phenomena that were and remain immediate in everyday experience. The theoretical accommodation of these newer phenomena demanded the formulation of new laws in which certain Dimensional Universal Constants first emerged in physical theory, as for example in the case of the Special Theory of Relativity and Quantum Mechanics.

To avert a philosophical catastrophe in the progressive development of physical theory, Bohr invoked 'The Correspondence Principle'. This principle demands that the formal statement of the laws conceived for the neo-classical and modern domains of physical experience correspond asymptotically to the laws conceived earlier for the classical domain in such cases where the phenomena being experimentally examined belong strictly to the domain of everyday experience. Bohr's Correspondence Principle thus leads to the result that the terms which in the neo-classical and modern physical theories include the Dimensional Universal Constants become asymptotically small, and thus negligible, in comparison with the terms which are retained in going over from the neo-classical to the classical domains of experience. Although Bohr's Correspondence Principle averts in this way a theoretical catastrophe, it does not however address itself to what might be a much more serious theoretical catastrophe, namely the violation of what is here called the 'Principle of Universal Correspondence'.

This principle, which stems directly from the above cited axiom and which asserts that the universal and irreducible aspects of nature are *imminent* in *all* natural phenomena, demands that in this specific sense *all* natural phenomena and their laws bear to each other universal correspondence. By this principle every aspect of nature is endowed with the universals, reflected in physics by the Dimensional Universal Constants, and are understood to be controlled by a universal law of evolution pertaining to the global action of these universals upon the space-time manifold in which they are conceived to be embedded, and which they internally drive, structure, and organize by a universal process of evolution. Every aspect of nature, including therefore such phenomena which by convention we refer to as classical, bears in this sense universal and complete correspondence to every other phenomenological aspect of nature, be it classical or non-classical according to the usual definitions.

By this thinking every natural phenomenon, classical as well as non-classical, is in effect a micro-cosmos of the universe, in the sense that they are all pre-eminently endowed with the universals which irreversibly drive, structure, and organize them in the space-time manifold, according to a universal evolutionary law, which operates in every phenomenological domain. Accordingly, all natural phenomena bear to each other, and to the universe as a whole, complete phenomenological correspondence, as do their laws which have been progressively forged out of experience in both the classical and non-classical domains.

The Principle of Universal Correspondence immediately presents a dilemma when it is invoked conceptually and formally in the domain of classical mechanics. From the above considerations, the symbols which label the Dimensional Universal Constants in the laws of physics are understood as designators, that is, as simply names with physical dimensions, that reflect the presence in the space-time manifold of immutable, and consequently ultimate processes called here universals. These symbols are accordingly the formal designators of these universals, within the formal statements of known physical laws.

According to Bohr's Correspondence Principle, which for obvious reasons I shall refer to as the principle of asymptotic correspondence, the terms in the formal statements of physical laws which include the symbols labelling the Dimensional Universal Constants are asymptotically eliminated when these laws are put into correspondence with the domain of classical mechanics. In Relativistic Mechanics, for example, the terms including the Universal Constant $C$, the speed of light, modify only the dynamical part of Newton's classical laws and do not and indeed cannot, for reasons given earlier, modify the unrestricted virtual covariance *a priori* ascribed to the symbol $F$ for all coordinate transformations.

The present dilemma concerns reconciling the Principle of Universal Correspondence with the elimination within the domain of classical mechanics, of the symbols which label the universals in the physical laws pertaining to the non-classical domains. The obvious question to answer is: wherein are the universals manifested in the domain of classical mechanics, as demanded by the Principle of Universal Correspondence? The answer is Force, that is, in the aspects of nature which are formally designated by the symbol $F$ in the propositions of classical mechanics. This is the symbol to which we earlier and inexorably ascribed on broad but incisive conceptual grounds an information content which dominates the information content designated by all the other symbols appearing in the formal proposition of classical mechanics. Now, however, we see from the Principle of Universal Correspondence that the universal aspects of the designator $F$ and its consequent dominant position in the formal statement of the laws of classical mechanics can and must be attributed to its rôle as a designator within the domain of classical mechanics, of the universals, as they are directly posited to experience within this domain. By this thinking, all forces in nature are endowed with the universals, and are manifestations in direct experience of a universal law of evolution which refers to the action of the universals on the space-time manifold to which I ascribe everywhere the ontological-geometrical-temporal property I call local impenetrability. Moreover, it follows from these considerations that a long-standing and prevalent view among mechanicians and scientists generally, which holds that Newton's laws define force, is untenable and indeed absurd. What these laws do is simply state certain connections between force and parameters that describe the behaviour of classical mechanical systems according to some preconceived ideas which are now in fact challenged by more recent developments in modern physics.

The known laws of classical mechanics do not define force, but instead give only limited expression to its information content, much of which in fact remains untapped by them. As a consequence, forces as they are directly posited to experience in the domain of classical mechanics accommodate a statement of a new fundamental and general law governing the performance of classical mechanical systems, which reports fundamental and general information which is not contained in the long-established laws of classical mechanics. The statement of this new law is made possible by the inherently high information content of force which derives from the universal aspect of all forces in nature, and from the fact that this information is only partially explicated by the known laws of classical mechanics.

*3. Concerning aspects of hierarchies of uniformity*

As shown earlier in this manuscript, we can interpret the resultant force posited to a non-free body as the vector sum of all non-uniform connections which exist between the body and universe. Each force individually contributing to this sum posits to the body a non-uniform aspect of the universe. In cases when the vector sum of these individually applied forces vanishes, we previously considered the body as free but not disjoined from the universe. In these cases the individual forces may be envisaged as existing in mirror symmetric pairs, the forces in each pair being consequently equal in magnitude. However, according to the usual laws of classical mechanics, the definition of a free body does not demand that the magnitudes of the individually applied forces be uniform for all pairs.

From these considerations we learn that there exist hierarchies of free bodies, all of which are equivalent according to the known laws of classical mechanics and which are therefore not discernible or identifiable by these laws. The hierarchies of free bodies may be identified and designated by either the degree of uniformity or non-uniformity of the magnitudes of the individual forces that are the immediate manifestations immanent in experience of particular aspects of non-uniformity existing between a body and the universe. Since all free bodies which belong to these various hierarchies ( of freedom ) are equivalent according to the presently established laws of classical mechanics, these laws cannot in principle render conditions which select among the many actual-possibilities these hierarchies afford at each instant a particular one that belongs to a particular hierarchy of freedom. The concept 'Hierarchies of Freedom' is a particular aspect of the concept 'Hierarchies of Uniformity'.

It is helpful to point out some other equivalent aspects of this concept, as it assumes a crucial role in the statement of a general principle of evolution which is in accord with the Principle of Universal Correspondence, and which is consequently understood to operate universally in all natural phenomena, including those which belong to the domain of classical mechanics. Some equivalent and related aspects of the concept 'Hierarchies of Uniformity' include 'Hierarchies of Symmetry', 'Hierarchies of Certainty', 'Hierarchies of Order', 'Hierarchies of Information', Hierarchies of Compatibility', 'Hierarchies of Harmony', 'Hierarchies of Forces', and 'Hierarchies of Consistency'. Moreover, in all of these cases it is important to distinguish between what in each case corresponds to the local aspects of uniformity and to its global spatial-temporal aspects. It is clear that the established propositions of classical mechanics do not and cannot make such a distinction, because the restrictions they impose on mechanical systems apply instantaneously and locally, everywhere as well as for all time. As the conditions they invoke, namely that forces be instantaneously in equilibrium everywhere and always, are uniform in space and time, they do not implicitly describe or define, nor do they condition the existence and the spatial-temporal evolution of local and global non-uniformity in their various hierarchies. This is the reason they are inherently devoid of historical thrust, causality, and evolutionary process.

It is the universality of all forces in nature, and therefore in particular of those forces which in the classical domain are designated by the symbol $\bar{F}$, that facilitates invoking and applying the principle of evolution cited above, in the domain of classical mechanics. The established laws of classical mechanics, in all of their equivalent formulations, express a particular and restricted aspect of the Principle of Maximum Uniformity, an aspect which, as explained earlier, is independent of location and independent of time. These laws consequently express universal propositions, that is, truths which are necessary in the strict logical sense, and are therefore not contingent upon space and time. These laws are in the sense of Liebniz logically universal, that is, necessary and analytic, as they are not contingent upon conditions that refer to particular places and to particular times. It is important to emphasize in this regard that these laws refer to a particular and restricted aspect of uniformity which is characterized and defined by the equilibrium of forces, and they assert that this particular aspect of Hierarchy of Uniformity is constantly maintained at all locations and is therefore not contingent upon space or time. In other words, the laws of classical mechanics as well as the particular Hierarchy of Uniformity to which they refer, namely the hierarchy characterized by the Equilibrium of Forces, and which as laws they report to be a general aspect of nature, are *both constant* in space and time, and are thus both free of contingency. If we follow this way of thinking, the usual laws of classical mechanics may be conceived as developing in two steps. The first consists of a definition of equilibrium, in which force is the aspect of nature to which the word equilibrium in the definition refers. The second uses this definition to express the universal law which asserts that equilibrium so defined is constantly maintained in nature, that is, everywhere and at all times. In other words, the particular Hierarchy of Uniformity which is characterized by the equilibrium of forces, as well as the laws of classical mechanics which ascribe this aspect of uniformity to all nature, are both constant with space and time, and thus free of contingency.

The existence in nature of Hierarchies of Uniformity, which as in the particular case of equilibrium are all directly revealed in experience by forces, leads here naturally to the identification of a universal law, which although free of contingencies in assertion, nevertheless conditions aspects of nature which are in fact contingent upon the evolution in space and time of distinct Hierarchies of Uniformity. The law does not in this case constantly refer to a particular hierarchy, but reports a universal proposition that governs a process of evolution which is contingent upon the emergence in space and time of the various Hierarchies of Uniformity. The usual laws of mechanics which are indeed embraced by this general law are a very special case of it, in so far as the particular Hierarchy of Uniformity in terms of which they are expressed is always *fixed* and therefore not contingent. This is precisely the reason why the established laws of mechanics are inherently and completely devoid of contingency in all aspects and consequently of historical thrust, causality, stability criteria, and evolution. This is, of course, also true for all of the so-called equivalent formulations of the laws of classical mechanics, and in particular therefore for their formulation in terms of the principle of least action. The reason I refer here particularly to the principle of least action is its power, and unifying role in physical theory. The power of this principle in the formulation given to it by Hamilton is seen by the fact that not only classical mechanics of particles and rigid bodies, but also elasticity and hydrodynamics, electromagnetism, and all modern field theories connected with ultimate particles (electron, proton, neutron) can be formulated with its help. All of the theories which are formulated with its help, therefore, share with Newton's formulation of the laws of classical mechanics the important feature of being devoid of historical commitment, causality, and inherent stability criteria. In other words, all of these theories are free of historical content, and consequently essentially devoid of an evolutionary principle.

*4. Concerning the principle of maximum uniformity and a general stability law*
We have shown earlier that the formulation of the laws of classical mechanics may be conceived in two essentially distinct steps, the first being a definition of equilibrium, whereas in the second the proposition is made that equilibrium as defined by the first step holds constantly everywhere and for all time. The notions of stability and equilibrium both developed by observing and critically examining the phenomenological behaviour of classical mechanical systems. As explained in the case of equilibrium, a general operational definition based on forces was established on the basis of experience, and then used in the formulation of the known laws of mechanics, which inherently report nothing about stability for reasons already gone into. Whereas the notion of stability has been described by many definitions, these have led to various stability criteria which are statements of convention rather than a general law—to a law that refers to stability as do the laws of mechanics to equilibrium. I shall now endeavour to formulate a statement of such a law, that is, of a general stability law, which refers to all of the Hierarchies of Uniformity and bears to them the same kind of relation as do the known laws of classical mechanics, to the particular hierarchy of uniformity characterized by the equilibrium of forces. For this purpose it is first necessary to identify and descriptively define the Hierarchies of Uniformity in terms of forces, which as explained above are interpreted here as the most fundamental, universal, and direct manifestation in experience of the non-uniform connections existing between the universe and bodies contained in it.

We start by considering in some detail the very special and fundamental Hierarchy of Uniformity to which the known laws of classical mechanics pertain. This special hierarchy is defined by the characteristics that the vector addition of all the non-uniform connections existing between a body and the universe which are posited in experience, and which we designate by the name force, sums to zero. It is clear that there can exist a conceivably infinite number of distinct configurations of forces impressed on a material point which individually designate the individual non-uniform connections between it and the universe, all of which equally belong to the very special hierarchy of force equilibrium. It is the *differences* between these distinct but otherwise equivalent force configurations which I define as the 'Hierarchies of Uniformity'. The figure below illustrates pictorially how we can conceive of an infinite number of distinct force configurations, all of which belong to the hierarchy of uniformity defined by the equilibrium of forces and which by their differences here define the 'Hierarchies of Uniformity'.

The 'Hierarchies of Uniformity', so descriptively defined in terms of force fields, are now used to formulate a Principle of Maximum Uniformity which virtually includes the known laws of classical mechanics as well as a general stability law. This stability law bears the same relation to stability which will be defined here in a particular way, as do the established laws of classical mechanics to the equilibrium of forces as it is defined in classical mechanics.

The Principle of Maximum Uniformity asserts that among all the force configurations which can be collectively and instantaneously accommodated in a finitely extended material domain, non-uniformly connected to the universe by maintained forces, and which individually belong to the special hierarchy of uniformity characterized by force equilibrium ; the particular set of force configurations which actually evolve and which satisfy the instantaneous and stringently exercised geometrical constraints, instantaneously maximizes a global positive definite scalar measure of uniformity obtained by summing local measures of uniformity that depend on the local force configurations over the entire domain.

This statement of the Principle of Maximum Uniformity differs essentially from the statements of the established laws of classical mechanics. Whereas, as explained above, the laws of classical mechanics are essentially atemporal, acausal, and consequently devoid of historical commitment and evolutionary process; the Principle of Maximum Uniformity, though conceived here as a universal proposition, nevertheless refers to essentially contingent aspects of nature expressed in terms of 'Hierarchies of Uniformity' which in general evolve non-uniformly in space-time. It is precisely because the universal and established laws of classical mechanics constantly refer to one, and only one, hierarchy of uniformity that they are free of contingency in all respects, and are consequently in principle amenable to mathematical formulation. This follows, as all mathematically stateable propositions are essentially free of contingencies which refer to space-time and therefore in principle devoid of historical content.

The Principle of Maximum Uniformity is indeed a *procedure* rather than a formally stateable proposition—it is the description of a process, of a universal process, that is, a process which is understood to operate universally. In this process the existence and operation in the space-time manifold of contingently stringent geometrical constraints, as well as absolutely stringent passive constraints, are among its essential features. Because (1) time, conceived as duration rather than the times of events ordered as points on the real time line, (2) the ontological-geometrical ground for passive-stringent geometrical constraints, is ascribed here to local impenetrability of matter, (3) force is the essential instrument in nature for effecting compatibility and excluding contradiction in nature by reconciling its universal and contingent aspects, and (4) the temporal and spatial contingencies expressed by the spatial-temporal evolution of various and distinct hierarchies of uniformity; are all involved in the operation in nature of the Principle of Maximum Uniformity, it follows that its description and statement cannot in principle be completely mathematically formulated.

This conclusion has direct bearing on the questions that concern the nature of biological theory and the kind of laws we can expect it to produce. It of course also bears on the nature of physical theory and the fundamental implications inherent in the formal statements of its laws. It is precisely because of the fact that they can be given mathematical expression that they are in principle devoid of all contingency, and consequently of historical content and thrust, inherent stability criteria, causality, and evolutionary process. Conversely, it is because the laws of physics are essentially ahistorical and acausal that they can be given mathematical formulation.

The second law of thermodynamics is indeed unique among the laws of physics. Whereas the other laws of physics do not know about ageing, and therefore about history, the second law does consider and compare earlier and later states of systems, but not how they evolve from the earlier to the later states.

We can sum up by saying that the physical laws as they are known are space-time invariant and thus not contingent, and that the aspects of nature to which they refer are devoid of ageing process. Laws of nature may however be space-time invariant and still refer to fundamental aspects of nature which are nevertheless contingent, and which therefore essentially include historical and evolutionary aspects. The Principle of Maximum Uniformity appears to be such a law, and laws which we may expect to emerge in biological theory will be essentially of this character. The Principle of Maximum Uniformity will be considered in a larger context and in much more detail from the biological side in a volume concerned with the constants of nature and biological theory, categories of information, and aspects of evolution, and in which it will assume a unifying role.

Stability according to the present definition is a characteristic of the instantaneous state of a system, just as is equilibrium; moreover, the stability so defined has both local and global aspects, which again correspond to the case of equilibrium. The instantaneously stable state is defined as the force configurations belonging to the highest Hierarchy of Uniformity which instantaneously satisfies all the conditions cited above in the statement of the Principle of Maximum Uniformity. According to this definition, instantaneous *global* stability is defined as the collection of instantaneous locally stable force configurations. The definitions given here to 'Hierarchies of Uniformity' and to stability are descriptive, pictorial, and conceptual, and not analytic or quantitative in a mathematical sense. For this purpose it is natural to consider continuously extended material domains, in which the forces joining an element to the universe are characterized by a stress tensor. The Principle of Maximum Uniformity and the general stability law that derives from it will be in part formulated in more analytical (terminology) language in another volume, in which it is planned to treat this subject in a more comprehensive manner, and particularly its biological ramifications.

The Principle of Maximum Uniformity is manifested in the domain of classical mechanics, as required by the Principle of Universal Correspondence, by the evolution in time at different locations of various and distinct *force* configurations. Each of these force configurations belongs to the Hierarchies of Uniformity, and have in common a particular member of the hierarchy, which is defined here by the equilibrium of forces. The progressive evolution in time of the Hierarchies of Uniformity is revealed in all experience and therefore in the classical domain, in particular, by the progressive evolution of different force configurations, each of which may also be interpreted as a hierarchy or order. As noted earlier, all forces are understood here to give direct expression in experience to the universals, which are reflected by the Dimensional Universal Constants and consequently to what is referred to in reference [6] as the domain of the universals. By this way of thinking, the operation in nature of the 'Principle of Maximum Uniformity', and the conception of its operation, demands the existence, and the consideration of the relation between, and interaction of, the domain of the universals and what I call in reference [6] the domain of the observables. This of course, applies equally to the operation in nature of the universal stability law manifested in every domain of experience, and which derives, as do the conventional laws of mechanics, from the Principle of Maximum Uniformity.

The Principle of Maximum Uniformity and the Universal Stability Law attendant upon it have been made operational, within the realm of classical mechanics, that is, exercised computationally in this realm by the development of an algorithm, by modelling certain aspects of the domain of universals by a potential theory. This model allows the formal description of the interaction between viscous flow fields which belong to the domain of the observables, and an ideal domain characterized by the potential theory from which, according to the algorithm, they emerge by what is analogous to a process of evolution. This has produced mathematical representations of viscous flow fields that evidently satisfy the fundamental partial differential equations of classical hydrodynamics and realistic boundary conditions.

The interaction between the domain of the universals and the observable domain brings necessarily under consideration multiple scales and the realization that they assume an essential role, especially their interrelationship, in the interaction between these domains. From the standpoint of classical mechanics, for example, such scales may be identified with temperature fluctuations in a heat bath which are related to the Universal Boltzmann Constant, and the production of inelastic deformations in a solid subjected to forces impressed by the universe from the outside. These points and considerations, as well as the relationships between the Principle of Maximum Uniformity, the stability law, the role of the Constants of Nature as the foundation of natural law and the development of biological theory, and the connection between these, and the existence in nature of Categories and Hierarchies of Information, all will be comprehensively examined and in concert in a volume more specifically directed at their ultimate biological aspects.

The considerations outlined above point to an ordering and designing principle in the universe, a universal evolutionary principle which acts constantly and immediately in accordance with the universe and the universals as they are immediately and everywhere presented in experience and in specific events. I now see the questions that have led in some cases to a search for teleological explanation as essentially clarified and possibly resolved. It is resolved by seeing in context that purpose and design in nature as well as history, and its thrust may be comprehended by considering the immediate relations between the universals and their constant operation in and upon the universe at each instant, and the fact that this is sufficient to give purpose, design, and direction toward a constant goal which may indeed conceptually be projected into the future. All action in the present includes goals directed at the future, in so far as they all inherently and necessarily include thrust toward the future.

*5. Concluding remarks*

A feature that strikingly and significantly distinguishes physical theory from biological theory is that its laws which have emerged from a description and critical examination of modes of experience which we by convention call physical are formulated in a very special language known as Mathematics. This language, its rules, its operations, and the primitive notions and abstract objects to which they refer are all essentially free of contingency and are consequently in principle atemporal, acausal, and ahistorical. It is a language which therefore cannot in principle describe or report on history, on evolution as a process, on selection, on identification, on force, on sensation, and on the nature of information, all conceived here as fundamental manifestations in experience of this process. All the laws of physics which have been given complete expression in this particular language are in fact devoid of historical content, causality, and evolution. There is nothing they report which distinguishes the past from the future or which refer to or condition the universal phenomena of birth, ageing, and death. Conversely, from the fact that the language called Mathematics is essentially atemporal, it follows immediately that a law formulated as a proposition in this language cannot in principle refer to or condition historical development and evolution.

A prevalent attitude among scientists and laymen as well holds that the rank of a scientific law, that is, its importance, depends upon whether or not it can be formulated in the language of Mathematics. The above considerations should dispel this misconception by pointing out some of the essential limitations of this language from the standpoint of describing, embracing, and conditioning certain fundamental aspects of human experience and scientific experience in particular. In my view, the most fundamental scientific law is a universal law which pertains to and conditions a universal evolutionary process and which therefore cannot be fully embraced by and be totally expressed in the mathematical language. This paper is in part committed to the identification of this law.

An equally significant and striking fact which concerns biological theory is that the ideas and laws which have emerged from observations and thinking based on the critical examination of biological materials have hardly been expressed, if at all, in the language which we call Mathematics. A careful examination of the ideas and laws which make up biological theory as we know it today reveals a profundity, an imagination, and a creative power no less penetrating than what has emerged in physical theory. This means, I believe, that what emerges as characteristic and prominent in a critical examination of biological materials and phenomena cannot be appropriately described, expressed, and communicated in the mathematical language. The reason is now becoming clear: it is simply that what is characteristic and eminent in biological phenomena is essentially the developmental and evolutionary aspects of nature, aspects which I have endeavoured to show here are in principle extra-mathematical, and that consequently the laws pertaining to them cannot in principle be given full expression in the language called Mathematics.

It is planned to amplify and extend the ideas and conclusions outlined here in subsequent essays to be presented under the same title in a separate volume. Direct contact will then be made with descriptions, experiments, and theories which have come from work with biological materials. Particular attention will be given to Professor C. H. Waddington's pioneering and experimentally revealing work concerned with the genetic assimilation of acquired characteristics [10–14] and the penetrating question he raised, namely: 'How does evolutionary adaptation *work*?' [15]. By so doing it is hoped to be able to present the design of a crucial experiment using biological materials, by which hypotheses and questions derived from the ideas presented here may be incisively tested.

## References

1. Lieber, Paul. 'A Principle of Maximum Uniformity obtained as a theorem on the Distribution of Internal Forces.' Institute of Engineering Research. University of California, Berkeley, Nonr-222 (87), No. MD-63-8, April 1963.

2. Lieber, Paul and Farmer, Arthur. 'Studies on Wave Propagation in Granular Media.' *Trans. American Geophysical Union*, Vol. 39, No. 2, April 1958.

3. Lieber, Paul. 'The Mechanical Evolution of Clusters of Binary Elastic Collisions and Conception of a Crucial Experiment on Turbulence.' Volume of the *Proceedings of the Symposium on Second-Order Effects in Elasticity, Plasticity, and Fluid Dynamics,* sponsored by the International Union of Theoretical and Applied Mechanics, April 1962.

4. Lieber, Paul and Wan, K. 'A Minimum Dissipation Principle for Real Fluids', *Proc. IX Int. Congress of Mechanics,* 1957.

5. Lieber, Paul. 'Categories of Information.' Office of Research Services, University of California, Berkeley. Nonr-222 (87) AM-65-13, October 1965.

6. Lieber, Paul. 'Constants of Nature; Biological Theory and Natural Law', in (C. H. Waddington, ed.) *Towards a Theoretical Biology, I* (University of Edinburgh Press, 1968).

7. Gauss, Carl F. 'On a New General Fundamental Principle of Mechanics', *Creele's Journal f. Math, 4* (1829) 232. Also appears in *Werke, 5,* 23.

8. Hertz, Heinrich. 'Collected Works, Vol. II', 'Principles of Mechanics' (Macmillan, New York, 1896).

9. Pattee, H. H. 'The Physical Basis of Coding and Reliability in Biological Evolution', in (C. H. Waddington, ed.) *Towards a Theoretical Biology, I* (University of Edinburgh Press, 1968).

10. Waddington, C. H. *Nature, 150* (1942) 563.

11. Waddington, C. H. *Evolution, 7* (1953) 118.

12. Waddington, C. H. *Evolution, 10* (1956) 1.

13. Waddington, C. H. *J. Genet., 55* (1957) 241.

14. Waddington, C. H. *Nature, 183* (1959) 1654.

15. Waddington, C. H. 'Evolutionary Adaptation', *Evolution after Darwin,* Vol. I (University of Chicago Press, 1960).

## Two poems by Mary Reynolds

*Conferenza di Bellagio*

('I'm nobody. Who are you? Are you nobody too?') –Hello!
At Berkeley, yes. Last summer, in a sem-
inar on nucleotide sequences. Not published yet.
–You did? I know the Michigan department,
Stronger than Harvard, and – Campari, grazie.
–Stochastic independence; I heard at last year's meetings,
Rigorous to a fault, an elegant presentation.
He has a Guggenheim.–Sorry, I didn't catch
The name? No, Sussex, since July.
–Behavioural approach; the larger context,
The influence of varying kinds of crisis. Did you see
My latest book?–Yes, yes, a lovely place,
So quiet. Well, a background paper only. The deter-
minants of allocative processes.–I think
Not tenure. Oh, he's sound enough, I know,
But there are other factors.–And I ques-
tion the equation too, he said, and is it cen-
tral to the argument would you say.–Hello, yes not
Until this morning. Plane delayed.–The food
Is great, their cellar must be something
Out of this world.–The Basle Club agreed.
Convertibility? Perhaps a hundred million.
–But not statistical.–Hello! I heard you, Friday,
The BBC, that panel; who arranged it?–Not compute
the coefficient of friction.–Earth is finite.
Standards require universal constants. The velo-
city of light is constant because I choose
To say it is.–The wind is rising.–Yes, you're right,
You're right, a lake trip might be even wiser,
Give them a chance to talk outside the sessions. If (–Hello!
'I'm nobody, who are you? Are you nobody too?
Then there's a pair of us. Don't tell,
They'd banish us, you know.')

*August 1967*

## Mary Reynolds

*Sestina*

(Conference on Highly Theoretical Biology and Machine Intelligence)

This ancient dwelling underneath the hill
Graced by the cypress and the singing bird,
Blessed for a thousand years by turning stars,
Has made us welcome, last upon this shore
Of all the long processional of man
Who comes and dwells and has his little day.

Now comes the breakfast bearer, as the day
Dawns with Italian voices down the hill;
Fiats will bring the staff (for every man
Cannot afford a Lancia); and the bird
Is on the wing above Pescallo shore
Searching abandoned garbage. Under the stars

Last night, the boat-borne songsters praised those stars,
The moon, the wine, their memories of the day;
And cruised, inebriate, closer to the shore
Than pleased the tired sleepers on the hill.
And Tom Cat, hunting, hunting for a bird
Invoked contempt for all the rule of man.

Noon. On the terrace, strident voice of man,
Who knows he soon will reach and pass the stars,
Considers how the essence of a bird,
Computerized and jet-propelled, one day
Will soar above the highest human hill
And come to rest on planetary shore.

Then will he find, upon that alien shore,
A simulated counterfeit of man?
Clanking to harsh command, on lunar hill,
Directed by equations based on stars,
And disregardful both of night and day?
Was it for this he dreamed himself a bird?

## Two poems

Will the computer, whirring like a bird,
Then make a dry martini on this shore
As now Vincenzo closes down our day;
All unaware that pre-computer man
Once knew high revelry beneath the stars,
Ten thousand years ago, upon this hill?

Then man-constructed bird, computer-man,
Will range Pescallo shore, under the stars,
And have their day, upon this ancient hill.

*August 1967*

## Appendix: notes on the second symposium by Michael A. Arbib

(Arbib was one of the very few who kept notes throughout most of the discussions. They contain a rather thorough survey of the discussions that went on between the longer presentations, and they are printed here for the stimuli they offer for further reflection on some of the points raised. Like most lecture notes, they are written in very condensed form; in fact, they can be taken as an informal examination – if you can understand them all, you have done your home-work properly in reading the rest of the book! – Editor).

1. OVERVIEW

Is there a theoretical biology? Is there a *special* mathematics for biology?

*Waddington* (i) General Processes in Biology:

A. Evolution

B. Epigenesis from phenotypes to ecosystem

(Note importance of selection for adaptive phenotypes)

C. Molecular biology–the study of mutable heritable algorithms

D. Origin of Life–is DNA:RNA:Protein the only system in which life could be produced?

E. General Structure of the Universe. General Physics. Metaphysics

(ii) The central nervous system is special both because it is an organ of thought, and because it affects the way we perceive, and thus the way we build theories.

*Kornacker's Scheme*

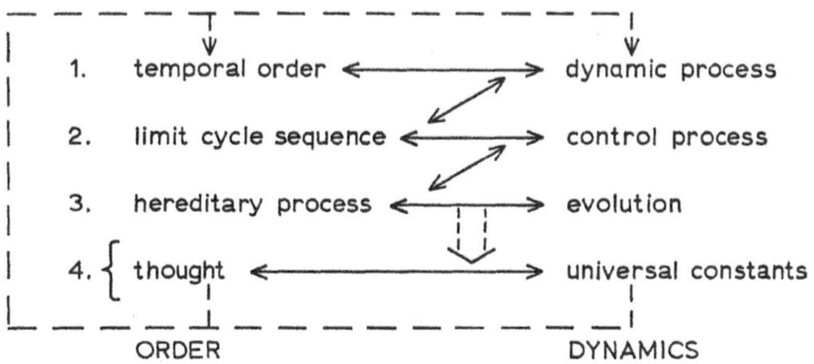


*Kornacker* Diagonals are obtained by calculation of correlations over the left-hand side of quantities on the right-hand side to get the next order of the hierarchy. In

Ted Bastin's model, matrix transformation cycles at one level are elements at the next level.

*Arbib* In what way do we have: temporal order :dynamic process : :thought: universal constants? I don't understand:

Correlation of dynamic processes over time yields limit cycle sequences?

Correlation of control processes over a limit cycle sequence yields hereditary processes?

Correlation of evolution (?) over hereditary processes yields thought?

Correlation of universal constants (what?) over thought (how?) yields time?

*Kornacker* Information transmission is the attempt to impose correlations between sender or receiver.

The states in the temporal order are moments of time.

The states in the limit cycle sequence are limit cycles. (*Arbib* So, in automata theory, we ignore transients between states – these must be reinserted if we desire a physical description.)

The states in the hereditary process are generations. (*Arbib* Thus the order, unlike the temporal order, is asynchronous – a fact we often choose to ignore in our theorizing.)

Are theories the states of the thought order?

*Pattee* In studying Brownian motion, the transition from a law of motion to noise marks a transition in your *design* or choice of *function*.

Coding and measurement take us from a dynamic process to a control process. Reading and writing then take us to the genome or program of the hereditary process.

*Kornacker* At the 2 to 3 level we obtain function for the first time, and with it:

| | | |
|---|---|---|
| code | | communication |
| memory | | learning |
| experience | and | maturing |
| perception | | insight |
| thought | | |

and with these, thought can study dynamics and time to close the loop.

2. EVOLUTION

The phenotype is the genotype's model of the environment.

*Maynard Smith* Neo-Darwinism seeks to explain the facts of adaptation, and the ascribability of functions to parts of organisms. Evolution is assumed to be the result of multiplication, variation, and heredity.

Heredity : says that there are at least 2 types of entity, A and B, such that A tends to beget A and B tends to beget B.

Variation : says that the above doesn't work completely reliably.

The theory requires that there exist differences in rate of multiplication of different entities. 'Fitness' of a type is a measure of this rate of multiplication – and is thus an ensemble concept, rather than a property of the individual. Unfortunately the theory cannot evaluate for you the fitness of a given type in a given environment.

One needs to distinguish the genotype ( a set of instructions ) and the phenotype ( their functioning embodiment ).

The only really strong assumption of the theory ( the Weissman assumption ) is that, whereas different environments produce different phenotypes from a given genotype, this change does not itself produce a change in the phenotype.

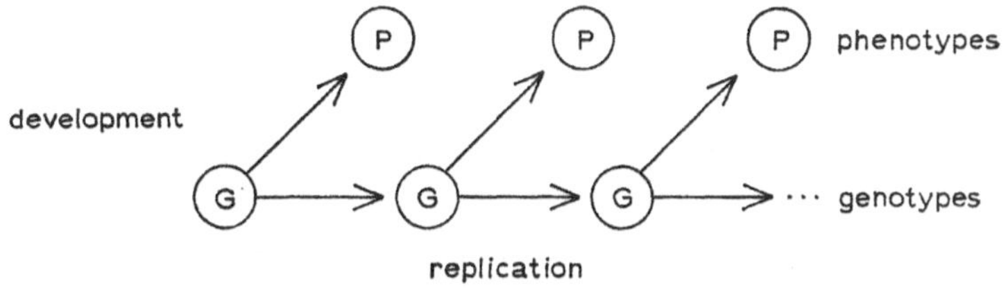


Genotype changes propagate, but phenotype changes do not. ( Lamarckianism is the contrary of this.)

Of course, a change in the environment, such as a focused beam of radiation, can change the genotype !

It does *not* seem necessary to posit randomness of variation, but one does assume that *most* changes of genotype lower fitness–or you wouldn't need natural selection !

In the beginning was the word WORD

WORE

GORE

GONE

and by the mutations came the gene GENE

One believes that the laws of physics and the earth environment permitted the sequence

$$G_1 \rightarrow G_2 \rightarrow \ldots \rightarrow G_N$$

such that $G_1$ could arise 'spontaneously', $G_N$ specifies a human, each step increased fitness, and no step needs to be regarded as a highly programmed change.

*Kornacker* To study this one needs to know how large is the range of phenotypes reachable from a genotype via one mutation. The theory of evolution would not be very interesting if we could take $N=1$ or 2.

*Waddington* Selection is of genotypes, but it acts on phenotypes. To specify the situation we need at least 2 genotypes and 2 environments. A genotype might develop in one environment but be selected in the other. We might then get selection for the most frequently occurring environment. We often get selection for ability to learn a skill rather than for the skill itself.

Each organism introduces new organismic relations – and then a new organism can evolve to exploit the new ecological niche provided by this relationship.

Evolution is not driven–there just exist metastable situations.

*Arbib* Consider the Volterra equations. They show that selection is not for simple maximization of species population, but involves time averages in an environment of multiple species.

What can we say about neutral changes of genotype which can accumulate until a combination may be reached which is advantageous or disadvantageous? Cf. a random walk with absorbing barrier.

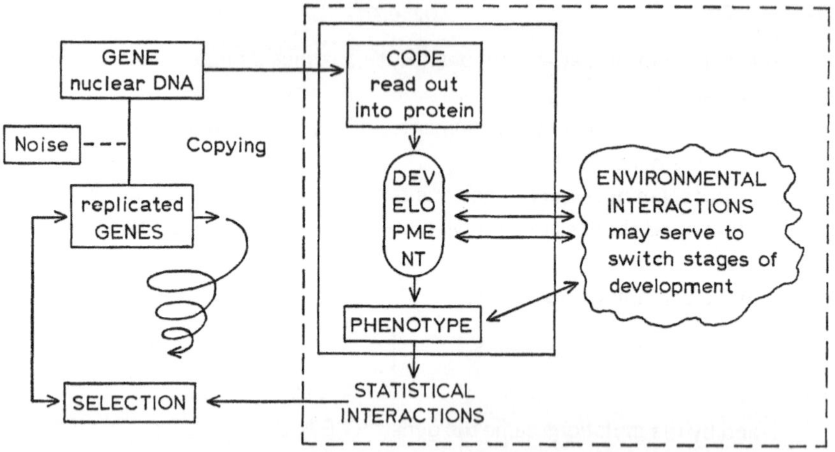


*Pattee* The essence of neo-Darwinism is that  is a black box; Waddington would have the theory only consider  as a black box.

Maynard Smith wants a theorem that species formation is more common than species extinction. It takes 10 generations to establish a gene change. (*Waddington.* Take care—this estimate comes from a theory with many unobservable parameters. A new species of fish was formed in Africa in 400 years.)
Lysogeny speeds up evolution.
Symbiosis acts as a higher-level store of genetic information:

lichen = fungus + moss; and each is the environment of the other.

What was the evolution of differentiation?

Note that evolution is *not* synonymous with increase of complexity, since the viruses are presumably simpler than their ancestors—but they could only evolve when there existed cells in the environment to support them.
*Pattee* Some bacteria have survived for billions of years without apparent dissatisfaction, yet man evolved too.

So perhaps one requires a theory of competitive and cooperative games which can throw light on coalition formation. (Cf. Waddington's comment on multiplication of niches.)
*Maynard Smith* We have ecotheories for non-evolving populations, and evolutionary theories for 'stable' ecosystems, but no stability criteria for general evolution. Is evolution a stable process?

If you evolve into an ecological niche, it may not still be there—you might be so efficient, that you would eat everybody else or die of starvation. (Cf. remarks on Volterra equations above.)

## 3. MORPHOGENESIS

*Waddington* What is a form? There seems to be some notion of regularity.
General categories:
1. Monotypic: e.g. crystal forms coming from one type of component;
vs. Polytypic: many biological forms come from many types of components.
*Drosophila* wing form can be altered by changes in 40 to 50 genes.
2. Synchronic: a form which is complete as soon as it appears;
vs. Diachronic: a form which is slowly elaborated and built up.
3. Element elaborated: like a jigsaw;
vs. Whole controlled: with feedback from the developing whole controlling each subsequent step.

In the biological case, the chreod usually corresponds to a polytypic or diachronic form which is whole controlled, corresponding to a morphogenetic field.
Various modes of form generation:

### Appendix: Notes on the second symposium

A. Unit generated: (a) Particles; (b) Fibres; (c) Sheets. How far up the scale can you go with chemistry? Perhaps (cf. July 1967, *Scientific American*) as far as virus head construction. Perhaps cells use specific attachment points (desmosomes as at synapses) to fix on to one another.

B. Instruction generated, of which a special case is

C. Template generated: contrast of simple and complex coding

D. Condition generated, which may involve stochastic conditions (cf. Turing's morphogenesis paper)

In intersusceptive template action, growth takes place all through the whole organ, not just at the edges as in apical growth.

*Scriven* Certain aspects of morphogenesis are to be studied in terms of temporal or spatial rhythms. Scriven has a treatment, developed from Turing's paper on morphogenesis, based on transport processes to move things from one place to another. (Robin Grands has a Turing manuscript for the non-linear case.)

Dynamic forms known to the chemical engineer:

1. Eddies on oil burners
   Blobs form to hold the flame under certain dynamic conditions.
2. Rotating the inner of two cylinders between which is liquid.
   Counter-rotating toroidal eddies form Taylor vortices.
   Increasing speed yields a switch to eddies – standing waves? As speed increases these start to progress.

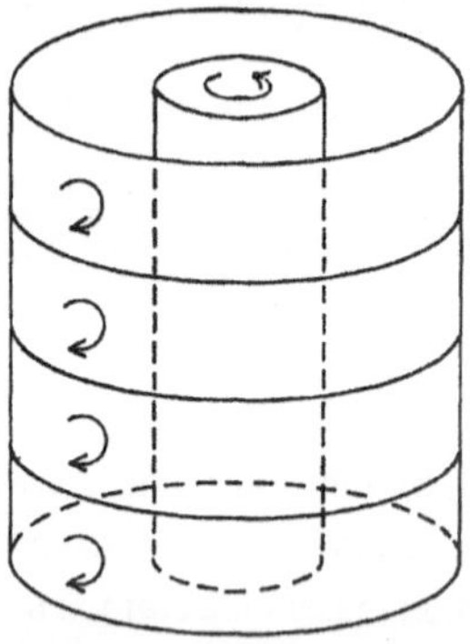

3. Bunsen burner

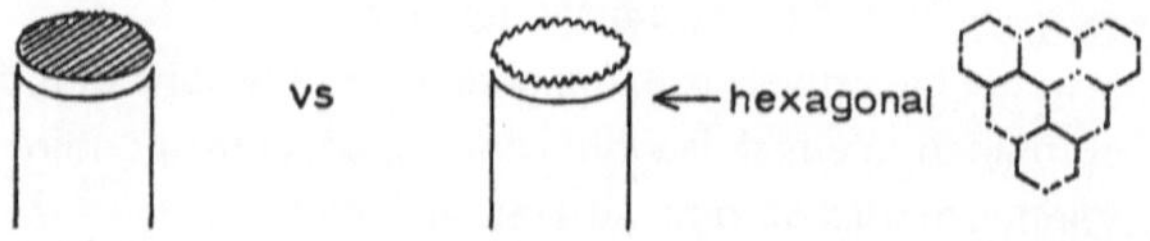

Can also get this structure in crystallization of binary melts, under certain conditions. Arctic tundra polygons from alternate thawing and freezing.

4. If a dishpan with a shallow (to avoid turbulence) layer of liquid is uniformly heated from below, a hexagonal pattern will form after convection is set up. In a circular dish, pattern formation starts from the perimeter inwards, with cylindrical units. In a small dish you never get hexagons, whereas in a large dish you eventually do. Perhaps the determination of the actual pattern depends on nonlinear effects, with 'noise' getting things started.

## 4. AUTOMATA THEORY AND THE ORIGIN OF LIFE

*Longuet-Higgins* A great unifying concept: mutable hereditary algorithms.

The 'secret of life' is the ability of living creatures to improve their programs.

1. Morphogenesis involves differentiation—change of cellular programs.
2. Evolution—change of organism's programs.
3. The brain may be regarded as a program for processing inputs, and is modifiable by learning.
4. Biogenesis—Where does the program come from? How is it modified?

To be appropriate automata should have an unbounded set of states, and be capable of rewriting their programs. Finite-state automata do not suffice for grammar—so why should they suffice for theoretical biology?

*Arbib* Rewriting programs is quicker (and safer!) and more compact than consulting a list of ancestral actions (or repeating ancestral mistakes!). How do we mix this up with evolutionary theory?

Note that my self-producing automata grow in tessellations – here there is no limit on the size of the automaton, though at any time a finite-state space will describe the automaton, in that temporal neighbourhood. Perhaps in formulating a proper theory of chreods we should consider them as moving through an expanding sequence of finite-dimensional subspaces of a 'Hilbert-Waddington biospace'!

*Longuet-Higgins* Gödel proved his incompleteness theorem via a self-referential 'I'm not provable' statement, which is still unprovable if you change the code. What (if anything) has this got to do with genetic coding?

*Pattee* What can be said about playing off the two principles of impotency: undecidability and uncertainty?

*Goodwin* We've been doing differential equations for years—then along came automata, which are irreversible. So they look good for biological irreversible processes—but don't let's get carried away.

## Appendix: Notes on the second symposium

*Arbib* Note that my 'Common Framework' paper indicates ways in which differential equations and automata can be combined.

*Kornacker* Groups $\xrightarrow{\text{damping}}$ Semigroups ? Do correlation terms destroy inverses but preserve the associative law ?

*Arbib* As soon as you partition a set, a group which acts on it loses its inverses :

$\begin{pmatrix}1\\2\\3\\4\end{pmatrix} \rightarrow \begin{pmatrix}4\\1\\3\\2\end{pmatrix}$ has inverse $\begin{pmatrix}1\\2\\3\\4\end{pmatrix} \rightarrow \begin{pmatrix}2\\4\\3\\1\end{pmatrix}$, but if we identify 3 and 4, labelling it $S$, we get problems ! Not only do we lose inverses, but we also lose determinism. We might approximate with $\begin{pmatrix}1\\2\\S\\S\end{pmatrix} \rightarrow \begin{pmatrix}S\\1\\S\end{pmatrix}$ which has no inverse, or with $\begin{pmatrix}1\\2\\S\end{pmatrix} \rightarrow \begin{pmatrix}S\\1\\2\end{pmatrix}$ which *does* have an inverse.

A large system is deterministically (Newtonian) describable in time and space – if you ask the right questions about it. A small system must be described at the wave function level. Can we talk of medium-scale systems (hopefully including RNA, etc.) which, with reference to the questions of interest to us, are describable as probabilistic automata ? i.e. they have enough stability, with reference to our questions, for the phase shifts of the wave functions to be of 'no' relevance.

Questions for the automata theory approach to tackle :

*Waddington* Bone growth

*Pattee* Think about virus construction (cf. July 1967 *Sci. Amer.*). Suggests one program this as a test. It's not a 'hydrodynamic instability' type of form generation.

*Pattee* The logico-mathematical approach to theory is the only practical one for complex systems – 'don't use the Hamiltonian to build a watch' – although the physical approach helps us with components, general concepts, etc.

[ ? ] The origin of life problem seems the appropriate place to make an interface between these two approaches in our attempt to link up the three great mysterious peaks, of development, evolution, and thought. (The components of self-reproducing automata are too complex.)

Michael A. Arbib

We must conform to the fact that part of the cell consists of tactic copolymerisations, with 20 subunits for amino acids, 4 for nucleotides, and several for the saccharides. With the closed matter cycle on the surface of the earth, fed by energy from the sun, how did we reach metastable 3D configurations with hereditary properties ? We also needed degradation mechanisms—to prevent the earth filling up with self-reproducing automata ! Building up and degrading down gave an ecocycle driven by the sun—the first cycle must ( ? cf. Marcus Goodall's idea on cell boundaries) have been non-living and outside the cell.

Fox has a preparation in which within 20 minutes of starting with amino acids and hot water, proteinoid micropheres arise which bud spontaneously. Pattee isn't happy with this as an origin of life contender—it's not necessarily hereditary. (*Maynard Smith* To be hereditary, more than one property must be propagatable. Thus, the flame is a poor model of reproduction because the offspring depends not on the 'parent' but on the substrate.)

So we're really interested in 'hereditary propagation which is modifiable by natural selection'.

We may think of death as an error-correction mechanism—which must be slow enough to allow living organisms to accumulate the information derived from natural selection.

What, then, is the simplest system which will operate at the origin-of-life level, with enough error-correction to have the possibility of evolution ? We're interested, then, in the occurrence of a specific reliable catalyst—it's not the permanence of the structure, it's the reliability of the catalyst that's important—the greater the capacity of replication, the less the demand for reliability.

The question of which chemical automata survive is not just a question of logic alone, but of how reliably a quantum-mechanical system can approximate the logic. ( Cf. Swanson & Landauer on reliability of components, in the I B M Journal.) *Monro* A single template can lead to the production of $10^{12}$ enzymes ( in a cell, or the whole organism ? ). Thus error-correction of templates is crucial—since $10^{12} \times$ a small error can be a whopper. Anyway, these figures show that D N A can control macrostates. A single error in the egg cell can cause death in the organism—it's incredible that anything survives !

There are two types of errors in the templates :

A letter gives a non-letter : there are enzymes to correct this.

A letter gives another letter : there's nothing you can do.

Errors in later stages of division—somatic mutations—are not as crucial, e.g. 1/10% errors in a haemoglobin molecule might not matter.

## Appendix: notes on the second symposium

Evolution of protein structure is related to evolution of organism—proteins are both structural and functional, cf. enzymes in active transport.

Typical metabolic patterns–protein synthesis, Krebs cycle, carbohydrate metabolism—recur in most species.

Another aspect of enzymes is the control of their synthesis and function. Control at point of enzyme synthesis through controller genes. The repressor is a protein. Presumably the repressor controls whether or not the operator can get going (in what sense?)

*Waddington* DNA amplification (in toad egg, not in bacteria): Ribosomal cistrons—for synthesizing ribosomes—are represented about 2,000 times in the toad. An interesting situation would occur if you could modify the amplification, e.g. to synthesize more actin and myosin in muscle cells (but it's not known whether or not this occurs).

In toad egg, 5,000 rings of DNA are thrown off to float around; this release is under the control of the cell as a whole. This presents the possibility of a rather flexible control beyond that we can discover in bacteria.

*Longuet-Higgins* In computer programs we can bring in subroutines, or skip them, on the basis of the computation state—surely a paradigm for repressors or derepressors.

*Waddington* In Arbib's self-reproducing automaton we may consider

V: copies the DNA.

U: gets the DNA to produce the new cytoplasm.

W: resets.

Question: Why are the genotype and phenotype separate?

*Longuet-Higgins* For 'sanitary' reasons to prevent overwriting of a 'master routine'.

*Arbib* If the alternative is that the objects are to copy themselves, we seem bound to get into problems of destructive readout with a 3D object in 3D space.

*Pattee* For reliability, one would want to make the genotype very stable. But then the dead time is too great for it to do anything. Solution: use ultra-reliable DNA to store the message, but build up a phenotype of other components to obtain a flexible organism. (We need a fast and accurate read-out mechanism.)

*Monro* There exist bacterial cells with different processors (sRNA) which instead of terminating will put in a new amino acid.

*Arbib* But even stronger, one could imagine a different cytoplasmic machine which would consistently assemble the DNA-decodings differently.

*Maynard Smith* A cell can 'accept' into its genes a useful 'subroutine' from an invading virus, thus speeding up evolution. Perhaps mitochondria were originally cell invaders which took up permanent residence, their division cycles became entrained, and we obtained an extra-nuclear genetic mechanism.

*Kornacker* For automata which model human thought, we need pattern recognition and generation with

input: spatio-temporal patterns in many sensory modalities.

output: patterns of muscular coordination.

*Kerner* thinks of biology as comprising *in vivo* biochemistry, genetics, and ecology and these are all amenable to theoretical treatment. The automata theory is 'overprinted'.

'What is the connection of software to hardware?'

*Arbib* In time-shared systems or parallel computations we have a hierarchy of cycles: bit times, word times, and job times. In choosing between synchronization (every operation takes the time of the slowest) and queuing (do a job as soon as the prerequisites are available) we may play off simplicity of operation and speed of operation.

An evolutionary theory will probably have to face up to a preponderance of breadth over depth in that fundamental principles will tend to rest on the accumulation of evolutionary 'accidents' or the importance of arbitrary decisions.

Note the different formations required to model different aspects of information processing.

decomposition of nets vs. semigroups vs. events.

$X$ for deterministic vs. $X \times Y$ for stochastic automata.

Hierarchies of computational complexity.

Contrast the questions:

Bellagio '66: Is random search necessary for evolution?

von Neumann: Is a minimum complexity needed for self-reproduction?

I suspect the answer to both is no – I can imagine a machine which computes according to some valuation function to see which of its features should best be modified. However, I might expect that there is a minimum complexity for a machine which can achieve this 'self-conscious evolution'.

5. NONLINEAR OSCILLATIONS

Iberall started teasing out a broad spectrum of nonlinear oscillators – macro-spectroscopy– from heart beat and below to circadian rhythms and beyond, and then realized that the oscillators *were* the system, with DC changes in the milieu changing operational points of the oscillators. This emphasized homeokinesis–the dynamic nature of regulation.

'Any compact system containing a complex of sustaining nonlinear oscillators and a series of algorithms to let it operate in a wide variety of ambient conditions . . .' is a living system, which may thus involve many types of successful mechanism.

We have dynamic regulating chains, be they stable, unstable, or marginally stable; self-activated motor activity; and, when time is adequate, entrainment of the oscillations.

Iberall would extend this scheme from involuntary to psychological systems: adolescence is an instability preparatory to reproduction; self-activation seems more important than pre-programmed activity.

*Kerner* Do the cycles matter, or are they just details of other things–e.g. in a factory, consider changes in the waste-paper-basket-emptying cycle.
*Goodwin* The periodic environment we live in imposes periodicities on the system, and once we've got a few it's easier to put the whole lot in.
*Elsasser* Poincaré's theory says that any stable system with one parameter changed will start to oscillate. All nonlinear systems oscillate – not just living systems.

*Goodwin* wants to describe, and then explain, the sequence of events within the cycle of cell division.

The naive view is that everything goes on in parallel, and when all cell concentrations are doubled, the cell splits. This is not so. Nor is there a causal chain, with completion of stage A triggering the start of stage B, etc. Rather, Brian found, we have phase-locked processes which interlock so that appropriate concentrations exist at various times.

There are then stability problems for temporal sequence of events (of macro-molecular synthesis) during cell division. Neoplastic growth (cancer) is an instability–one wants a theory which can also describe this instability.

He assumes the relevant variables are enzyme concentrations. The functional units are then taken to be control circuits based on a functionally linked set of genes whose end-product metabolites can feed back to inhibit the genes.

There are 200–300 such functional units, roughly, per bacterial cell—two units are judged distinct if you can turn one off without affecting the other. The units are *not* localized—messages can diffuse from one place to another, and one doesn't want to only consider operons controlled by a single gene.

One assumes variables are exclusive to one of the units (?). Of course, there are non-specific interactions, due to sharing a common pool of ribosomes. Glutamate appears in two cycles and yields strong interactions, while pH gives weak interactions.

Studies of 10 of these circuits showed them operating in an oscillatory mode, with the oscillations giving the basis for sequential production of enzymes—the systems being phase locked with one another, yielding different points for maximal concentration of various enzymes. This presumably serves to minimize the time required for cell division.

*Scriven* Chemical reactors are non-isothermal and thus highly non-linear. There are *many* steady states, both stable and unstable, and some of the latter can be stabilized by appropriate control to give higher yields than the stable steady states. There exist limit cycles more efficient than steady states.

They are attempting to study this at the molecular level, and to join statistical mechanics and continuum mechanics, as well as to use optimization theory. They are also concerned with improving the design of interlinkages in chemical plants.

One needs constitutive relations to define the system, as well as boundary conditions in time and space to define the rest of the world for the system,

$$\text{biology} \rightleftharpoons \text{chemical engineering},$$

e.g. catalysis and optimization of complex chemical systems.

Relation of this work on chemical process optimization to Goodwin's work.

1 Complex reaction kinetic schemes.

Nature of steady-state hypothesis for low concentrations of steady-state intermediates.

2 Multiple chemical reactions of greater than first-order nonlinear, consider relations between phases, e.g. in polymerisation.

Characterizing a chemical plant: (i) what's it for?; (ii) inventory; (iii) flow diagram; (iv) material balances—check they're closed.

*Maynard Smith* Surely the strange thing about the cell is that it has so few steady states.

*Arbib* Goodwin only distinguishes the growing state and the dividing state–but surely these are too few?

6. STATISTICAL MECHANICS

*Kerner* Gibbs ensemble method is not restricted to use in mechanics. Biological systems are so complex that you may need an analysis which gives an overview.

We know more about $10^{23}$ molecules in a gas than we do about the moon–earth–sun system. Temperature is a useful measure for the former, not the latter.

Gibbs gave a statistical theory of differential equations. D.E.s are useful in biology, e.g. in ecology (the dirtiest form of biology, but with a long mathematical history going back to Malthus & Benjamin Franklin). An ecosystem may be defined by $10^6$, say, variables, one for each species number.

The Gibbs strategy says you must know

(i) one conserved quantity of the observed system

(e.g. energy in classical physics, something similar in Volterra-Lotka dynamics, mass in a closed chemical kinetic system).

(ii) a good way to introduce statistics

e.g. a probability distribution on the ergodic surface (i.e. a surface of constant value for the conserved quantity).

(iii) nothing else–so all you know is which constant surface you are on.

What probability distribution should be placed on the surface? The uniform distribution is O.K. if you have a Lionville theorem to assure that it's preserved by the dynamics–and you pray that the ergodic hypothesis is valid. You have to find the appropriate time scale, and check that there are no isolated 'pockets' in the motion.

You don't need a Hamiltonian for a Gibbsian analysis, just a conserved quantity. Let $N_r$ be the number of animals in the $r$th species. Volterra set up the equations

$$N_r = e_r N_r + \frac{1}{\beta_r} \sum_s \alpha_{sr} N_s N_r \quad \text{with} \quad \alpha_{sr} = -\alpha_{rs}$$

With $V_r = \log \frac{N_r}{q_r}$, ($q_r$ being a stationary value) you get equations for which Lionville's theorem holds. The conserved quantity is then of the form

$$G = \Sigma \tau_r (e^{V_r} - V_r) \equiv \text{const.}$$

What are the observables of the system? Suppose we could think classically and observe the motion of one of the $10^{23}$ particles in a gas (cf. Cowan's one neuron and the EEG). A Brownian particle is a good thermometer. We want to know averages telling us amplitudes of fluctuations, mean levels of crossings, etc. All these are expressed in terms of temperature. Using an ensemble $e^{-G/\theta}$, where $\theta$, the modulus of the distribution, is the temperature, we get a thermostatics. Here is the idea of a Gibbs ensemble—a description of complex phenomena with many averages expressed in terms of a single parameter.

*Kornacker* To get from mechanics $\left(\frac{dE}{dt}=F_e \, . \, V\right)$ to *heat* is to introduce averaging over all values consistent with our partial state of knowledge.

$$\text{If } \bar{E}=\int Ep\,dx, \text{ then } \frac{d\bar{E}}{dt}=\int\left(E\frac{dp}{dt}\right)dx+\int\left(p\frac{dE}{dt}\right)dx.$$

The central dogma of statistical mechanics: you gotta be ignorant. Who you, the observer, are is crucial. In classical mechanics you can ignore the nature of the measuring devices. Measuring devices perform averaging—a pressure measure is an integration of molecular forces. A time average looks like

$$\frac{1}{\tau}\int_{t-\tau}^{t} g(x(t))dt=\frac{1}{\tau}\int_{t-\tau}^{t}\left[\int g(x)\delta(x-x(t))dx\right]dt=\int g(x)\overline{(x-x(t))}\,dx$$

and this is where the ensemble comes from. This requires the time constant in the measuring device to be long enough so that . . . (*Arbib*, what?)

By this criterion, one molecule could be macroscopic for suitable time intervals. The less the number of particles, the lower the frequency, and so the longer the time interval required—in general.

A working hypothesis could be that $\overline{\delta(x-x(t,x_0))}$ is the microcanonical distribution. ($x_0$: initial condition)

Of course, with probability 0 (or greater?) the average could be degenerate, e.g. if all particles are initially moving in parallel.

Let $\langle\ \rangle$ denote a time average on a single system over a time which is several relaxation times for the measuring device, but is short compared with the relaxation time of the system measured.

$\frac{d\langle E\rangle}{dt}=\langle F_e . V\rangle$, whereas work is given by $\frac{dW}{dt}=\langle F_e\rangle . \langle V\rangle$, and the difference between work and energy is

$$\frac{dQ}{dt}=\frac{d\langle E\rangle}{dt}-\frac{dw}{dt}=\langle F_e . V\rangle-\langle F_e\rangle . \langle V\rangle$$

and so heat is a macroscopic measure of the degree of *correlation* of force and displacement.

There are many ways of measuring the strength of correlation:

linear $\overline{xy}-\overline{x}\,\overline{y}$

probabilistic Does $p(x,y)=p(x)\,p(y)$?

If so, they are independent, which implies uncorrelated, but the converse is not true: $\overline{\sin . \cos}=\overline{\sin} . \overline{\cos}$

? A more powerful way of approaching correlation is via an entropy measure:

$$S(p_x)=\int p_x \ln p_x \text{ and ask, does } S(p_x)+S(p_y)=S(p_{xy})\text{ ?}$$

Heat is only sensitive to linear correlations, and Kornacker claims the trouble with entropy is that it picks up too many correlations, and this causes many problems.

Heat is intrinsically indirectly observable.

The second law of thermodynamics will be set up to within fluctuations, but Kornacker claims that it is irrelevant to biology.

Temperature depends on momentum fluctuations $\langle (p-\langle p \rangle)^2 \rangle$.

## 7. METAPHYSICS

Elsasser believes biology is not the kind of subject you can unify, but that there exist unifying ideas like those of Darwin, or Watson and Crick.

Longuet-Higgins thinks biological systems are physico-chemical, but that the questions we ask about cells are not those for which we have operators in quantum mechanics. There is an energy operator but not a kidney operator!

It's more expedient to build a theory in terms of the concepts we're really interested in – since it is inexpedient to do everything in terms of quantum mechanics, need we worry whether it's adequate?

*Elsasser* Starting from quantum mechanics, you'd like to make inferences. But, as Bohr said, you can't measure something as complex as an organism without destroying it. So there are enough loopholes to give it a living character or *autonomy*: a class is autonomous if, as a matter of principle, you can't deduce its regularities (the essence of scientific predictability) from quantum mechanics.

*Arbib* Quantum measurements which suffice to convince one an earth thing is a cat might not suffice to determine an extraterrestial creature? How does this relate to Elsasser's terminology?

*Bohm* Nobody has reduced statistical mechanics to quantum mechanics, and they may be inconsistent. Perhaps there is a contradiction between biology and quantum mechanics–and this would demand the creation of a new system embracing both.

A wave function does not describe an object in time or space.

Bohm (*Rev. Mod. Phys.*, 1966) showed that von Neumann had built his conclusions on observable classes into his axioms–if you take Euclidean geometry as a fact, then non-Euclidean geometry yields a contradiction!

Gregory notes that an engineer succeeds by, e.g., dividing a radio into components – functional units. This should be our lead to overcoming the quantum mechanical problem of immense numbers.

How much should the biologist be an engineer, how much a physicist?

*Goodwin* The discovery of appropriate variables for biology is itself an act of creation.
*Bohm* Compare Piaget's study of the evolution of object perception in children.
*Kerner* The only question of creation he allows is: 'Where does the next instant of time come from?'
A model is a dynamic map–what is a dynamic model?
*What is organization?*
*Kornacker* would reserve the term for functional organization (crystals having order rather than organization)–parts related to a whole geometrically, perhaps, but with the emphasis on action.

Phy sics says irreversible processes, of which one is scalar and one is vector, cannot be coupled isotropically – an anisotropic membrane which couples two such processes is organized, but one which is poisoned is not.
*Waddington* Surely the organization is in the whole system of which the membrane is part.
*Iberall* Kornacker would accept a laser as an organized system of which a crystal is *part.*
*Maynard Smith* At least in biology, an organization is something of which you can ask 'What is it for; to what does it contribute?'
*Longuet-Higgins* There's a hierarchy. Subsystems have function in relation to systems having functions in relation to . . .
*Bohm* Distinguish the organization and that which organizes it.
*Arbib* We are interested in systems whose dynamics can be described using far

fewer parameters than are required to describe the dynamics of its individual parts (non-holonomic constraints?). This implies maintenance of a certain structure which permits these few parameters to remain an adequate description. When they cease to be adequate, the organization has changed–cf. ageing and death. It all depends on the level of observation and questioning we apply.

*Grene* This notion doesn't exclude gases. For biology we concentrate on organizations with properties of not just survival, but hereditary and developmental properties. A simpler characterization of biological organization may follow.

*Waddington* Not just homeostasis but homeorhesis–stabilized paths, not stabilized parts.

*Pattee* The problem still remains of reconciliation with physics. An electron has no function. What is the simplest system with a function? What is the property (is there such a property?) of a thing that marks it as alive, or as an artefact of a living system?

*Elsasser* Incredible–a scientific conference where people can talk metaphysics without being shot down!

## List of Participants

*(Second Symposium. 3–12 August 1967)*

| | |
|---|---|
| Michael A. **Arbib** | Automata Theorist. Stanford University. |
| E.W. ('Ted') **Bastin** | Physicist, Computer Scientist. Language Research Unit, Cambridge. |
| David **Bohm** | Theoretical Physicist. Birkbeck College, University of London. |
| Jack **Cowan** | Neuroscientist. Imperial College, London (now Dept. of Mathematical Biology, Chicago University). |
| W. M. **Elsasser** | Physicist. Princeton University (now University of Maryland). |
| Martin A. **Garstens** | Physicist. Office of Naval Research, Washington (now University of Maryland). |
| Brian **Goodwin** | Theoretical Biologist. University of Sussex, Brighton. |
| Richard L. **Gregory** | Neuroscientist. University of Edinburgh. |
| Marjorie **Grene** | Philosopher. University of California (now University of Texas). |
| A. S. ('Art') **Iberall** | Systems Analyst. General Technical Services Inc., Pennsylvania. |
| Edward H. **Kerner** | Physicist. University of Delaware. |
| Karl **Kornacker** | Neuroscientist. Massachusetts Institute of Technology. |
| Paul **Lieber** | Physicist. University of California, Berkeley. |
| Christopher **Longuet-Higgins** | Theoretical Chemist and Computer Scientist. University of Edinburgh. |
| John **Maynard Smith** | Geneticist. University of Sussex, Brighton. |
| Robin E. **Monro** | Molecular Biologist. M R C Unit for Molecular Biology, Cambridge, England. |
| Howard H. **Pattee** | Physicist. Stanford University. |

L. E. ('Skip') **Scriven** Chemical Engineer. University of Minnesota, Minneapolis.

C. H. ('Wad') **Waddington** Biologist. University of Edinburgh.

*Secretary*

Miss D. **Manning** University of Edinburgh (now Internat. Inst. Genetics and Biophysics, Naples).

# Author Index

*References to extended treatments are given in italics*

## Author Index

## Author Index

## Author Index

# Subject Index

## Subject Index

## Subject Index

## Subject Index

## Subject Index

## Subject Index

## Subject Index